THE BIEBERBACH CONJECTURE
**Proceedings of the Symposium
on the Occasion of the Proof**

P. Koebe

L. Bieberbach

C. Loewner

G. M. Goluzin

H. Grunsky

M. M. Schiffer

M. S. Robertson

I. E. Bazilevich

I. M. Milin

N. A. Lebedev

L. de Branges

W. K. Hayman

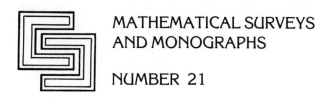

MATHEMATICAL SURVEYS
AND MONOGRAPHS

NUMBER 21

THE BIEBERBACH CONJECTURE
Proceedings of the Symposium on the Occasion of the Proof

Albert Baernstein II, David Drasin,
Peter Duren, and Albert Marden, Editors

American Mathematical Society
Providence, Rhode Island

1980 *Mathematics Subject Classification* (1985 *Revision*). Primary 30C50; Secondary 30C70, 30C75, 30C85, 30D55, 30E20, 30E25, 33A65, 33A70, 42A50, 47A45.

Library of Congress Cataloging-in-Publication Data
The Bieberbach conjecture.
    (Mathematical surveys and monographs, ISSN 0076-5376; no. 21)
    Includes bibliographies.
    1. Geometric function theory—Congresses. 2. Bieberbach, Ludwig, 1886– —Congresses. I. Baernstein, A. (Albert), 1941–   . II. American Mathematical Society. III. Series.
QA331.B584   1986   515   86-10843
ISBN 0-8218-1521-0 (alk. paper)

    Copying and reprinting. Individual readers of this publication, and nonprofit libraries acting for them, are permitted to make fair use of the material, such as to copy an article for use in teaching or research. Permission is granted to quote brief passages from this publication in reviews, provided the customary acknowledgment of the source is given.
    Republication, systematic copying, or multiple reproduction of any material in this publication (including abstracts) is permitted only under license from the American Mathematical Society. Requests for such permission should be addressed to the Executive Director, American Mathematical Society, P.O. Box 6248, Providence, Rhode Island 02940.
    The owner consents to copying beyond that permitted by Sections 107 or 108 of the U.S. Copyright Law, provided that a fee of $1.00 plus $.25 per page for each copy be paid directly to the Copyright Clearance Center, Inc., 21 Congress Street, Salem, Massachusetts 01970. When paying this fee please use the code 0076-5376/86 to refer to this publication. This consent does not extend to other kinds of copying, such as copying for general distribution, for advertising or promotion purposes, for creating new collective works, or for resale.

    Copyright ©1986 by the American Mathematical Society. All rights reserved.
Printed in the United States of America
The American Mathematical Society retains all rights
except those granted to the United States Government.
The paper used in this book is acid-free and falls within the guidelines
established to ensure permanence and durability.

Contents

Preface   .   vii

List of Contributors   xv

Mathematical Papers

Classical analysis: present and future
    LARS V. AHLFORS   1

Inequalities for polynomials
    RICHARD ASKEY AND GEORGE GASPER   7

On interpolation, Blaschke products, and balayage of measures
    ARNE BEURLING   33

Powers of Riemann mapping functions
    LOUIS DE BRANGES   51

300 years of analyticity
    JEAN DIEUDONNÉ   69

Problems in mathematical physics connected with the Bieberbach conjecture
    PAUL R. GARABEDIAN   79

Extremal methods
    D. H. HAMILTON   85

The method of the extremal metric
    JAMES A. JENKINS   95

Some problems in complex analysis
    PETER W. JONES   105

Comments on the proof of the conjecture on logarithmic coefficients
    I. M. MILIN   109

Notes on two function models
    N. K. NIKOL′SKIĬ AND V. I. VASYUNIN   113

The growth of the derivative of a univalent function
CHRISTIAN POMMERENKE   143

Shift-invariant subspaces from the Brangesian point of view
DONALD SARASON   153

The Cauchy integral, chord-arc curves, and quasiconformal mappings
STEPHEN W. SEMMES   167

Zippers and univalent functions
WILLIAM P. THURSTON   185

Personal Accounts

The story of the verification of the Bieberbach conjecture
LOUIS DE BRANGES   199

Reminiscences of my involvement in de Branges's proof of the Bieberbach conjecture
WALTER GAUTSCHI   205

My reaction to de Branges's proof of the Bieberbach conjecture
RICHARD ASKEY   213

Poem
WOLFGANG FUCHS   217

Preface

In March of 1984 the message began to travel. Louis de Branges was claiming a proof of the Bieberbach conjecture. And his method had come from totally unexpected sources: operator theory and special functions. The story seemed fantastic at the time, but it turned out to be true.

This achievement provided the impetus for an international conference at Purdue University during the week of March 11–14, 1985: the *Symposium on the Occasion of the Proof of the Bieberbach Conjecture*. Its purpose was to consider the impact of de Branges's work and to make a broad survey of some topics of current study in complex function theory. It was not possible to cover all of the interesting work now taking place, but we hoped our selection would provide some illumination of the past as well as some indications as to where the subject might be headed.

Fifteen mathematicians spoke on various aspects of complex analysis and related areas. Thirteen of their lectures are recorded here, most in expanded form. Unfortunately, the lectures by E. Bombieri and W. K. Hayman are not included. However, two additional papers are presented by Soviet mathematicians who were unable to attend the conference: I. M. Milin and N. K. Nikol′skiĭ (the latter in collaboration with V. I. Vasyunin). At the end of the volume are some personal accounts of the story of de Branges's proof, its evolution and eventual confirmation by the group of geometric function theorists in Leningrad, as told by some of the principal characters in the drama. Milin's discussion provides additional information about these events. (Another contemporary account by O. M. Fomenko and G. V. Kuz′mina [**24**] has appeared elsewhere. Further personal accounts have been given by C. H. FitzGerald [**22**], Ch. Pommerenke [**69**], and others.)

Ludwig Bieberbach proposed his famous conjecture in 1916 [**9**, p. 946]: If $f(z) = z + a_2 z^2 + a_3 z^3 + \cdots$ is analytic and univalent in the unit disk, then $|a_n| \leq n$ for all n, with equality occurring (only) for rotations of the "Koebe function"

$$k(z) = z(1-z)^{-2} = z + 2z^2 + 3z^3 + \cdots.$$

Its simple elegance attracted the efforts of many mathematicians over the course of 68 years, and they devised a number of different techniques to obtain partial solutions. The Bieberbach conjecture became one of the most famous unsolved problems of mathematics. Some of the people who contributed to its solution are represented in photographs on the frontispiece.

Riemann founded the field of geometric function theory around 1850 with his fundamental mapping theorem. Only after the turn of the century, however, did Koebe and others take the first steps toward a theory of univalent (=schlicht) functions. Koebe [44, 45, 46] formalized the study of normalized families of univalent functions (such as the class S) and obtained primitive forms of the basic distortion and covering theorems. At about the same time, Montel [63] introduced the concept of a normal family of analytic functions, and Carathéodory [14] established his important kernel theorem on sequences of univalent functions.

Bieberbach [9] proved his conjecture for $n = 2$, using the area principle which had just been established by T. H. Gronwall [29]. The inequality $|a_2| \leq 2$ led easily to sharp forms of Koebe's distortion and covering theorems.

Charles Loewner (Karl Löwner) [54] brought a new level of sophistication to the subject with his 1923 paper representing slit mappings in terms of a differential equation. The Carathéodory convergence theorem shows that the slit mappings are dense in S. As an application of his method, Loewner proved the Bieberbach conjecture for the third coefficient. It was this method, as interpreted by de Branges, which finally gave the full solution.

G. M. Goluzin, the founder of the present school of geometric function theory in Leningrad, refined Loewner's method and used it to obtain the sharp form of the rotation theorem, as well as the "Goluzin inequalities" on the values of a function $f \in S$ at prescribed points of the disk. His monograph [27] has been very influential.

Around the time of Loewner's work, Littlewood [51] considered integral means of functions in S and proved the uniform bound $|a_n| < en$. Some years later, Bazilevich [8] made a further study of these means and proved $|a_n| < en/2 + C$, where C is an absolute constant. Hayman [32, 33] showed that $\alpha = \lim |a_n|/n$ exists for each $f \in S$ and that $\alpha < 1$ unless f is a rotation of the Koebe function. This result does not follow from de Branges's theorem, nor does Hayman's related result [36] that $||a_{n+1}| - |a_n||$ is bounded by an absolute constant. In addition, Hayman's analysis applies to more general notions of univalence, where many other methods seem to fail. More recently, Baernstein [3] completed Littlewood's program by showing that $\int |f(re^{i\theta})|^p \, d\theta \leq \int |k(re^{i\theta})|^p \, d\theta$ for all $f \in S$ ($0 < r < 1$, $0 < p < \infty$), where k is the Koebe function.

Littlewood and Paley [52] proved that the coefficients of odd functions in S are bounded by a universal constant. They conjectured that the best bound is 1 because the square-root transform of the Koebe function is $z/(1-z^2) = 1 + z^3 + z^5 + \cdots$. However, Fekete and Szegö [20] promptly used Loewner's method to disprove this for a_5. Shortly afterwards, in 1936, M. S. Robertson [**70**] conjectured that the Littlewood-Paley conjecture holds on the average:

$$1 + |a_3|^2 + |a_5|^2 + \cdots + |a_{2n-1}|^2 \leq n$$

for all odd functions in S. He observed that this easily implies the Bieberbach conjecture. Robertson used Loewner's method to prove his conjecture for $n = 3$. Much later, Robertson [**71**] observed that his conjecture implies a conjecture of Rogosinski [**72**] on the coefficients of subordinate functions.

During the same period, Schiffer developed a calculus of variations for families of univalent functions. His method of boundary variation [**73, 74**], applicable to very general extremal problems, showed that any function in S which maximizes $|a_n|$ must map the disk onto the complement of a single analytic arc [**76**] which lies on a trajectory of a certain quadratic differential. The omitted arc was found to have monotonic modulus and other nice properties. Bieberbach's conjecture asserted that it must be a radial half-line. In 1955, Garabedian and Schiffer [**25**] finally succeeded in using this approach, in combination with Loewner's method, to prove $|a_4| \leq 4$. Garabedian discusses variational methods in his present article.

Beginning about 1950, Jenkins [**40, 41, 42, 43**] developed his general coefficient theorem, which has had notable success in solving a variety of extremal problems for univalent functions. Rooted in ideas of Grötzsch and Teichmüller, the method applies to domains slit along trajectories of quadratic differentials; its setting thus complements that of the variational method. Jenkins discusses one aspect of the method in his article.

While Schiffer was perfecting the variational method, Grunsky was working on generalizations of the area theorem. The Grunsky inequalities [**30**] give restrictions on the coefficients of

$$\log \frac{f(z) - f(\varsigma)}{z - \varsigma} = -\sum_{n=0}^{\infty} \sum_{m=0}^{\infty} c_{nm} z^n \varsigma^m$$

for $n, m \geq 1$ which are necessary and sufficient for the univalence of f. The Grunsky inequalities received relatively little attention until 1960, when Charzyński and Schiffer [**15**] used them to give an *elementary* proof of $|a_4| \leq 4$. This brought Grunsky's inequalities into the limelight. Pederson [**65**] and Ozawa [**64**] independently used them to prove $|a_6| \leq 6$. Garabedian and Schiffer [**26**] strengthened the Grunsky inequalities, and Pederson and Schiffer [**66**] used the stronger result to prove $|a_5| \leq 5$.

During this period there were many other applications of area methods to obtain extensions of the Grunsky inequalities and related results. The book of N. A. Lebedev [**48**] describes these developments.

The Grunsky inequalities essentially give information on the coefficients of the *logarithm* of a function $f \in S$. During the 1960's, I. M. Milin [**57, 58, 59, 60, 61, 62**] systematically developed a technique for exponentiating the Grunsky inequalities to get more direct access to the coefficients of f. One consequence was the general bound $|a_n| < 1.243\,n$. His method made important use of some general inequalities on the coefficients of exponentiated power series which he obtained jointly with Lebedev [**49, 59, 62**]. Milin considered the logarithmic coefficients of f, defined by

$$\log \frac{f(z)}{z} = 2 \sum_{n=1}^{\infty} \gamma_n z^n,$$

and obtained good estimates on their average growth. These results led him to conjecture that

$$\sum_{k=1}^{n} (n-k+1)k|\gamma_k|^2 \leq \sum_{k=1}^{n}(n-k+1)\frac{1}{k}$$

for all n. On the basis of one of the Lebedev-Milin inequalities, Milin's conjecture is easily seen to imply Robertson's conjecture, and so to imply the Bieberbach conjecture. Elsewhere in this volume, Milin describes the evolution of these ideas.

The Goluzin inequalities are easily derived from the Grunsky inequalities. In 1972, FitzGerald [**21**] found a way to exponentiate the Goluzin inequalities, removing the logarithms. This led him to a proof that $|a_n| < \sqrt{7/6}\,n < 1.081\,n$.

Finally came the work of de Branges [**13**]. Developing some of his earlier ideas [**10, 11, 12**], he used a variant of the Loewner method, together with a careful analysis of related composition operators, to prove Milin's conjecture. His argument depended on proving the monotonicity of a set of solutions to an initial-value problem. By a remarkable coincidence, the required result, equivalent to the positivity of certain sums of Jacobi polynomials, was already contained in a recent paper of Askey and Gasper [**2**]. The proof by de Branges was reformulated during his visit to Leningrad in collaboration with I. M. Milin, E. G. Emel'yanov, A. Z. Grinshpan, and others. FitzGerald and Pommerenke [**23**] then gave another version of the proof.

While the resolution of the Bieberbach conjecture removes the most famous unsolved problem on univalent functions, many interesting problems remain and the field continues to develop in many directions. Detailed accounts of work up to 1983 may be found in the books of Hayman [**33**], Jenkins [**41**], Goluzin [**27**], Milin [**62**], Lebedev [**48**], Pommerenke [**68**], Schober [**77**], Aleksandrov [**1**], Duren [**18**], and Goodman [**28**], and in the survey articles of Hayman [**37**], Duren [**17**], Pfluger [**67**], and Baernstein [**4**]. We wish to mention some very recent work.

First of all, there is the work of N. G. Makarov [**55**] on the relation between the harmonic measure and the Hausdorff measure of sets on a Jordan curve. Along the way, he found a remarkable "log log log" law on the radial growth of the derivative of a univalent function. This gives very precise bounds on the pointwise behavior (a.e.) of $|f'(re^{i\theta})|$, but there remain interesting open questions about the growth of the integral means $\int |f'(re^{i\theta})|^p\,d\theta$, especially for negative p. Makarov [**56**] used ergodic theory to disprove a previous conjecture that the Koebe function maximizes these integral means, at least in order of magnitude, for $p > \frac{1}{3}$. All of this is discussed in Pommerenke's article in this volume.

Hamilton [**31**] has discovered significant results on extreme points and support points of the class S. These are discussed in his article in this volume. The main idea is to consider linear functionals continuous on S but not necessarily continuous on the full space of analytic functions. Duren and Leung [**19**] present an alternate approach to Hamilton's construction of extreme points which are not support points, and they give additional information on generalized support points. Leung and Schober [**78, 50**] have studied support points of the class Σ and have found new evidence in favor of the conjecture that the omitted arcs have maximum branching.

Much recent effort has been devoted to the related study of *harmonic* mappings in the unit disk. Harmonic mappings have long been investigated by differential geometers in particular because of their role in the theory of minimal surfaces. The paper by Clunie and Sheil-Small [**16**] shows that despite many obstacles part of the classical (Koebe-Bieberbach) theory of analytic mappings does extend to harmonic mappings. More recent developments appear, for example, in the work of Hengartner and Schober [**38, 39**].

We mention also the ingenious work of J. L. Lewis [**53**] in which methods used in the study of free-boundary problems in partial differential equations are adapted to give a partial solution of A. W. Goodman's "minimum area" problem. Further information is given in the papers of Barnard [**5**] and Barnard and Lewis [**6**].

Finally, because of the role of special functions in de Branges's proof, there has been an active search for other connections of this type. For instance, Koornwinder [**47**] has used a stronger form of the Askey-Gasper inequality to generalize the Milin inequality. These developments are discussed in the articles of Askey and Gasper and of de Branges in this volume.

Less than six months' time separated the first planning for the Symposium and its realization. The organizing committee consisted of the four undersigned editors and Glenn Schober. Financial support was provided by the National Science Foundation, the Institute for Mathematics and its Applications (University of Minnesota), and Purdue University. We are grateful to the sponsors for their prompt and vigorous support, and to the staff of Purdue University for their hospitality. We thank all of the speakers for their enthusiastic response to an unusual challenge. The care and effort they devoted to their talks and manuscripts is clearly evident. Finally, we express our appreciation to the 170 mathematicians who joined us to celebrate the passing of one milestone in function theory, and to anticipate a promising future.

Albert Baernstein II   David Drasin
Peter Duren   Albert Marden

REFERENCES

1. I. A. Aleksandrov, *Parametric extensions in the theory of univalent functions*, Izdat. "Nauka", Moscow, 1976. (Russian)
2. R. Askey and G. Gasper, *Positive Jacobi polynomial sums.* II, Amer. J. Math. **98** (1976), 709–737.
3. A. Baernstein, *Integral means, univalent functions and circular symmetrization*, Acta Math. **133** (1974), 139–169.
4. ____, "Bieberbach's conjecture for tourists," in *Harmonic analysis (Minneapolis, Minn., 1981)*, Lecture Notes in Math., vol. 908, Springer-Verlag, 1982, pp. 48–73.
5. R. W. Barnard, "The omitted area problem for univalent functions," in *Topics in complex analysis*, Contemp. Math., vol. 38, Amer. Math. Soc., Providence, R. I., 1985, pp. 53–60.
6. R. W. Barnard and J. L. Lewis, *Note on the omitted area problem*, Michigan Math. J. (to appear).
7. I. E. Bazilevich, *On distortion theorems and coefficients of univalent functions*, Mat. Sb. **28(70)** (1951), 147–164. (Russian)
8. ____, *On distortion theorems in the theory of univalent functions*, Mat. Sb. **28(70)** (1951), 283–292. (Russian)
9. L. Bieberbach, *Über die Koeffizienten derjenigen Potenzreihen, welche eine schlichte Abbildung des Einheitskreises vermitteln*, S.-B. Preuss. Akad. Wiss. (1916), 940–955.
10. L. de Branges, *Coefficient estimates*, J. Math. Anal. Appl. **82** (1981), 420–450.
11. ____, *Grunsky spaces of analytic functions*, Bull. Sci. Math. **105** (1981), 401–406.
12. ____, *Löwner expansions*, J. Math. Anal. Appl. **100** (1984), 323–337.
13. ____, *A proof of the Bieberbach conjecture*, Acta Math. **154** (1985), 137–152.
14. C. Carathéodory, *Untersuchungen über die konformen Abbildungen von festen und veränderlichen Gebieten*, Math. Ann. **72** (1912), 107–144.
15. Z. Charzyński and M. Schiffer, *A new proof of the Bieberbach conjecture for the fourth coefficient*, Arch. Rational Mech. Anal. **5** (1960), 187–193.
16. J. Clunie and T. Sheil-Small, *Harmonic univalent functions*, Ann. Acad. Sci. Fenn. Ser. A I **9** (1984), 3–25.
17. P. L. Duren, *Coefficients of univalent functions*, Bull. Amer. Math. Soc. **83** (1977), 891–911.
18. ____, *Univalent functions*, Springer-Verlag, Heidelberg and New York, 1983.
19. P. L. Duren and Y. J. Leung, *Generalized support points of the set of univalent functions*, J. Analyse Math. (to appear).
20. M. Fekete and G. Szegö, *Eine Bemerkung über ungerade schlichte Funktionen*, J. London Math. Soc. **8** (1933), 85–89.
21. C. H. FitzGerald, *Quadratic inequalities and coefficient estimates for schlicht functions*, Arch. Rational Mech. Anal. **46** (1972), 356–368.
22. ____, *The Bieberbach conjecture: retrospective*, Notices Amer. Math. Soc. **32** (1985), 2–6.
23. C. H. FitzGerald and Ch. Pommerenke, *The de Branges theorem on univalent functions*, Trans. Amer. Math. Soc. **290** (1985), 683–690.
24. O. M. Fomenko and G. V. Kuz'mina, *The last 100 days of the Bieberbach conjecture*, Math. Intelligencer **8** (1986), No. 1, pp. 40–47.
25. P. R. Garabedian and M. Schiffer, *A proof of the Bieberbach conjecture for the fourth coefficient*, J. Rational Mech. Anal. **4** (1955), 427–465.
26. ____, *The local maximum theorem for the coefficients of univalent functions*, Arch. Rational Mech. Anal. **26** (1967), 1–32.
27. G. M. Goluzin, *Geometric theory of functions of a complex variable*, Gosudarst. Izdat., Moscow, 1952; German transl., Deutscher Verlag, Berlin, 1957; 2nd ed., Izdat. "Nauka", Moscow, 1966; English transl., Amer. Math. Soc., Providence, R. I., 1969.
28. A. W. Goodman, *Univalent functions*. Vols. I and II, Mariner Publishing Co., Tampa, Florida, 1983.
29. T. H. Gronwall, *Some remarks on conformal representation*, Ann. of Math. **16** (1914–1915), 72–76.

30. H. Grunsky, *Koeffizientenbedingungen für schlicht abbildende meromorphe Funktionen*, Math. Z. **45** (1939), 29–61.

31. D. H. Hamilton, *Extremal boundary problems for schlicht functions*, Proc. London Math. Soc. (to appear).

32. W. K. Hayman, *The asymptotic behaviour of p-valent functions*, Proc. London Math. Soc. **5** (1955), 257–284.

33. ____, *Multivalent functions*, Cambridge University Press, London and New York, 1958.

34. ____, *Bounds for the large coefficients of univalent functions*, Ann. Acad. Sci. Fenn. Ser. A I, no. 250 (1958), 13 pp.

35. ____, *On the coefficients of univalent functions*, Proc. Cambridge Philos. Soc. **55** (1959), 373–374.

36. ____, *On successive coefficients of univalent functions*, J. London Math. Soc. **38** (1963), 228–243.

37. ____, *Coefficient problems for univalent functions and related function classes*, J. London Math. Soc. **40** (1965), 385–406.

38. W. Hengartner and G. Schober, *Univalent harmonic mappings*, Trans. Amer. Math. Soc. (to appear).

39. ____, *Harmonic mappings with given dilatation*, J. London Math. Soc. (to appear).

40. J. A. Jenkins, *A general coefficient theorem*, Trans. Amer. Math. Soc. **77** (1954), 262–280.

41. ____, *Univalent functions and conformal mapping*, Springer-Verlag, Berlin, 1958.

42. ____, "On certain coefficients of univalent functions," in *Analytic functions*, Princeton Univ. Press, Princeton, N. J., 1960, pp. 159–194.

43. ____, *On certain coefficients of univalent functions*. II, Trans. Amer. Math. Soc. **96** (1960), 534–535.

44. P. Koebe, *Über die Uniformisierung beliebiger analytischer Kurven*, Nachr. Akad. Wiss. Göttingen Math.-Phys. Kl. (1907), 191–210.

45. ____, *Über die Uniformisierung der algebraischen Kurven durch automorphe Funktionen mit imaginärer Substitutionsgruppe*, Nachr. Akad. Wiss. Göttingen Math.-Phys. Kl. (1909), 68–76.

46. ____, *Ränderzuordnung bei konformer Abbildung*, Nachr. Königl. Ges. Wiss. Göttingen, Math.-Phys. Kl. (1913), 286–288.

47. T. Koornwinder, *Squares of Gegenbauer polynomials and Milin-type inequalities* (to appear).

48. N. A. Lebedev, *The area principle in the theory of univalent functions*, Izdat. "Nauka", Moscow, 1975. (Russian)

49. N. A. Lebedev and I. M. Milin, *An inequality*, Vestnik Leningrad. Univ. **20** (1965), No. 19, 157–158. (Russian)

50. Y. J. Leung and G. Schober, *On the structure of support points in the class* Σ, J. Analyse Math. (to appear).

51. J. E. Littlewood, *On inequalities in the theory of functions*, Proc. London Math. Soc. **23** (1925), 481–519.

52. J. E. Littlewood and R. E. A. C. Paley, *A proof that an odd schlicht function has bounded coefficients*, J. London Math. Soc. **7** (1932), 167–169.

53. J. L. Lewis, *On the minimum area problem*, Indiana Univ. Math. J. **34** (1985), 631–661.

54. C. Loewner (K. Löwner), *Untersuchungen über schlichte konforme Abbildungen des Einheitskreises*. I, Math. Ann. **89** (1923), 103–121.

55. N. G. Makarov, *On the distortion of boundary sets under conformal mappings*, Proc. London Math. Soc. **51** (1985), 369–384.

56. ____, *A note on integral means of the derivative in conformal mapping*, Proc. Amer. Math. Soc. **96** (1986), 233–236.

57. I. M. Milin, *The area method in the theory of univalent functions*, Dokl. Akad. Nauk SSSR **154** (1964), 264–267; English transl. in Soviet Math. Dokl. **5** (1964), 78–81.

58. ____, *Estimation of coefficients of univalent functions*, Dokl. Akad. Nauk SSSR **160** (1965), 769–771; English transl. in Soviet Math. Dokl. **6** (1965), 196–198.

59. ___, *On the coefficients of univalent functions*, Dokl. Akad. Nauk SSSR **176** (1967), 1015–1018; English transl. in Soviet Math. Dokl. **8** (1967), 1255–1258.

60. ___, *Adjacent coefficients of univalent functions*, Dokl. Akad. Nauk SSSR **180** (1968), 1294–1297; English transl. in Soviet Math. Dokl. **9** (1968), 762–765.

61. ___, *Hayman's regularity theorem for the coefficients of univalent functions*, Dokl. Akad. Nauk SSSR **192** (1970), 738–741; English transl. in Soviet Math. Dokl. **11** (1970), 724–728.

62. ___, *Univalent functions and orthonormal systems*, Izdat. "Nauka", Moscow, 1971; English transl., Amer. Math. Soc., Providence, R. I., 1977.

63. P. Montel, *Sur les suites infinies de fonctions*, Ann. Sci. École Norm. Sup. **24** (1907), 233–334.

64. M. Ozawa, *On the Bieberbach conjecture for the sixth coefficient*, Kōdai Math. Sem. Rep. **21** (1969), 97–128.

65. R. N. Pederson, *A proof of the Bieberbach conjecture for the sixth coefficient*, Arch. Rational Mech. Anal. **31** (1968), 331–351.

66. R. Pederson and M. Schiffer, *A proof of the Bieberbach conjecture for the fifth coefficient*, Arch. Rational Mech. Anal. **45** (1972), 161–193.

67. A. Pfluger, *Über die Koeffizienten schlichter Funktionen*, Bonner Math. Schriften, Sonderband Nr. 121 (Bonn, 1980), 41–61.

68. Ch. Pommerenke, *Univalent functions* (with a chapter on quadratic differentials by G. Jensen), Vandenhoeck and Ruprecht, Göttingen, 1975.

69. ___, *The Bieberbach conjecture*, Math. Intelligencer **7** (1985), No. 2, pp. 23–25; 32.

70. M. S. Robertson, *A remark on the odd schlicht functions*, Bull. Amer. Math. Soc. **42** (1936), 366–370.

71. ___, "Quasi-subordinate functions," in *Mathematical essays dedicated to A. J. Macintyre*, Ohio Univ. Press, Athens, Ohio, 1970, pp. 311–330.

72. W. Rogosinski, *On the coefficients of subordinate functions*, Proc. London Math. Soc. **48** (1943), 48–82.

73. M. Schiffer, *A method of variation within the family of simple functions*, Proc. London Math. Soc. **44** (1938), 432–449.

74. ___, *On the coefficients of simple functions*, Proc. London Math. Soc. **44** (1938), 450–452.

75. ___, *Variation of the Green function and theory of the p-valued functions*, Amer. J. Math. **65** (1943), 341–360.

76. ___, *On the coefficient problem for univalent functions*, Trans. Amer. Math. Soc. **134** (1968), 95–101.

77. G. Schober, *Univalent functions—selected topics*, Lecture Notes in Math., vol. 478, Springer-Verlag, 1975.

78. ___, "Some conjectures for the class Σ," in *Topics in complex analysis*, Contemp. Math., vol. 38, Amer. Math. Soc., Providence, R. I., 1985, pp. 13–21.

Contributors

LARS V. AHLFORS, Department of Mathematics, Harvard University, Cambridge, MA 02138

RICHARD ASKEY, Department of Mathematics, University of Wisconsin, Madison, WI 53706

ARNE BEURLING, School of Mathematics, Institute for Advanced Study, Princeton, NJ 08540

LOUIS DE BRANGES, Department of Mathematics, Purdue University, West Lafayette, IN 47907

JEAN DIEUDONNÉ, Académie des Sciences, Paris, France

WOLFGANG FUCHS, Department of Mathematics, Cornell University, Ithaca, NY 14853

PAUL R. GARABEDIAN, Courant Institute of Mathematical Sciences, New York University, New York, NY 10012

GEORGE GASPER, Department of Mathematics, Northwestern University, Evanston, IL 60201

WALTER GAUTSCHI, Department of Computer Science, Purdue University, West Lafayette, IN 47907

DAVID HAMILTON, Department of Mathematics, University of Maryland, College Park, MD 20742

JAMES A. JENKINS, Department of Mathematics, Washington University, St. Louis, MO 63130

PETER W. JONES, Department of Mathematics, Yale University, New Haven, CT 06520

CONTRIBUTORS

I. M. MILIN, Steklov Mathematical Institute (LOMI), 191011 Leningrad, U.S.S.R.

N. K. NIKOL′SKIĬ, Steklov Mathematical Institute (LOMI), 191011 Leningrad, U.S.S.R.

CHRISTIAN POMMERENKE, Technische Universität Berlin, Fachbereich Mathematik, 1 Berlin 12, Germany

DONALD SARASON, Department of Mathematics, University of California, Berkeley, CA 94720

STEPHEN W. SEMMES, Department of Mathematics, Yale University, New Haven, CT 06520

WILLIAM P. THURSTON, Department of Mathematics, Princeton University, Princeton, NJ 08544

V. I. VASYUNIN, Steklov Mathematical Institute (LOMI), 191011 Leningrad, U.S.S.R.

Classical Analysis: Present and Future

LARS V. AHLFORS

The National Science Foundation, in its wisdom, calls this group the Classical Analysts. This is a classification that we should accept with pride, and we should also be thankful for this opportunity to celebrate a momentous event that has recently taken place. It is perhaps a good reason to reflect on where we stand and where we are headed.

We live in exciting times. Great progress is made in all the sciences. In astronomy, the grandfather of them all, radio astronomy is not yet old, and satellite observatories scan the skies. The mind shudders when we read about dark holes or clouds of galaxies.

In physics the situation is the same. We read about new progress in Unified Field Theory, perhaps without knowing exactly how far it has advanced. There are yearly discoveries of new particles, accompanied by discoveries of new Nobel prize winners.

Biologists are splicing the genes and putting bacteria to work. In chemistry and medicine sensational reports follow each other at an accelerated rate. Closer to our own science, computers have revolutionized much of the commercial world, and supercomputers are being built. Finally, artificial intelligence may be a threat to our livelihood, but also a great promise for the future.

The excitement has spilled over to mathematics as well. In recent years several conjectures of long standing have suddenly been solved or at least partially solved. The four-color problem was solved by two mathematicians and a computer, a remarkable first in the history of mathematics. For a while all mail from a well-known university, or at least from its mathematics department, was stamped with the magic words: Four is enough. The human part of the team deserves most of the credit, but there are skeptics who are waiting for a less technological proof.

Startling mathematical discoveries have become so common that they are even reported in the daily press. Some time ago I learned from *The New York Times* that the Mordell conjecture in number theory had been proved by a young German. This came very close to proving Fermat's last theorem, a truly celebrated conjecture from 300 years ago which can be understood by any literate adult. It is typical that Faltings' proof had been preceded by extremely intense work by the world's leading algebraic geometers. This explains, but does not take away from, Faltings' success story.

Classical analysis has its share of famous conjectures. The Prime Number Theorem was one such conjecture, and it stirred up a great deal of interest when, near the end of the last century, it was proved simultaneously by Hadamard and de la Vallée-Poussin. Today its proof has become so simplified that it can be taught to advanced undergraduates.

The granddaddy of all conjectures is the Riemann hypothesis. It is a safe guess that most analysts have tried to prove it at least once. Many have come close and some have believed that they had a proof, but it has collapsed under closer scrutiny. Who knows, maybe it will be proved in the near future. If so, it would create a sensation among mathematicians, and it would rate a headline in *The New York Times*.

Classical analysis has spawned topology, and the Poincaré conjecture in topology is justly famous. I know that there has been much progress on this problem. I must confess that I don't even know the present status of the problem, but if it had been solved for the three-dimensional case I would undoubtedly know it.

It is time to pass to the topic of the day, the Bieberbach conjecture. Everybody in this room knows that it concerns the inequality $|a_n| \leq n$ for normalized univalent, or schlicht, functions, and that it has recently been proved by Louis de Branges of this university.

In many respects the story connected with this conjecture is most unusual. Bieberbach was a respected mathematician, but his famous conjecture which dates back to 1916 was based on very flimsy evidence, to say the least. In spite of the almost complete lack of significant clues, wishful thinking in favor of simple and elegant solutions is so strong that most mathematicians would almost automatically believe the conjecture to be true. Only a few skeptics have continued to look for counterexamples. de Branges has shown that the believers were right.

The most remarkable thing about the solution is that it was found by a known mature mathematician, not by an unknown precocious youngster. On second thought, it is perhaps not so astonishing. It is absolutely certain that the conjecture could not have been proved within a few years, neither by Bieberbach nor by anybody else. For one thing, the mathematical techniques were not nearly sufficient at the time. The fact is that an enormous amount of preliminary work had to be done before any single person could have arrived at a proof. As it is, it took 68 years of related research before the proof was found. Whether it could have been done in 50 years, or whether it would have taken 100 years without de Branges, is a question which cannot be answered. But it is hardly surprising that the solution had to wait for the computer age.

Nobody could or would like to take away from de Branges's triumph. After all, he is the one who will go down in history, or at least in the history of mathematics, as the one who proved Bieberbach's conjecture. On the other hand, it would not be fair to forget his predecessors who paved the way. Above all, one has to mention Karl Löwner who took a giant step beyond Bieberbach by proving $|a_3| \leq 3$. He did more than that by inventing a new variational method which made it possible to approximate all univalent functions by a subclass which could be expressed in semi-explicit form. It became much easier to come to grips with the problem, but at the time the known techniques were not strong enough to prove more than $|a_3| \leq 3$. In more recent times the method has been used successfully in various connections, and it still plays an important role, at least implicitly, in de Branges's proof.

The next step was the proof of $|a_4| \leq 4$, in 1955, by Garabedian and Schiffer. Their first proof was an incredibly difficult tour de force, which was later replaced by a much simpler proof using the Grunsky inequality (Charzynski and Schiffer). It was followed by proofs of $|a_6| \leq 6$ (Pederson and Ozawa) and $|a_5| \leq 5$ (Pederson and Schiffer). Thus, when de Branges entered the scene, the Bieberbach conjecture had been proved for $n \leq 6$. According to hearsay de Branges was encouraged to continue his search for a proof when a computer seemed to verify the conjecture up to $n = 26$.

All these early proofs had to overcome great difficulties. For this purpose the Grunsky-Goluzin inequalities were a very versatile tool. Although not directly involved in de Branges's proof, these inequalities have played a crucial role for a deeper understanding of all problems connected with univalent functions. There can be no doubt that they have served as an inspiration for the important work of Lebedev and Milin on the logarithmic coefficients of a univalent function, that is to say, the coefficients of $\log(f(z)/z)$. Through this work Milin was led to formulate what is known as the Milin conjecture, the proof of which would immediately imply the Bieberbach conjecture. Because of Milin, de Branges needed to prove only the Milin conjecture, and with hindsight one can see that Milin's conjecture was not only a great help on the way to Bieberbach, but actually an indispensable step.

de Branges's proof of Milin's conjecture is not easy, but still surprisingly simple in comparison with the enormous efforts that had been needed for the cases $n \leq 6$. In the popular press it was said that de Branges's chances were of the order of one in a million. Besides being an absurd use of probability, this is of course also pure nonsense. All mathematicians need a certain amount of luck, but de Branges did not succeed because he was lucky. He succeeded because he is a good mathematician, because he was perseverant, and because he had faith in special functions. This faith in the power of special functions is apparent already in many of his earlier publications.

Yes, he was lucky that Askey and Gasper had proved, for completely different purposes, a deep theorem on special functions which happened to fit in with de Branges's proof. Whether he or somebody else discovered this connection is irrelevant. What matters is that de Branges's proof serves as a strong reminder that mathematics is one; what is proved in one field of mathematics may be equally important in another.

To my knowledge, de Branges may be the first who has tapped the rich reservoir of knowledge hidden in the volumes on special functions and sometimes relegated to a corner of the library, and applied it to the coefficient problem. In this case, what he used was original work by contemporary mathematicians, but it raises the question whether special functions, as a powerful tool, have not been unduly neglected.

What has been the reaction to the fact that Bieberbach's conjecture has acquired the status of a theorem? Due to the somewhat unusual circumstances of the initial publication, the immediate reaction was not free from disbelief, but it was soon followed by complete acceptance and recognition of the author's remarkable feat. Some have lamented the passing of the Bieberbach conjecture, I hope more in jest than in earnest. In a way, the conjecture was a beacon which, when still far away, served as a target for the research. Now when the aim has been reached, a search for new directions may cause temporary difficulties, but it could also be an inspiration for entirely new ideas. It is understandable that some mathematicians may feel at a loss, because some of their papers have become partly obsolete. They should realize instead, as I have tried to emphasize, that their work has helped pave the way for the final solution. It can be predicted that in the near future many simplifications and other variants of the proof will see the light of day. Whatever their merits, it will not be forgotten that they would not be there if it were not for de Branges. More important, it seems quite likely that the leading ideas of the proof will find other applications of equal value.

The title of my talk implies a foolish promise to say something about the future of classical analysis. As a warning I cannot resist quoting my favorite *New York Times* book critic who chides an author for his "clumsy attempts to portentously foreshadow the future." I like the phrase, but I doubt its accuracy as a general principle. The present, and especially the past, do indeed foreshadow the future, although it is necessary to exercise great caution when making predictions. For better or for worse, present and past are the only available facts for assessing the future.

My unaided memory goes back to about 1930. It does not include the original announcement of the Bieberbach conjecture, but it covers many of its reverberations and, above all, its impressions on a young mathematician. When looking back, the most striking observation is that mathematics has grown exponentially. The number of mathematicians and the volume of the periodic literature are now at least tenfold of what they were before 1930. It can be predicted with absolute certainty that this spectacular growth will continue to accelerate. But volume alone is a poor measure. Whether the quality has also grown or will continue to grow is a difficult question. My personal feeling is that potentially first-rate mathematicians have a better chance to develop now than fifty years ago, but I have no statistics to base it on.

A safer basis for a prognosis is to look for trends that are likely to continue. This is a very subjective business, and I do not expect everybody to agree. To me the greatest change that has taken place in my own special field, the one that in Germany used to be called simply "Funktionentheorie," is its increasing geometrization. To be sure, Klein and Poincaré had already introduced a geometric element by their use of noneuclidean and projective geometry. This was the classical use of geometry. The picture changed radically with the introduction of quasiconformal mappings as part of function theory.

I remember vividly the first time I heard about quasiconformal mappings. It was in 1930. I was visiting Carathéodory in Munich, where I also met a young Frenchman by the name of de Possel. During a conversation he asked if we had heard of a new kind of extremal problem considered by somebody in Leipzig. He was referring to Grötzsch, a student of Koebe's, and a problem now known as Grötzsch's box problem. I had never heard of Grötzsch, although it turned out that our work had much in common. The problem is ridiculously simple, almost a joke, and nobody had an idea that it would mushroom to become a fruitful part of the well-established theory of conformal mapping. Even the name did not exist. Actually, the theory developed from two different seeds, one planted by Grötzsch and one by Lavrentiev in the Soviet Union who recognized its importance for elliptic differential equations.

To tell the truth, the idea caught on very slowly. This was partly due to the personality of Grötzsch, but later on also to the political turmoil in Germany and finally to the total lack of communications during the war. Ironically, because much was published in the antisemitic *Deutsche Mathematik*, many papers are still difficult to find.

Quasiconformal mappings enter the theory of classical function theory in three ways:

(i) As a simplification of proofs. This was the least important way, although historically it was the first to demonstrate their usefulness.

(ii) As a generalization of conformal mappings which was interesting in its own right. It was also important because it could be generalized to several dimensions in a way that differs completely from holomorphic functions of several complex variables.

(iii) Through the problem of extremal quasiconformal mappings, which leads to a beautiful and unexpected connection with a class of holomorphic quadratic differentials on Riemann surfaces.

This third connection is based on a brilliant discovery by Teichmüller. It forms the core of his famous paper of 1939, full of proven and unproven conjectures, ideas, and theorems, difficult to separate from each other, but all highly unconventional. Whatever feelings the name of Teichmüller evokes, the depth of his ideas cannot be disputed. It spawned the theory of Teichmüller Spaces, which is still pursued by an active group of devoted mathematicians.

The geometrization of function theory found other manifestations as well. It was realized that the relations between conformal mappings and differential geometry, although known to exist, had been badly neglected. In particular, it became clear that the whole class of Riemannian metrics conformal with the euclidean metric could convey a great deal of information by very simple means. This development culminated in Beurling's theory of extremal length, which to this day remains one of the most powerful tools, not only in traditional function theory, but also for the study of multidimensional quasiconformal and quasiregular mappings.

I realize that I have talked too much about events that started long ago. My obvious excuse is that I was personally more involved at that time. From a more detached point of view, the fact is that the geometric approach is still going strong, as seen for instance from the continued interest in univalent functions, which will go on despite the triumph over Bieberbach's conjecture. Among recent additions to the arsenal of classical analysis, I should not fail to mention the direct and indirect influence of functional analysis, which is visible in the lively interest in such varied topics as the corona problem, the fascinating properties hidden behind the letters BMO, and the recent prevalence of Carleson measures. The progress in these topics interests me greatly, but I find it wiser to let my younger colleagues worry about their future implications.

All told, and this is the main point I want to make, there is ample evidence that classical analysis draws increasingly on neighboring branches of mathematics. de Branges's proof is itself a convincing example, and I feel confident that this healthy trend will continue.

I was supposed to speak about the future, and here I have been talking mostly about the past. My approach has been to look upon the future as history in reverse. An acceleration of the pace is inevitable, but barring catastrophes the human mind does not change suddenly and mathematics will probably go on essentially as before. The future is not a mirror image of the past, but I like to think of it as a mirage. The Bieberbach conjecture was a mirage, but it has become solid reality. Who knows when the next mirage will become an oasis of heavenly delight?

Inequalities for Polynomials

RICHARD ASKEY AND GEORGE GASPER

ABSTRACT. The polynomial inequality that de Branges used in his solution of the Bieberbach conjecture is equivalent to the positivity of the sum of certain Jacobi polynomials. There are many other positive sums of Jacobi polynomials, including some that are fractional derivatives of order one-half of the ones de Branges used. A historical summary of some of these inequalities is given, Koornwinder's extension of the de Branges-Milin type inequality is outlined as an example of a refinement suggested by inequalities for hypergeometric functions, and a few places to look for other refinements are mentioned.

1. Introduction and early trigonometric polynomial inequalities. de Branges's solution of the Bieberbach conjecture was a surprise to both of us. Askey's comments on his disbelief and surprise are given elsewhere [7], so they do not need to be repeated. However de Branges's use of a polynomial inequality of the type he used [18] is not without precedent in the study of special classes of univalent functions. Some of the earlier inequalities will be summarized, since the inequality de Branges used was discovered while trying to understand and generalize some of the older inequalities.

The first inequality of the type we will consider is due to Fejér:

$$(1.1) \quad \sum_{k=0}^{n} \sin\left(k + \frac{1}{2}\right)\theta = [1 - \cos(n+1)\theta]/2\sin\frac{\theta}{2} \geq 0, \qquad 0 \leq \theta \leq \pi.$$

This is an extension of the obvious fact that

$$(1.2) \quad \int_0^x \sin t \, dt = 1 - \cos x \geq 0.$$

Fejér was led to (1.1) by the desire of having a positive sum that comes from the formal delta function

$$(1.3) \quad 1 + 2\sum_{n=1}^{\infty} \cos n\theta.$$

Askey's work was supported in part by NSF grant MCS 840071, and Gasper's by NSF grant DMS 8403566.

The earliest positive approximation to (1.3) is

$$(1.4) \quad 1 + 2\sum_{n=1}^{\infty} r^n \cos n\theta = \frac{1-r^2}{1 - 2r\cos\theta + r^2} > 0, \quad -1 < r < 1.$$

This sum is usually attributed to Poisson. It was also found (but not published) by Gauss [46]. For some applications it is useful to have a polynomial approximation of (1.3). Fejér's polynomial [24]

$$(1.5) \quad 1 + 2\sum_{k=1}^{n} \left(1 - \frac{k}{n+1}\right) \cos k\theta$$

is probably the first known polynomial approximation of (1.3) which is nonnegative for $n = 0, 1, \ldots$, $0 \le \theta \le \pi$. Fejér [24] used a sum equivalent to (1.1) and

$$(1.6) \quad 1 + 2\sum_{k=1}^{n} \cos k\theta = \frac{\sin(n+\frac{1}{2})\theta}{\sin(\theta/2)} = \frac{\cos n\theta - \cos(n+1)\theta}{1 - \cos\theta}$$

to prove the nonnegativity of (1.5).

Another natural extension of the integral in (1.2) is the sum

$$(1.7) \quad g_n(\theta) = \sum_{k=1}^{n} \sin k\theta.$$

Since

$$g_n(\theta) = \left[\cos\frac{\theta}{2} - \cos\left(n + \frac{1}{2}\right)\theta\right] \Big/ 2\sin\frac{\theta}{2},$$

this function changes sign on any subinterval of $[0, \pi]$ for large n. There are a number of ways to modify (1.7) to obtain a positive sum. One easy way is to observe that

$$(1.8) \quad g_n(\theta) + g_{n-1}(\theta) = \left[2\cos\frac{\theta}{2}\right] \sum_{k=1}^{n} \sin\left(k - \frac{1}{2}\right)\theta = 2\sum_{k=1}^{n-1} \sin k\theta + \sin n\theta,$$

and this is nonnegative for $0 \le \theta \le \pi$ because of (1.1). A second way is to consider

$$\sum_{k=1}^{n} g_k(\theta) = \sum_{k=1}^{n} (n+1-k)\sin k\theta.$$

The inequality

$$(1.9) \quad \sum_{k=1}^{n} (n+1-k)\sin k\theta > 0, \quad 0 < \theta < \pi,$$

was found by Lukács, communicated to Fejér, and published in [28, Satz 23]. Fejér wrote that Lukács's proof was by direct calculation. Fejér gave a more sophisticated proof which will be mentioned in §2.

A third change of (1.7) was discovered by Fejér. He considered the Fourier series

$$\frac{\pi - \theta}{2} = \sum_{k=1}^{\infty} \frac{\sin k\theta}{k}, \qquad 0 < \theta < \pi,$$

in a number of papers. For example, in [**26**] he showed that

$$\left|\sum_{k=1}^{n} \frac{\sin k\theta}{k}\right| \leq \frac{\pi}{2} + 1.$$

This series has been studied in connection with Gibbs' phenomenon, and graphs of the partial sums have been published. See [**49**] for a historical survey. Gibbs' phenomenon is of interest near $\theta = 0$, and the graphs suggest that the partial sums are positive for $0 < \theta < \pi$. The interesting points is $\theta = \pi$. In 1910 Fejér conjectured that

$$(1.10) \qquad \sum_{k=1}^{n} \frac{\sin k\theta}{k} > 0, \qquad 0 < \theta < \pi.$$

See [**28**, §22]. This conjecture was proven by Jackson [**51**] and Gronwall [**48**]. While there have been few applications of (1.10), it is important because of the methods developed to prove it and even more for extensions that were suggested by it.

Shortly after the proofs of Jackson and Gronwall, Alexander [**3**] proved the following closely related result.

THEOREM 1.1. *If $a_1 \geq \cdots \geq a_n > 0$, then*

$$\sum_{k=1}^{n} \frac{a_k z^k}{k}$$

is univalent for $|z| < 1$.

The connection between Theorem 1.1 and (1.10) was made explicit in a later result of Dieudonné [**21**].

THEOREM 1.2. *The polynomial*

$$f(z) = \sum_{k=1}^{n} a_k z^k$$

is univalent for $|z| < 1$ if and only if

$$(1.11) \qquad \sum_{k=1}^{n} \frac{a_k z^{k-1} \sin k\theta}{\sin \theta} \neq 0, \qquad |z| < 1, \ 0 \leq \theta \leq \pi.$$

When $a_k = k^{-1}$ in (1.11), Theorems 1.1 and 1.2 give

$$(1.12) \qquad \sum_{k=1}^{n} \frac{z^{k-1} \sin k\theta}{k \sin \theta} \neq 0, \qquad |z| < 1, \ 0 \leq \theta \leq \pi.$$

Then for z real, letting $z \to 1$ in (1.12) gives
$$\sum_{k=1}^{n} \frac{\sin k\theta}{k} \geq 0, \quad 0 \leq \theta \leq \pi,\ n = 1, 2, \ldots.$$

One early application of (1.1) and (1.5) was found by W. H. Young [74] and rediscovered by Fejér [27].

THEOREM 1.3. *If $a_n \to 0$ and $\Delta^2 a_n := a_n - 2a_{n+1} + a_{n+2} \geq 0$, then*
$$f(\theta) = \frac{a_0}{2} + \sum_{n=1}^{\infty} a_n \cos n\theta$$
converges for $0 < \theta \leq \pi$ to a continuous function on $(0, \pi]$ and $f(\theta) \geq 0$, $0 < \theta \leq \pi$.

The proof of Theorem 1.3 uses a couple of summations by parts.

The nonnegativity of Fejér's polynomial was used by Sheil-Small [65] to give short proofs of Bernstein's inequality
$$|t'_n(\theta)| \leq n\|t_n\|_\infty, \quad -\infty < \theta < \infty,$$
where $t_n(\theta)$ is a trigonometric polynomial of degree n, of Szegö's inequality [66]
$$\max_{|z|\leq 1} |p'_n(z)| \leq n \max_{|z|\leq 1} |\operatorname{Re} p_n(z)|,$$
where $p_n(z)$ is a polynomial of degree n, and of some related inequalities.

Young [73] obtained a result for cosine series that is analogous to (1.10). He proved that

(1.13) $$1 + \sum_{k=1}^{n} \frac{\cos k\theta}{k} \geq 0, \quad 0 \leq \theta \leq \pi.$$

Rogosinski and Szegö [60] extended this to

(1.14) $$\frac{1}{1+\alpha} + \sum_{k=1}^{n} \frac{\cos k\theta}{k+\alpha} \geq 0, \quad 0 \leq \theta \leq \pi,\ -1 < \alpha \leq 1.$$

Gasper [40] showed that (1.14) continues to hold for $-1 < \alpha \leq \bar{\alpha} \sim 4.567$, where $\bar{\alpha}$ is the positive zero of a seventh-degree polynomial, and that this is best possible.

It is natural to look for sine series results like Theorem 1.3. Two obvious ones follow from (1.1) and (1.9) and summation by parts once and twice respectively. Fejér found two deeper results which are more important.

In [29] Fejér proved that

(1.15) $$\sum_{k=0}^{n} \frac{(4)_{n-k}}{(n-k)!}(k+1)\sin(k+1)\theta > 0, \quad 0 < \theta < \pi,$$

and in [30] he proved that

(1.16) $$\sum_{k=0}^{n} \frac{(3)_{n-k}}{(n-k)!}\left(k+\frac{1}{2}\right)\sin\left(k+\frac{1}{2}\right)\theta > 0, \quad 0 < \theta < \pi.$$

The shifted factorial $(a)_k$ is defined by
$$(a)_k = a(a+1)\cdots(a+k-1) = \Gamma(k+a)/\Gamma(a).$$
These imply the following theorem.

THEOREM 1.4. *If $a_k \to 0$ and $\Delta a_k = a_k - a_{k+1} \geq 0$, $\Delta^2 a_k \geq 0$ and $\Delta^3 a_k = \Delta(\Delta^2 a_k) \geq 0$, then*
$$f_r(\theta) = \sum_{k=0}^{\infty} a_k r^k \left(k + \frac{1}{2}\right) \sin\left(k + \frac{1}{2}\right)\theta \geq 0, \qquad 0 \leq \theta \leq \pi,\ 0 \leq r < 1,$$
and there is a function $f(\theta)$ with
$$\lim_{r \to 1} f_r(\theta) = f(\theta) \geq 0, \qquad 0 \leq \theta \leq \pi,$$
and
$$\int_0^\pi f(\theta) \sin\frac{\theta}{2}\,d\theta < \infty.$$
If, in addition, $\Delta^4 a_k = \Delta(\Delta^3 a_k) \geq 0$, then a similar result holds for
$$g_r(\theta) = \sum_{k=0}^{\infty} a_k r^k (k+1) \sin(k+1)\theta,$$
and now
$$\int_0^\pi g(\theta) \sin\theta\,d\theta < \infty.$$

Fejér [31, §IX] used (1.15) to prove that
$$f(z) = z + \sum_{n=2}^{\infty} a_n z^n$$
is univalent for $|z| < 1$ if $\Delta^j a_n \geq 0$, $n = 0, 1, \ldots$, when $j = 0, 1, 2, 3, 4$. Szegö [68] reduced these conditions to $\{a_n\}$ being triply monotonic, i.e., $\Delta^j a_n \geq 0$, $j = 0, 1, 2, 3$. These conditions cannot be replaced by similar inequalities when $j < 3$ (see [36]). Robertson [57] proved an equivalent form of the inequality

(1.17) $$\sum_{k=0}^{n} \frac{(3)_{n-k}}{(n-k)!} (k+1) \sin\left(k + \frac{1}{2}\right)\theta > 0, \qquad 0 < \theta < \pi,$$

and used it to improve Szegö's result. He showed that $f(z)$ is close-to-convex in $|z| < 1$ relative to $-\log(1-z)$ when its coefficients are triply monotonic. The inequality (1.17) follows immediately from (1.16) and (1.1). Fuchs [37] proved an inequality equivalent to (1.17) and used it to show that if $\{a_n\}$ is triply monotonic, then the function
$$f_k(z) = z^k \sum_{n=0}^{\infty} a_n z^n$$
is k-valent for $k = 1, 2, \ldots$, and
$$\operatorname{Re}\left(\frac{1-z}{z^{k-1}} f_k'(z)\right) > \frac{a}{2}, \qquad a = \lim_{n \to \infty} a_n.$$

To try to understand these inequalities better, observe that inequality (1.16) implies

(1.18) $$\int_0^x (x-t)^2 t \sin t \, dt \geq 0, \qquad x > 0,$$

and (1.5) implies

(1.19) $$\int_0^x (x-t) \cos t \, dt \geq 0, \qquad x > 0.$$

Williamson [72] proved (1.18) and used it to give a short proof of Royall's result [61] that the Laplace transform $f(s)$ of a three-times monotone function is univalent for $\mathrm{Re}(s) > 0$.

In the three integral inequalities (1.2), (1.18), and (1.19) there is a trigonometric function as part of the integrand, and the oscillations of this function are uniform. In (1.2) each arch has the same area, and the first arch is positive, so (1.2) is obvious on geometric grounds. In (1.19) the first arch is only half as large as each of the other arches, so it must be counted more. One way to do this is to integrate a second time, or more often if necessary. In this case

$$\int_0^x \int_0^s \cos t \, dt \, ds = \int_0^x (x-t) \cos t \, dt$$

gives a nonnegative function, since

$$\int_0^x \int_0^s \cos t \, dt \, ds = \int_0^x \sin s \, ds = 1 - \cos x.$$

In the case of (1.18) the oscillating function $\sin t$ is multiplied by the increasing function t, so it may not be possible to integrate sufficiently many times to obtain a nonnegative function. It is in this case, and (1.18) is best possible in the sense that

(1.20) $$\int_0^x (x-t)^a t \sin t \, dt \geq 0, \qquad x \geq 0,$$

holds for $a \geq 2$ but not for any $a < 2$ (see [44, Theorem 8]). It is also best possible in the sense that

(1.21) $$\int_0^x (x-t)^a t^{1+\varepsilon} \sin t \, dt \geq 0, \qquad x \geq 0,$$

fails for all $a > 0$ when $\varepsilon > 0$. To see this, divide by x^a, replace t by yt, and set $a = cx/y$. Then

(1.22) $$\int_0^\infty e^{-ct} t^{1+\varepsilon} \sin yt \, dt$$

is the limit as $x \to \infty$. If (1.21) held, then the function in (1.22) would be nonnegative for $y > 0$ and all $c > 0$. But that is not true (see [4]).

It seems clear that one wants inequalities to be best possible and often best possible in a number of ways. Many of the above inequalities are best possible in a number of ways. For example, the 3 in (1.16) cannot be replaced by c with

$c < 3$, and the $k + \frac{1}{2}$ cannot be replaced by a function that grows like k^α for any $\alpha > 1$ or replaced by $k + a$ with $a < \frac{1}{2}$, even if the 3 is replaced by c, no matter how large c is. See the argument below. The same is true of (1.15), but here 4 can be replaced by 3 if the inequality is only required to hold for $0 < \theta < \theta_0$. The largest θ_0 that works is $2\pi/3$ (see Schweitzer [64]). If $(k+1)\sin(k+1)\theta$ in (1.15) is replaced by $(k+2)\sin(k+1)\theta$, then 4 can be replaced by 3 and still have an inequality which holds for $0 < \theta < \pi$ (see [4]).

So far we have just dealt with iterated partial sums. It will be very useful to consider fractional sums. Given a series

$$\sum_{n=0}^{\infty} a_n, \tag{1.23}$$

form the generating function

$$f(r) = \sum_{n=0}^{\infty} a_n r^n. \tag{1.24}$$

In all our examples, $a_n = O(n^\alpha)$ for some α, so (1.24) converges for $|r| < 1$.

From the binomial theorem we have

$$(1-r)^{-\alpha} f(r) = \sum_{n=0}^{\infty} \left[\sum_{k=0}^{n} \frac{(\alpha)_{n-k}}{(n-k)!} a_k \right] r^n.$$

When $\alpha = 1$ the coefficient of r^n is a partial sum of (1.23). The sums (1.15) and (1.16) arise when $\alpha = 4$ and $\alpha = 3$ respectively. A necessary condition that

$$\sum_{k=0}^{n} \frac{(\alpha)_{n-k}}{(n-k)!} a_k \geq 0, \quad n = 0, 1, \ldots,$$

is that $f(r) \geq 0$, $0 \leq r < 1$. For the series (1.15) the generating function is

$$\sum_{n=0}^{\infty} r^n (n+1) \sin(n+1)\theta = \frac{(1-r^2)\sin\theta}{(1 - 2r\cos\theta + r^2)^2},$$

and this is nonnegative for $0 \leq r < 1$, but approaches 0 when $r \to 1$, $0 < \theta \leq \pi$. This is the real reason $n+1$ cannot be replaced by n^α, $\alpha > 1$, or $n + a$, $a < 1$. Also

$$\sum_{n=0}^{\infty} r^n \sum_{k=0}^{n} \frac{(c)_{n-k}}{(n-k)!} (k+1) \frac{\sin(k+1)\theta}{\sin\theta} = \frac{1}{(1-r)^c} \cdot \frac{(1-r^2)}{(1 - 2r\cos\theta + r^2)^2},$$

and when $\theta = \pi$ this becomes

$$\frac{1}{(1-r)^{c-1}} \cdot \frac{1}{(1+r)^3}.$$

When $c < 4$ the coefficient of r^n will be negative when n is one. When $c = 4$ the coefficient of r^n is zero when n is odd. This is another way in which (1.15) is best possible. The same is true for (1.1), (1.10), and (1.16), but not (1.13). However for (1.10) there is a refinement due to Vietoris [71] which shows that this particular "best possible" can be improved. See [11] for another proof of Vietoris's inequalities and a general problem that leads to them, and [19] for some interesting refinements.

2. Inequalities for Jacobi polynomial sums. To really understand the inequalities of Fejér and others mentioned in the first section we need to introduce Jacobi polynomials, $P_n^{(\alpha,\beta)}(x)$. These polynomials can be defined by the Rodrigues formula:

$$(2.1) \quad (1-x)^\alpha (1+x)^\beta P_n^{(\alpha,\beta)}(x) = \frac{(-1)^n}{2^n n!} \frac{d^n}{dx^n}[(1-x)^{n+\alpha}(1+x)^{n+\beta}],$$

or by the hypergeometric representation:

$$(2.2) \quad P_n^{(\alpha,\beta)}(x) = \frac{(\alpha+1)_n}{n!} {}_2F_1\left(\begin{array}{c} -n,\ n+\alpha+\beta+1 \\ \alpha+1 \end{array}; \frac{1-x}{2}\right).$$

The generalized hypergeometric series is defined by

$$(2.3) \quad {}_pF_q\left(\begin{array}{c} a_1,\ldots,a_p \\ b_1,\ldots,b_q \end{array}; x\right) = \sum_{n=0}^\infty \frac{(a_1)_n \cdots (a_p)_n}{(b_1)_n \cdots (b_q)_n} \frac{x^n}{n!}.$$

Jacobi polynomials are orthogonal with respect to the beta distribution, put on $[-1,1]$ to give symmetry about $x=0$ when $\alpha=\beta$. The orthogonality is

$$(2.4) \quad \int_{-1}^1 P_m^{(\alpha,\beta)}(x) P_n^{(\alpha,\beta)}(x)(1-x)^\alpha(1+x)^\beta\,dx = \begin{cases} 0, & m \neq n,\ \alpha,\beta > -1, \\ h_n^{(\alpha,\beta)}, & m = n. \end{cases}$$

They contain the functions of the first section, since

$$(2.5) \quad \frac{P_n^{(-1/2,-1/2)}(\cos\theta)}{P_n^{(-1/2,-1/2)}(1)} = \cos n\theta,$$

$$(2.6) \quad \frac{P_n^{(1/2,1/2)}(\cos\theta)}{P_n^{(1/2,1/2)}(1)} = \frac{\sin(n+1)\theta}{(n+1)\sin\theta},$$

and

$$(2.7) \quad \frac{P_n^{(1/2,-1/2)}(\cos\theta)}{P_n^{(1/2,-1/2)}(1)} = \frac{\sin(n+\frac{1}{2})\theta}{(2n+1)\sin(\theta/2)}.$$

Since

$$(2.8) \quad P_n^{(\alpha,\beta)}(-x) = (-1)^n P_n^{(\beta,\alpha)}(x), \quad \text{(from (2.1))},$$

formula (2.7) can be rewritten as

$$(2.9) \quad \frac{P_n^{(1/2,-1/2)}(\cos\theta)}{P_n^{(-1/2,1/2)}(1)} = \frac{\sin(n+\frac{1}{2})\theta}{\sin(\theta/2)}.$$

The first positive sum of the type in §1 that involves a different special case of Jacobi polynomials is Fejér's sum

$$(2.10) \qquad \sum_{k=0}^{n} P_k(x) > 0, \qquad -1 < x \leq 1,$$

where the Legendre polynomial $P_n(x)$ is the special case $\alpha = \beta = 0$ of $P_n^{(\alpha,\beta)}(x)$. Fejér [25] obtained (2.10) from (1.1) via a formula of Mehler:

$$(2.11) \qquad P_n(\cos\theta) = \frac{2}{\pi} \int_\theta^\pi \frac{\sin(n+\tfrac{1}{2})\varphi\, d\varphi}{[2\cos\theta - 2\cos\varphi]^{1/2}}.$$

Both the result and the proof are important.

All formulas for Jacobi polynomials which are stated without reference can be found in [69, Chapter 4].

The next result is a theorem of Kogbetliantz [52]. His proof does not seem to be complete, but simple proofs now exist. Kogbetliantz used a different normalization since he only considered Jacobi polynomials when $\alpha = \beta$. He used

$$C_n^\lambda(x) = \frac{(2\lambda)_n}{(\lambda+\tfrac{1}{2})_n} P_n^{(\lambda-1/2,\lambda-1/2)}(x)$$

and the generating function

$$(2.12) \qquad \sum_{n=0}^{\infty} C_n^\lambda(x) r^n = (1 - 2xr + r^2)^{-\lambda}.$$

However we will continue to use Jacobi polynomials and their standard normalization since they are necessary for the inequalities de Branges used. A slight variant of (2.12) is

$$\frac{1-r^2}{(1-2xr+r^2)^{\lambda+1}} = \sum_{k=0}^{\infty} \frac{(k+\lambda)(2\lambda)_k P_k^{(\lambda-1/2,\lambda-1/2)}(x)}{\lambda k! P_k^{(\lambda-1/2,\lambda-1/2)}(1)} r^k.$$

Kogbetliantz showed that if $\lambda > 0$, then

$$(2.13) \qquad \sum_{k=0}^{n} \frac{(2\lambda+2)_{n-k}(2\lambda)_k(k+\lambda)P_k^{(\lambda-1/2,\lambda-1/2)}(x)}{(n-k)!k!\lambda P_k^{(\lambda-1/2,\lambda-1/2)}(1)} > 0, \qquad -1 < x \leq 1.$$

When $\lambda \to 0$, (2.13) reduces to a multiple of (1.5), and when $\lambda = 1$, it becomes (1.15). Inequality (2.13) is best possible in a number of ways. When $x = -1$ and n is odd, the sum vanishes. The factor $(2\lambda+2)_{n-k}$ cannot be replaced by $(a)_{n-k}$ for any $a < 2\lambda+2$, for the series is then negative when $x = -1$ and n is odd.

Szegö [67] generalized (2.10) to

$$(2.14) \qquad \sum_{k=0}^{n} P_k(x) z^k \neq 0, \qquad -1 < x < 1, \ |z| \leq 1, \ n = 0, 1, \ldots.$$

This is analogous to (1.12).

Turán [**70**] found a proof of (1.10) which is very important. He proved

THEOREM 2.1. *If*

$$\sum_{k=0}^{\infty} a_k \sin\left(k + \frac{1}{2}\right)\theta \geq 0, \qquad 0 \leq \theta \leq \pi,$$

then

$$\sum_{k=0}^{\infty} \frac{a_k \sin(k+1)\varphi}{k+1} > 0, \qquad 0 < \varphi < \pi,$$

unless $a_k = 0$, $k = 0, 1, \ldots$.

Thus (1.1) implies (1.10). One of the first people to appreciate the importance of Fejér's proof of (2.10) was Feldheim. In a letter to Fejér dated March 12, 1944, he showed that

$$(2.15) \quad \frac{P_n^{(\alpha,\alpha)}(\cos\theta)}{P_n^{(\alpha,\alpha)}(1)} = \frac{2\Gamma(\alpha+1)}{\Gamma(\alpha-\beta)\Gamma(\beta+1)} \int_0^{\pi/2} \sin^{2\beta+1}\varphi \cos^{2\alpha-2\beta-1}\varphi$$

$$\times (1 - \sin^2\theta\cos^2\varphi)^{n/2} \frac{P_n^{(\beta,\beta)}(\cos\theta(1-\sin^2\theta\cos^2\varphi)^{-1/2})}{P_n^{(\beta,\beta)}(1)} d\varphi,$$

$\alpha > \beta > -1$, $0 \leq \theta \leq \pi$. This was published much later [**34**].

Bailey [**14**] showed that

$$(2.16) \quad \sum_{n=0}^{\infty} [h_n^{(\alpha,\beta)}]^{-1} P_n^{(\alpha,\beta)}(\cos 2\theta) P_n^{(\alpha,\beta)}(\cos 2\varphi) r^n$$

$$= \frac{\Gamma(\alpha+\beta+2)(1-r)2^{-\alpha-\beta-1}}{\Gamma(\alpha+1)\Gamma(\beta+1)(1+r)^{\alpha+\beta+2}} F\left(\begin{array}{c}\frac{\alpha+\beta+2}{2}, \frac{\alpha+\beta+3}{2}; \frac{a^2}{K^2}, \frac{b^2}{K^2}\\ -; \alpha+1; \beta+1\end{array}\right),$$

where

$$a = \sin\theta\sin\varphi, \qquad b = \cos\theta\cos\varphi, \qquad K = (r^{1/2} + r^{-1/2})/2,$$

$$F\left(\begin{array}{c}a_1,\ldots,a_k; b_1,\ldots,b_j; c_1,\ldots,c_l\\ d_1,\ldots,d_p; e_1,\ldots,e_q; f_1,\ldots,f_r\end{array}; x,y\right) = \sum_{m,n \geq 0} \frac{(a_i)_{m+n}(b_i)_m(c_i)_n}{(d_i)_{m+n}(e_i)_m(f_i)_n} \frac{x^m y^n}{m!n!},$$

and $(a_i)_{m+n} = (a_1)_{m+n}\cdots(a_k)_{m+n}$, etc. The right-hand side of (2.16) is obviously positive when $\alpha, \beta > -1$. Feldheim's integral (2.15) and the positivity of the Poisson kernel (2.16) combine to give

THEOREM 2.2. *If* $\alpha > \beta > -1$ *and*

$$\sum_{n=0}^{\infty} a_n \frac{P_n^{(\beta,\beta)}(x)}{P_n^{(\beta,\beta)}(1)} \geq 0, \qquad -1 \leq x \leq 1,$$

then

$$\sum_{n=0}^{\infty} a_n \frac{P_n^{(\alpha,\alpha)}(y)}{P_n^{(\alpha,\alpha)}(1)} > 0, \qquad -1 < y < 1,$$

unless $a_n = 0$, $n = 0, 1, \ldots$.

As a corollary of Theorem 2.2 and (2.10) we have

$$(2.17) \qquad \sum_{k=0}^{n} \frac{P_k^{(\alpha,\alpha)}(x)}{P_k^{(\alpha,\alpha)}(1)} > 0, \qquad -1 < x < 1,\ \alpha \geq 0,\ n = 0, 1, \ldots.$$

The case $\alpha = \frac{1}{2}$ is (1.10).

Fejér [28, footnote 34] came very close to finding a very easy proof of Kogbetliantz's inequality (2.13). If the sum in (2.13) is denoted by $K_n^\lambda(x)$, then Fejér observed that

$$\sum_{n=0}^{\infty} K_n^\lambda(x) r^n = \frac{1-r^2}{(1-r)^{2\lambda+2}(1-2xr+r^2)^{\lambda+1}}$$

$$= \frac{1-r^2}{(1-r)^2(1-2xr+r^2)} \cdot \frac{1}{(1-r)^{2\lambda}(1-2xr+r^2)^\lambda}.$$

When $\lambda = 0$, the resulting series has nonnegative power series coefficients since it is just the second sum of (1.4), and Fejér proved the nonnegativity of these power series coefficients.

When $\lambda = \frac{1}{2}$, the remaining factor is just

$$\sum_{n=0}^{\infty} \left[\sum_{k=0}^{n} P_k(x) \right] r^n,$$

and the coefficient of r^n is positive $-1 < x \leq 1$. The product of absolutely monotonic functions (those with nonnegative power series coefficients) is absolutely monotonic, so Fejér had a proof of Kogbetliantz's result when $\lambda = \frac{1}{2}, 1, \frac{3}{2}, \ldots$. To find an easier proof that works for $\lambda > 0$, use the infinitesimal generator of the semigroup $(1-r)^{-2\lambda}(1-2xr+r^2)^{-\lambda}$ (see [10]).

The result on the absolute monotonicity of $(1-r)^{-2\lambda}(1-2xr+r^2)^{-\lambda}$ for $\lambda > 0$, $-1 \leq x \leq 1$, yields

$$(2.18) \qquad \sum_{k=0}^{n} \frac{(2\alpha+1)_{n-k}}{(n-k)!} \frac{(2\alpha+1)_k}{k!} \frac{P_k^{(\alpha,\alpha)}(x)}{P_k^{(\alpha,\alpha)}(1)} > 0, \qquad -1 < x \leq 1,\ \alpha > -\tfrac{1}{2}.$$

Theorem 2.2 then gives

$$(2.19) \qquad \sum_{k=0}^{n} \frac{(\lambda+1)_{n-k}(\lambda+1)_k}{(n-k)!k!} \frac{P_k^{(\alpha,\alpha)}(x)}{P_k^{(\alpha,\alpha)}(1)} > 0, \qquad -1 < x \leq 1,\ 2\alpha \geq \lambda > -1.$$

When $\alpha = \frac{1}{2}$, $\lambda = 1$, this is just Lukács's inequality (1.9).

The part of [34] written by Feldheim closes with a cryptic sentence: "The inequalities (2) and (5) can be generalized to certain sums of Jacobi polynomials ...." This is unclear for a number of reasons. First, (5) is an identity rather than an inequality. Second, there are a number of ways to extend the inequalities in [34], and the most important extension is of (2.17) above. This is (3) in this paper. The first time this was extended it was not extended in a useful way [8]. The extension there was to

$$\text{(2.20)} \qquad \sum_{k=0}^{n} \frac{P_k^{(\alpha,\beta)}(x)}{P_k^{(\alpha,\beta)}(1)} > 0, \qquad -1 < x \leq 1,$$

when $\alpha \geq |\beta|$ and to some other cases. Later when one of us tried to use (2.20) to prove the positivity of some quadrature schemes, it became clear that (2.20) is the wrong extension of (2.17). The right one is to

$$\text{(2.21)} \qquad \sum_{k=0}^{n} \frac{P_k^{(\alpha,\beta)}(x)}{P_k^{(\beta,\alpha)}(1)} > 0, \qquad -1 < x \leq 1,$$

for suitable α, β.

There is an old result of Bateman [15] that allows one to obtain (2.21) for $\alpha + \beta \geq 0$, $\beta \geq \alpha$, from (2.17). For Jacobi polynomials it is

$$\text{(2.22)} \quad (1+x)^{\beta+\mu} \frac{P_n^{(\alpha-\mu,\beta+\mu)}(x)}{P_n^{(\beta+\mu,\alpha-\mu)}(1)} = \frac{\Gamma(\beta+\mu+1)}{\Gamma(\beta+1)\Gamma(\mu)} \int_{-1}^{x} (1+y)^{\beta} \frac{P_n^{(\alpha,\beta)}(y)}{P_n^{(\beta,\alpha)}(1)} (x-y)^{\mu-1} \, dy,$$

$\mu > 0$, $\beta > -1$. When $\alpha = \frac{1}{2}$, $\beta = -\frac{1}{2}$, $\mu = \frac{1}{2}$, this is another way of writing (2.11). When used on (2.19) this gives

$$\text{(2.23)} \qquad \sum_{k=0}^{n} \frac{(\lambda+1)_{n-k}(\lambda+1)_k}{(n-k)!k!} \frac{P_k^{(\alpha,\beta)}(x)}{P_k^{(\beta,\alpha)}(1)} > 0, \qquad -1 < x \leq 1,$$

when $\beta \geq \alpha$ and $\alpha + \beta \geq \lambda > -1$.

There are two strong indications that (2.23) can be extended to a large set of (α, β, λ) with $\alpha > \beta$. When $\alpha = \frac{1}{2}$, $\beta = -\frac{1}{2}$, the case $\lambda = 0$ is equivalent to (1.1). A summation by parts shows that

$$\text{(2.24)} \quad \sum_{k=0}^{n} \frac{(\alpha+\beta+3)_{n-k}(2k+\alpha+\beta+1)(\alpha+\beta+1)_k}{(n-k)!(\alpha+\beta+1)k!} \frac{P_k^{(\alpha,\beta)}(x)}{P_k^{(\beta,\alpha)}(1)}$$
$$= \sum_{k=0}^{n} \frac{(\alpha+\beta+2)_{n-k}(\alpha+\beta+2)_k}{(n-k)!k!} \frac{P_k^{(\alpha+1,\beta)}(x)}{P_k^{(\beta,\alpha+1)}(1)}.$$

It is not obvious, but it is true that the right-hand side of (2.24) is positive for $\alpha = -\beta = \frac{1}{2}$ since the left-hand side of (2.24) is a positive multiple of the sum in (1.16). Thus (2.23) is true for $\lambda = 1$ when $\alpha = \frac{3}{2}$, $\beta = -\frac{1}{2}$. The case $\lambda = 0$, $\alpha = \frac{5}{2}$, $\beta = -\frac{1}{2}$ of (2.23) was proved in [9] and shown to be equivalent to

$$(n+1)\frac{\sin(n-1)\theta}{\sin\theta} - (n-1)\frac{\sin(n+1)\theta}{\sin\theta} \leq (3+\cos\theta)\left(n - \frac{\sin n\theta}{\sin\theta}\right)$$

for $0 < \theta < \pi$. This is stronger than the inequality

$$(n+1)\frac{\sin(n-1)\theta}{\sin\theta} - (n-1)\frac{\sin(n+1)\theta}{\sin\theta} \leq 4\left(n - \frac{\sin n\theta}{\sin\theta}\right),$$

which Robertson [56] proved and used to derive inequalities between some coefficients of analytic functions which have real power series coefficients and are convex in the direction of the imaginary axis for $|z| < 1$.

All of these facts and a few more led to the conjecture that (2.23) holds for $0 \leq \lambda \leq \alpha + \beta$ when $\alpha \geq \beta \geq -\frac{1}{2}$. After some more isolated cases were done, the following proof was found [9] for the special case $\lambda = 0$, $\beta = 0$. Start by using (2.2) to obtain

$$\sum_{k=0}^{n} P_k^{(\alpha,0)}(x) = \sum_{k=0}^{n} \frac{(\alpha+1)_k}{k!} \sum_{j=0}^{k} \frac{(-k)_j (k+\alpha+1)_j}{(\alpha+1)_j j!} \left(\frac{1-x}{2}\right)^j$$

$$= \sum_{j=0}^{n} \frac{(\frac{1-x}{2})^j (-1)^j}{(\alpha+1)_j j!} \sum_{k=j}^{n} \frac{(\alpha+1)_{k+j}}{(k-j)!}$$

$$= \sum_{j=0}^{n} \frac{(\alpha+1)_{2j}}{(\alpha+1)_j j!} \left(\frac{x-1}{2}\right)^j \sum_{k=0}^{n-j} \frac{(2j+\alpha+1)_k}{k!}.$$

Then use $(2a)_{2j} = 2^{2j}(a)_j(a+\frac{1}{2})_j$ and $\sum_{k=0}^{n}(a)_k/k! = (a+1)_n/n!$. The first of these is just a simple sieve, and the second is the natural sum analogue of $\int_0^x t^{a-1} dt = x^a/a$ and is easily proven by induction. The result is

$$(2.25) \quad \sum_{k=0}^{n} P_k^{(\alpha,0)}(x) = \sum_{j=0}^{n} \frac{(\frac{\alpha+1}{2})_j (\frac{\alpha+2}{2})_j (2j+\alpha+2)_{n-j}}{j!(n-j)!(\alpha+1)_j} [2(x-1)]^j$$

$$= \frac{(\alpha+2)_n}{n!} {}_3F_2 \left(\begin{array}{c} -n, \, n+\alpha+2, \, (\alpha+1)/2 \\ \alpha+1, \, (\alpha+3)/2 \end{array} ; \frac{1-x}{2} \right),$$

where $(a)_{n-j} = (a)_n(-1)^j/(1-a-n)_j$ has been used.

This is the first direct connection with de Branges's proof, for the $_3F_2$ in (2.25) contains the one he needed to show was positive. There are sign changes from one term of this $_3F_2$ to the next, so one needs to look for another representation to prove the positivity. If this were a $_2F_1$ or a $_3F_2$ with power series variable equal to one, there might be a transformation of the series which would work. In the present case that would not be the first thing to look for, since a transformation would involve a double sum at best. Another possibility is to see if it is a square or a sum of squares with positive coefficients. One could try to see if it is positive at both end points and monotone or convex in between. The last two are too much to hope for, and they are not true. There is a general theorem that says that a polynomial which is nonnegative on $[-1, 1]$ has the form $[q(x)]^2 + (1-x^2)[r(x)]^2$ or $(1-x)[q(x)]^2 + (1+x)[r(x)]^2$ for polynomials $q(x)$ and $r(x)$ (see [69, Theorem 1.21.1]). General theorems of this type are almost never useful, and it is not useful here. However when one looks in handbooks to see if there are any $_3F_2$'s which are squares, one rapidly discovers an identity of Clausen [20] (see [23,

4.3(1)] or [**13**, p. 86])

$$(2.26) \quad \left[{}_2F_1\left(\begin{matrix}a,\ b\\a+b+\frac{1}{2}\end{matrix};t\right)\right]^2 = {}_3F_2\left(\begin{matrix}2a,\ 2b,\ a+b\\a+b+\frac{1}{2},\ 2a+2b\end{matrix};t\right).$$

The only thing wrong with this ${}_3F_2$ is that one numerator parameter is $n+\alpha+1$ rather than $n+\alpha+2$, and one denominator parameter is $(\alpha+2)/2$ rather than $(\alpha+1)/2$. The other parameters agree. The first thing to do is to reduce the ${}_3F_2$ to a ${}_2F_1$ using the beta integral. This is just

$${}_3F_2\left(\begin{matrix}a,b,c\\d,e\end{matrix};t\right) = \frac{\Gamma(d)}{\Gamma(c)\Gamma(d-c)}\int_0^1 {}_2F_1\left(\begin{matrix}a,b\\e\end{matrix};tu\right)u^{c-1}(1-u)^{d-c-1}\,du$$

for $d > c > 0$. This gives

$$(2.27) \quad {}_3F_2\left(\begin{matrix}-n,\ n+\alpha+2,\ (\alpha+1)/2\\(\alpha+3)/2,\ \alpha+1\end{matrix};\frac{1-x}{2}\right)$$

$$= \frac{\Gamma(\alpha+1)}{[\Gamma(\frac{\alpha+1}{2})]^2}\int_0^1 {}_2F_1\left(\begin{matrix}-n,\ n+\alpha+2\\(\alpha+3)/2\end{matrix};u\left(\frac{1-x}{2}\right)\right)$$

$$\times u^{(\alpha-1)/2}(1-u)^{(\alpha-1)/2}\,du.$$

The ${}_2F_1$ in the integrand is a symmetric Jacobi polynomial, since

$$P_n^{(\beta,\beta)}(t) = \frac{(\beta+1)_n}{n!}\,{}_2F_1\left(\begin{matrix}-n,\ n+2\beta+1\\\beta+1\end{matrix};\frac{1-t}{2}\right).$$

When $\beta = (\alpha+1)/2$, this is the ${}_2F_1$ in (2.27). The ${}_3F_2$ in Clausen's formula can be represented in the same way, so all we need to be able to do is obtain $P_n^{((\alpha+1)/2,(\alpha+1)/2)}(x)$ from $P_n^{(\alpha/2,\alpha/2)}(x)$ with a positive kernel. Feldheim's integral (2.15) does this, but it is not useful in this setting because we have to integrate to get the ${}_3F_2$'s.

However there is a strong duality for the classical orthogonal polynomials of Jacobi, Laguerre, and Hermite as functions of x and as functions of n. They satisfy second-order differential equations in x, and second-order difference equations in n. This suggests that $P_n^{(\gamma,\gamma)}(x)$ can be given as

$$(2.28) \quad P_n^{(\gamma,\gamma)}(x) = \sum_{j=0}^{[n/2]} c_{j,n} P_{n-2j}^{(\beta,\beta)}(x)$$

with $c_{k,n} \geq 0$. Gegenbauer [**47**] found these coefficients explicitly. They are

$$(2.29) \quad c_{j,n} = \frac{(\gamma+1)_n}{(2\gamma+1)_n}\frac{(\gamma-\beta)_j}{j!}\frac{(2\beta+1)_{n-2j}(\gamma+\frac{1}{2})_{n-j}(\beta+\frac{1}{2}+n-2j)}{(\beta+1)_{n-2j}(\beta+\frac{3}{2})_{n-j}(\beta+\frac{1}{2})}.$$

Clearly $c_{j,n} > 0$ when $\gamma > \beta > -\frac{1}{2}$. See [**50**] for a proof. It is worth remarking that L. J. Rogers found a q-extension of this formula and used a special case in his derivation and first proof of the Rogers-Ramanujan identities (see [**58**, **59**]). Another simple derivation of (2.28)–(2.29) is given at the end of this section.

When these formulas are combined using (2.28) with $\beta = \alpha/2$ and $\gamma = (\alpha+1)/2$, the result is

$$(2.30) \quad \sum_{k=0}^{n} P_k^{(\alpha,0)}(x) = \sum_{j=0}^{[n/2]} \frac{(\frac{1}{2})_j (\frac{\alpha+2}{2})_{n-j} (\frac{\alpha+3}{2})_{n-2j} (\alpha+1)_{n-2j}}{j! (\frac{\alpha+3}{2})_{n-j} (\frac{\alpha+1}{2})_{n-2j} (n-2j)!}$$

$$\times \left[{}_2F_1 \left(\begin{matrix} (-n+2j)/2, \ (n-2j+\alpha+1)/2 \\ (\alpha+2)/2 \end{matrix} ; \frac{1-x}{2} \right) \right]^2.$$

Clearly the sum is nonnegative when $\alpha > -1$, and a little work gives the positivity for $-1 < x \le 1$ when $\alpha > -2$. See [9] for details. Other proofs of this positivity are given in [42, pp. 414–415], [45], and in [44], where (2.23) is proved for $\alpha + \beta \ge \lambda \ge 0$, $\beta \ge -\frac{1}{2}$, by using a sum of squares of Jacobi polynomials.

This gives

$$(2.31) \quad {}_3F_2 \left(\begin{matrix} -n, \ n+\alpha+2, \ (\alpha+1)/2 \\ (\alpha+3)/2, \ \alpha+1 \end{matrix} ; t \right) > 0, \qquad 0 \le t < 1, \ \alpha > -2.$$

A number of people have lamented the fact that there is not an easier proof of (2.31), at least for the special cases $\alpha = 2, 4, 6, \ldots$ used by de Branges. It is possible to prove the nonnegativity for these special cases without using Gegenbauer's explicit expansion. It suffices to show that

$$(2.32) \quad P_n^{(\alpha+\frac{1}{2},\alpha+\frac{1}{2})}(x) = \sum_{k=0}^{[n/2]} a_{k,n} P_{n-2k}^{(\alpha,\alpha)}(x),$$

with $a_{k,n} \ge 0$ when $2\alpha = -1, 0, 1, \ldots$. This follows from a result of Schoenberg [63]. He defined positive definite functions on a bounded metric space X with metric ρ as those real continuous functions $f(t)$, $0 \le t \le \text{diameter } X$, which satisfy

$$\sum_{i,j=1}^{n} c_i c_j f(\rho(x_i, x_j)) \ge 0$$

for all $n = 1, 2, \ldots$, real numbers c_i, and all points $x_i \in X$. For the unit sphere in R^N the positive definite functions are

$$f(t) = \sum_{n=0}^{\infty} a_n \frac{P_n^{(\alpha,\alpha)}(\cos t)}{P_n^{(\alpha,\alpha)}(1)}, \qquad \alpha = (N-3)/2$$

with $a_n \ge 0$, $\sum_{n=0}^{\infty} a_n$ finite. Since the unit sphere in R^N can be totally geodesically imbedded in the unit sphere in R^{N+1}, we have

$$P_n^{(\alpha+\frac{1}{2},\alpha+\frac{1}{2})}(x) = \sum_{k=0}^{n} b_{k,n} P_k^{(\alpha,\alpha)}(x), \qquad b_{k,n} \ge 0.$$

This gives (2.32) since $P_n^{(\alpha,\alpha)}(x)$ is even when n is even and odd when n is odd. If the reader still wants an easier proof, there are two alternatives. One is to find one. The other alternative is to read one of the other proofs, say the proofs in [42] or [44]. After reading one of these, the above proof will seem simpler.

There are many other inequalities which could be mentioned. One found by Ruscheweyh [62] and partly extended by Lewis [55] is very interesting. Ruscheweyh [62] showed that

$$(2.33) \qquad \sum_{k=0}^{n} a_k \frac{P_k^{(\alpha,\alpha)}(x)}{P_k^{(\alpha,\alpha)}(1)} z^k \neq 0, \qquad |z| \leq 1, \ -1 < x < 1,$$

when $1 = a_0 \geq a_1 \geq \cdots \geq a_n \geq 0$ and $\alpha \geq 0$.

When $\alpha = 0$, this contains Szegö's result (2.14), and when $|z| < 1$ and $\alpha = \frac{1}{2}$, it reduces to Theorem 1.1 when Theorem 1.2 is used.

Lewis [55] extended the case $a_k = 1$ to

$$\sum_{k=0}^{n} \frac{(\lambda+1)_{n-k}}{(n-k)!} \frac{(\lambda+1)_k}{k!} \frac{P_k^{(\alpha,\beta)}(x)}{P_k^{(\beta,\alpha)}(1)} z^k \neq 0$$

for $-1 \leq x \leq 1$ and $|z| < 1$ when $0 \leq \lambda \leq \alpha + \beta$ and $\beta \geq \alpha$.

Ruscheweyh [62] also proved that

$$\sum_{k=0}^{n} \frac{P_{mk}^{(\alpha,\alpha)}(x)}{P_{mk}^{(\alpha,\alpha)}(1)} z^k \neq 0$$

for $-1 < x < 1$, $|z| \leq 1$ when $1 = a_0 \geq a_1 \geq \cdots \geq a_n \geq 0$ and $\alpha \geq (m-1)/2$, $m = 1, 2, \ldots$. This is a different type of inequality, and it is not clear how to use it yet, or how close it is to being best possible. It is very interesting.

Two general surveys of positivity and the classical orthogonal polynomials are given in [5] and [42]. A number of open problems are given in [6].

The treatment above was historical, and as history often is, the results were not found in the most systematic way possible. To get a better idea of what is happening, consider the following multiplier problem. Find all sequences $\{t_n\}$ having the property that

$$(2.34) \qquad f(x) = \sum_{n=0}^{\infty} a_n P_n^{(\alpha,\beta)}(x) \geq 0, \qquad -1 \leq x \leq 1,$$

implies

$$Tf(x) = \sum_{n=0}^{\infty} t_n a_n P_n^{(\alpha,\beta)}(x) \geq 0, \qquad -1 \leq x \leq 1.$$

When $\alpha \geq \beta > -1$ and either $\alpha + \beta \geq 0$ or $\beta \geq -\frac{1}{2}$, the answer is

$$t_n = \int_{-1}^{1} \frac{P_n^{(\alpha,\beta)}(x)}{P_n^{(\alpha,\beta)}(1)} d\mu(y)$$

for a nonnegative measure $d\mu(y)$ (see Gasper [41]). All of this should be done for distributions, and since there is no problem in doing this, we will use divergent series without worrying about them. Unfortunately this answer is as useful as general answers often are. It cannot be used to prove any of the results mentioned above. However it does suggest that the sum

$$K(x,y,z) = \sum_{n=0}^{\infty} \frac{P_n^{(\alpha,\beta)}(x) P_n^{(\alpha,\beta)}(y) P_n^{(\alpha,\beta)}(z)}{h_n^{(\alpha,\beta)} P_n^{(\alpha,\beta)}(1)}$$

plays a central role. If $f(x)$ is in L^1 with respect to $(1-x)^\alpha(1-x)^\beta$ and

$$f(x) = \sum_{n=0}^{\infty} a_n P_n^{(\alpha,\beta)}(x), \qquad -1 \leq x \leq 1,$$

then

$$Tf(x;y) = \sum_{n=0}^{\infty} a_n P_n^{(\alpha,\beta)}(x) P_n^{(\alpha,\beta)}(y) / P_n^{(\alpha,\beta)}(1)$$

$$= \int_{-1}^{1} K(x,y,z) f(z)(1-z)^\alpha (1+z)^\beta \, dz$$

when $\alpha \geq \beta > -1$, $\alpha + \beta > -1$. When $y = 1$, the kernel $K(x, y, z)$ becomes the delta function

$$K(x,1,z) = \delta(x,z) = \sum_{n=0}^{\infty} P_n^{(\alpha,\beta)}(x) P_n^{(\alpha,\beta)}(z) / h_n^{\alpha,\beta}.$$

When $z = 1$, we have

$$K(x,1,1) = \sum_{n=0}^{\infty} P_n^{(\alpha,\beta)}(x) P_n^{(\alpha,\beta)}(1) / h_n^{\alpha,\beta}$$

$$= 2^{-\alpha-\beta-1} \sum_{n=0}^{\infty} \frac{(2n+\alpha+\beta+1)\Gamma(n+\alpha+\beta+1)}{\Gamma(\alpha+1)\Gamma(n+\beta+1)} P_n^{(\alpha,\beta)}(x).$$

The partial sums of this series are a positive multiple of

$$\sum_{k=0}^{n} \frac{(2k+\alpha+\beta+1)\Gamma(k+\alpha+\beta+1)}{\Gamma(k+\beta+1)} P_k^{(\alpha,\beta)}(x)$$

$$= \frac{\Gamma(n+\alpha+\beta+2)}{\Gamma(n+\beta+1)} P_n^{(\alpha+1,\beta)}(x) = A \frac{(\alpha+\beta+2)_n}{n!} \frac{P_n^{(\alpha+1,\beta)}(x)}{P_n^{(\beta,\alpha+1)}(1)}.$$

See [69, (4.5.3)]. Further partial sums lead to

$$\sum_{k=0}^{n} \frac{(c)_{n-k}}{(n-k)!} \frac{(\alpha+\beta+2)_k}{k!} \frac{P_k^{(\alpha+1,\beta)}(x)}{P_k^{(\beta,\alpha+1)}(1)},$$

and when $c = \alpha + \beta + 2$, this is just (2.24). It is then a natural question to consider the general expansion (2.34) with $f(x) \geq 0$ and ask if there is a c for which

(2.35) $$\sum_{k=0}^{n} \frac{(c)_{n-k}}{(n-k)!} a_k P_k^{(\alpha,\beta)}(x) \geq 0, \qquad -1 \leq x \leq 1, \; n = 0, 1, \ldots.$$

We have been considering the case

(2.36) $$f(x) = \sum_{n=0}^{\infty} \frac{(\lambda)_n}{n!} \frac{P_n^{(\alpha,\beta)}(x)}{P_n^{(\beta,\alpha)}(1)}, \qquad 0 < \lambda < \alpha + \beta + 1, \ \beta > -1.$$

To see that $f(x)$ defined by (2.36) is nonnegative for $-1 \le x \le 1$, observe that Example 19 in Bailey [**13**, p. 102] gives the positivity of

$$\sum_{n=0}^{\infty} r^n \frac{(\alpha+\beta+1)_n}{n!} \frac{P_n^{(\alpha,\beta)}(x)}{P_n^{(\beta,\alpha)}(1)}$$

for $0 \le r < 1$, $\alpha + \beta > -1$, $\beta > -1$. Then integrate this with respect to $r^{\lambda-1}(1-r)^{\alpha+\beta-\lambda}$ to obtain a positive multiple of $f(x)$.

Another interesting case is $f(x) = (1-x)^{-\alpha}(1+x)^{-\beta}$. This arises when considering the positivity of some quadrature schemes. The more general function $(1-x)^{-\gamma}(1+x)^{-\delta}$ contains the inequalities of Vietoris mentioned in the first section (see [**11**] for this). Although we probably will never be able to solve these problems completely, it is clear that they should be looked at every so often to see if other important results can be found.

There are integral analogues that have been considered. For Bessel functions

(2.37) $$\int_0^x (x-t)^\lambda t^{\lambda+\frac{1}{2}} J_\alpha(t) \, dt > 0,$$

$0 < \lambda \le \alpha - \frac{1}{2}$, $x > 0$, is an analogue of (2.23). See [**43**] for a proof. An analogous problem for Jacobi polynomials of when

(2.38) $$\int_x^1 (t-x)^\lambda P_n^{(\alpha,\beta)}(t)(1-t)^\gamma (1+t)^\delta \, dt \ge 0, \qquad -1 \le x \le 1, \ n = 0, 1, \ldots,$$

has not been considered seriously. The special case $\beta = 1$, $\lambda = \delta = 0$, $\gamma = (\alpha - 1)/2$ of (2.38) is equivalent to the inequality de Branges used. It is probably important to look at (2.38) in detail, and also to consider the two integration problems that arise when $P_n^{(\alpha,\beta)}(x)$ is replaced by

$$P_\lambda^{(\alpha,\beta)}(\cosh u) = {}_2F_1 \left(\begin{array}{c} (\alpha+\beta+1+i\lambda)/2, \ (\alpha+\beta+1-i\lambda)/2 \\ \alpha+1 \end{array} ; \frac{1-\cosh u}{2} \right).$$

See Koornwinder [**53**] for a survey of these Jacobi functions.

There is another way to look at these questions, and it was started by Stieltjes and Hilbert. They considered the hypergeometric polynomial

$$g_n(t) = \frac{(\alpha+1)_n}{n!} {}_2F_1 \left(\begin{array}{c} -n, \ n+\alpha+\beta+1 \\ \alpha+1 \end{array} ; t \right)$$

and for fixed real values of α and β found how many zeros are on the interval $(0,1)$, how many are on $(1,\infty)$, how many are on $(-\infty, 0)$, and thus how many nonreal zeros there are. See [**69**, Theorem 6.72] for a statement and proof. The next case to consider is

$$h_n(t) = (\alpha+1)_n (\gamma+1)_n {}_3F_2 \left(\begin{array}{c} -n, \ n+\alpha+\beta+1, \ t \\ \alpha+1, \ \gamma+1 \end{array} ; 1 \right).$$

It may be possible to determine the number of real zeros and the number of complex zeros of $h_n(t)$. A lot can be said because of the existence of two sets of orthogonal polynomials

(2.39) $$Q_n(x; \alpha, \beta, N) = {}_3F_2\left(\begin{matrix} -n,\ n+\alpha+\beta+1,\ -x \\ \alpha+1,\ -N \end{matrix}; 1\right),$$

for $n = 0, 1, 2, \ldots, N$, and

(2.40) $$p_n(x; a, b) = i^n {}_3F_2\left(\begin{matrix} -n,\ n+2a+2b-1,\ a+ix \\ a+b,\ 2a \end{matrix}; 1\right),$$

which are orthogonal with respect to

$$\binom{x+\alpha}{x}\binom{N-x+\beta}{N-x}, \quad x = 0,1,\ldots,N, \quad \text{and} \quad |\Gamma(a+ix)\Gamma(b+ix)|^2$$

on $(-\infty, \infty)$ respectively. See the chart at the end of [**12**] for references to these and other classical hypergeometric orthogonal polynomials. To treat

$${}_3F_2\left(\begin{matrix} -n,\ n+\alpha+\beta+1,\ t \\ \alpha+1,\ \gamma+1 \end{matrix}; u\right)$$

as a function of u or a function of t for fixed u, $0 < u < 1$, will be much harder.

One minor observation is that the polynomial $p_n(x) = p_n(x; a, b)$ defined by (2.40) satisfies

$$p_n(-x) = (-1)^n p_n(x).$$

Thus $p_{2n+1}(0) = 0$ and $p_{2n}(0) \neq 0$ since adjacent orthogonal polynomials cannot vanish at the same point. It is easy to see that for $a, b > 0$

$$(-1)^n p_{2n}(0) > 0,$$

and thus that

$${}_3F_2\left(\begin{matrix} -2n,\ 2n+2a+2b-1,\ a \\ a+b,\ 2a \end{matrix}; 1\right) > 0.$$

When $a = (\alpha+1)/2$ and $b = 1$, this gives the end-point results de Branges used. Unfortunately we have not been able to use this set of orthogonal polynomials to give another proof of (2.31).

We close this section with a new derivation of Gegenbauer's solution to the connection coefficient problem for ultraspherical polynomials. Here it will be better to use the notation for ultraspherical polynomials given by the generating function (2.12)

(2.12) $$\sum_{n=0}^{\infty} C_n^\lambda(x) r^n = (1 - 2xr + r^2)^{-\lambda}.$$

Differentiating with respect to x gives

$$\sum_{n=1}^{\infty} \frac{d}{dx} C_n^\lambda(x) r^n = 2\lambda r (1 - 2xr + r^2)^{-\lambda-1},$$

so
(2.41) $$\frac{d}{dx}C_n^\lambda(x) = 2\lambda C_{n-1}^{\lambda+1}(x).$$

When $x = \cos\theta$ in (2.12), the right-hand side factors to
$$(1 - re^{i\theta})^{-\lambda}(1 - re^{-i\theta})^{-\lambda},$$
and the binomial theorem and the Cauchy product of power series gives
$$C_n^\lambda(\cos\theta) = \sum_{k=0}^{n} \frac{(\lambda)_{n-k}(\lambda)_k}{(n-k)!k!} e^{i(n-2k)\theta},$$
or
$$C_n^\lambda(\cos\theta) = \sum_{k=0}^{n} \frac{(\lambda)_{n-k}(\lambda)_k}{(n-k)!k!} \cos(n-2k)\theta.$$

Differentiating with respect to θ and dividing by $-\sin\theta$ gives
$$2\lambda C_{n-1}^{\lambda+1}(\cos\theta) = \sum_{k=0}^{n} \frac{(\lambda)_{n-k}(\lambda)_k(n-2k)}{(n-k)!k!} \frac{\sin(n-2k)\theta}{\sin\theta},$$
and since $(n-2k)\sin(n-2k)\theta$ is an even function of k about the point $2k = n$, this is the same as

(2.42) $$\lambda C_{n-1}^{\lambda+1}(\cos\theta) = \sum_{k=0}^{[(n-1)/2]} \frac{(\lambda)_{n-k}(\lambda)_k}{(n-k)!k!}(n-2k)\frac{\sin(n-2k)\theta}{\sin\theta}.$$

Recall that
$$\frac{\sin(n+1)\theta}{\sin\theta} = C_n^1(\cos\theta).$$
This allows us to rewrite (2.42) as
$$\lambda C_{n-1}^{\lambda+1}(x) = \sum_{k=0}^{[(n-1)/2]} \frac{(\lambda)_{n-k}(\lambda)_k}{(n-k)!k!}(n-2k)C_{n-1-2k}^1(x).$$

Differentiate $(j-1)$ times to obtain
$$2^{j-1}(\lambda)_j C_{n-j}^{\lambda+j}(x) = \sum_{k=0}^{[(n-j)/2]} \frac{(\lambda)_{n-k}(\lambda)_k}{(n-k)!k!}(n-2k)2^{j-1}(1)_{j-1} C_{n-j-2k}^j(x).$$

Now replace n by $n+j$ and λ by $\lambda-j$. The result is

(2.43) $$(\lambda-j)_j C_n^\lambda(x) = \sum_{k=0}^{[n/2]} \frac{(\lambda-j)_{n+j-k}(\lambda-j)_k}{(n+j-k)!\,k!}(n+j-2k)(1)_{j-1} C_{n-2k}^j(x).$$

We can rewrite (2.43) as

(2.44) $$C_n^\lambda(x) = \sum_{k=0}^{[n/2]} \frac{(\lambda)_{n-k}(\lambda-j)_k}{(j+1)_{n-k}k!}\frac{(n+j-2k)}{j} C_{n-2k}^j(x).$$

If the nonnegative integer j is replaced by the real number μ, (2.44) becomes

$$(2.45) \qquad C_n^\lambda(x) = \sum_{k=0}^{[n/2]} \frac{(\lambda)_{n-k}(\lambda - \mu)_k(n+\mu-2k)}{(\mu+1)_{n-k}k!\mu} C_{n-2k}^\mu(x),$$

which is Gegenbauer's formula. The right-hand side of (2.45) is a rational function of μ, the left-hand side is constant in μ, and they agree for infinitely many values. So they are identically equal.

3. Extensions of de Branges's inequality and some speculations. Let S denote the class of functions

$$f(z) = z + \sum_{n=2}^{\infty} a_n z^n$$

which are analytic and univalent in $|z| < 1$, and set

$$\log \frac{f(z)}{z} = 2 \sum_{n=1}^{\infty} \gamma_n z^n.$$

The main reason we were surprised by de Branges's use of the inequalities (2.31) for $\alpha = 2, 4, 6, \ldots$ to prove the Bieberbach conjecture

$$(3.1) \qquad |a_n| \leq n, \qquad n = 2, 3, \ldots, \; f \in S,$$

and the Milin conjecture

$$(3.2) \qquad \sum_{k=1}^{n}(n+1-k)(k|\gamma_k|^2 - k^{-1}) \leq 0, \qquad n = 1, 2, \ldots, \; f \in S,$$

was that while (3.1) and (3.2) are sharp in the sense that equality holds for the Koebe function, inequality (2.31) is not sharp in the following sense. When $\alpha \geq 0$, it is equivalent to the inequality (2.23) for $\lambda = \beta = 0$, and this inequality for $(\alpha, \beta, \lambda) = (\alpha, 0, 0)$ is a fractional integral of order one half (using Bateman's integral (2.22)) of the sum for $(\alpha + \frac{1}{2}, -\frac{1}{2}, 0)$. This last sum is nonnegative when $\alpha + \frac{1}{2} \geq \frac{1}{2}$. This led to the question of whether de Branges's inequalities (3.2) could be extended using other cases of (2.23) or another nonnegative polynomial sum.

With this in mind, recall that in his proof of (3.2) de Branges used a general form of the Löwner differential equation to obtain the equations

$$(3.3) \qquad \sigma_k(t) + \frac{t}{k}\sigma_k'(t) = \sigma_{k+1}(t) - \frac{t}{k+1}\sigma_{k+1}'(t), \qquad k = 1, 2, \ldots, n,$$

and showed that
(3.4)

$$\sigma_k(t) = \frac{k(n+k+1)!}{(n-k)!(2k+1)!} \int_t^\infty {}_3F_2\left(\begin{matrix} k-n,\; n+k+2,\; k+\frac{1}{2} \\ k+\frac{3}{2},\; 2k+1 \end{matrix}; \frac{1}{s}\right) s^{-k-1}\, ds,$$

$k = 1, 2, \ldots, n$, $\sigma_{n+1}(t) = 0$, satisfies (3.3). Then he used $\sigma_k'(t) \leq 0$ for $t \geq 1$, and this follows from (2.31).

de Branges (in an August 1984 letter to one of the authors) and Koornwinder [54] independently observed that when formula (2.30) is used to write the $_3F_2$ in (3.4) as a sum of squares of $_2F_1$ series, these squares of $_2F_1$ series can also be used to obtain nonincreasing solutions of the differential equations (3.3). Explicitly, de Branges observed that

$$\sigma_k^*(t) = \frac{(n+k)!}{2^{2k}k!(k-1)!(n-k)!}$$

(3.5)
$$\times \int_t^\infty \left[{}_2F_1\left(\frac{k-n}{2}, \frac{n+k+1}{2}; k+1; s^{-1}\right) \right]^2 s^{-k-1}\, ds$$

$$= \frac{(n+k)! t^{-k}}{2^{2k}k!k!(n-k)!} {}_4F_3\left(\begin{matrix} k-n,\ n+k+1,\ k,\ k+1/2 \\ k+1,\ k+1,\ 2k+1 \end{matrix}; \frac{1}{t}\right),$$

$k = 1, 2, \ldots, n$, $\sigma_{n+1}^*(t) \equiv 0$, are nonincreasing solutions of the differential equations (3.3). Koornwinder [54], following the notation in FitzGerald and Pommerenke [35], observed that $\tau_k(t) = \tau_k^n(t) = \sigma_k^*(e^t)$ are nonincreasing solutions of

(3.6) $\quad \tau_k(t) + k^{-1}\tau_k'(t) = \tau_{k+1}(t) - (k+1)^{-1}\tau_{k+1}'(t), \quad k = 1, 2, \ldots, n,$

(these follow by replacing t by e^t in (3.3)). This leads to

(3.7) $\quad \displaystyle\sum_{k=1}^n \tau_k^n(0)[k|\gamma_k|^2 - k^{-1}] \leq 0, \quad n = 1, 2, \ldots, f \in S.$

This can be rewritten as

(3.8) $\quad \displaystyle\sum_{j=1}^n d_j^n \sum_{k=1}^j (k|\gamma_k|^2 - k^{-1}) \leq 0,$

with

$$d_j^n = d_{j-1}^n = \frac{(\frac{1}{2})_{(n+j)/2}(\frac{1}{2})_{(n-j)/2}}{(\frac{n+j}{2})!(\frac{n-j}{2})!}, \quad n-j \text{ even}.$$

The expansion formula (2.30) shows that the inequalities (3.2) are in the convex hull of the inequalities (3.7).

Since the differential equations (3.6) are invariant under translations, it is easily seen that (3.7) can be extended to

(3.9) $\quad \displaystyle\sum_{k=1}^n \tau_k^n(t)(k|\gamma_k|^2 - k^{-1}) \leq 0, \quad t \geq 0,\ n = 1, 2, \ldots, f \in S.$

Observe that, since

$${}_2F_1\left(\frac{k-n}{2}, \frac{n+k+1}{2}; k+1; s^{-1}\right) = \frac{(n-k)!}{(2k+1)_{n-k}} C_{n-k}^{k+\frac{1}{2}}((1-s^{-1})^{1/2})$$

and the zeros of ultraspherical polynomials are contained in the interval $(-1, 1)$, it follows from (3.5) that the functions $\tau_k^n(t+t_0)$ cannot be written as linear combinations with nonnegative coefficients of the functions $\tau_k^m(t)$ for any $t_0 > 0$.

There are other integrals of hypergeometric functions which satisfy (3.6), but we have not had time to consider the consequences for univalent functions.

In [45] it was observed that for $f \in S$ the inequalities (3.2) can be written in the form

$$\text{(3.10)} \qquad \sum_{k=0}^{n} \frac{(2)_k (2)_{n-k}}{k!(n-k)!} (|\gamma_{k+1}|^2 - (k+1)^{-2}) \leq 0, \qquad n = 0, 1, \ldots.$$

This was extended to

$$\text{(3.11)} \qquad \sum_{k=0}^{n} \frac{(\alpha)_k (\beta)_{n-k}}{k!(n-k)!} (|\gamma_{k+1}|^2 - (k+1)^{-2}) \leq 0, \qquad n = 0, 1, \ldots,$$

for $0 < \alpha \leq 2$, $\beta \geq 2$, by showing these inequalities are in the convex hull of the inequalities (3.2). This is not the case for all n when $\alpha > 2$ or $0 < \beta < 2$. The functions $\sigma_k(t)$, which satisfy the differential equations (3.3) with the endpoint conditions $\sigma_k(1) = (\alpha)_{k-1}(\beta)_{n-k}/k!(n-k)!$, are not nonincreasing for all k and n when $\alpha > 2$ or $0 < \beta < 2$. Hence the arguments in [18] and [35] cannot be used to prove (3.11) when $\alpha > 2$ or $0 < \beta < 2$. Since the Milin constant is positive (see [22, §5.4]), the inequality (3.11) fails for $\alpha = 2$, $\beta = 1$. A few other special cases are mentioned in [45]. It would be interesting to see when (3.11) holds.

There are a number of indications that the even coefficients of a function in S are related, and also that the odd coefficients are related. Recall the inequalities of Garabedian, Ross, and Schiffer [38, 39], and of Bombieri [16] which proved the Bieberbach conjecture for functions sufficiently close to the Koebe function. In addition Ahlfors [1, 2] proved that

$$16 - |a_4|^2 \geq (16/15)(4 - |a_2|^2).$$

These inequalities suggest that

$$(2n)^2 - |a_{2n}|^2 \geq A_n(4 - |a_2|^2)$$

and

$$(2n+1)^2 - |a_{2n+1}|^2 \geq B_n(9 - |a_3|^2)$$

for positive constants A_n and B_n. To prove this by a method similar to de Branges's proof, one would like to have a polynomial inequality depending on a parameter that can be proven for large values of the parameter from the inequality for small values of the parameter. It should also distinguish between polynomials of even and odd degree. The ultraspherical polynomials (or symmetric Jacobi polynomials) are the obvious polynomials to consider because of Feldheim's integral (2.15) and Gegenbauer's expansion (2.28).

REFERENCES

1. L. Ahlfors, "An inequality between the coefficients a_2 and a_4 of a univalent function," in *Some problems of mathematics and mechanics*, Izdat. "Nauka" Leningrad, 1970, pp. 71–74; English transl. in Amer. Math. Soc. Transl. (2) **104** (1976), 57–60; also in *Collected papers*. Vol. 2, Birkhäuser, Boston, 1982, pp. 480–483.

2. ____, *Conformal invariants: Topics in geometric function theory*, McGraw Hill, New York, 1973.

3. J. W. Alexander, *Functions which map the interior of the unit circle upon simple regions*, Ann. of Math. (2) **17** (1915–16), 12–22.

4. R. Askey, *Some absolutely monotonic functions*, Studia Sci. Math. Hungar. **9** (1974), 51–56.

5. ____, *Orthogonal polynomials and special functions*, Regional Conf. Lect. Appl. Math. vol. 21, SIAM, Philadelphia, Pa., 1975.

6. ____, *Some problems about special functions and computations*, Rend. Sem. Mat. Univ. Politec. Torino, Special Volume, 1985, 1–22.

7. ____, *My reaction to de Branges's proof of the Bieberbach conjecture*, (in this volume).

8. R. Askey and J. Fitch, *Integral representations for Jacobi polynomials and some applications*, J. Math. Anal. Appl. **26** (1969), 411–437.

9. R. Askey and G. Gasper, *Positive Jacobi polynomial sums. II*, Amer. J. Math. **98** (1976), 709–737.

10. R. Askey and H. Pollard, *Some absolutely monotonic and completely monotonic functions*, SIAM J. Math. Anal. **5** (1974), 58–63.

11. R. Askey and J. Steinig, *Some positive trigonometric sums*, Trans. Amer. Math. Soc. **187** (1974), 295–307.

12. R. Askey and J. Wilson, *Some basic hypergeometric orthogonal polynomials that generalize Jacobi polynomials*, Mem. Amer. Math. Soc. No. 319 (1985).

13. W. N. Bailey, *Generalized hypergeometric series*, Cambridge Univ. Press, Cambridge, 1935; reprinted, Hafner, New York, 1964.

14. ____, *The generating function of Jacobi polynomials*, J. London Math. Soc. **13** (1938), 8–12.

15. H. Bateman, *The solution of linear differential equations by means of definite integrals*, Trans. Camb. Phil. Soc. **21** (1909), 171–196.

16. E. Bombieri, *On the local maximum property of the Koebe function*, Invent. Math. **4** (1967), 26–67.

17. L. de Branges, *Coefficient estimates*, J. Math. Anal. Appl. **82** (1981), 420–450.

18. ____, *A proof of the Bieberbach conjecture*, Acta Math. **154** (1985), 137–152.

19. G. Brown and E. Hewitt, *A class of positive trigonometric sums*, Math. Ann. **268** (1984), 91–122.

20. T. Clausen, *Ueber die Fälle, wenn dei Reihe von der Form...ein Quadrat von der Form...hat*, J. Reine Angew. Math. **3** (1828), 89–91.

21. J. Dieudonné, *Recherches sur quelques problèmes relatifs aux polynômes et aux fonctions bornées d'une variable complexe*, Ann. Sci. École Norm. Sup. **48** (1931), 247–358.

22. P. L. Duren, *Univalent functions*, Springer-Verlag, New York, 1983.

23. A. Erdélyi, *Higher transcendental functions*. Vol. 1, McGraw Hill, New York, 1952; reprinted, Krieger, Malabar, Florida, 1981.

24. L. Fejér, *Sur les fonctions bornées et intégrables*, C. R. Acad. Sci. Paris **131** (1900), 984–987; in [**32**], pp. 37–41.

25. ____, *Sur le développement d'une fonction arbitraire suivant les fonctions de Laplace*, C.R. Acad. Sci. Paris **146** (1908), 224–227; in [**32**], pp. 319–322.

26. ____, *Sur les singularités de la série de Fourier des fonctions continues*, Ann. Ecole Norm. Supér. **28** (1911), 64–103; in [**32**], pp. 654–688.

27. ____, *Über die Positivität von Summen, die nach trigonometrischen oder Legendreschen Funktionen fortschreiten (erste Mitt.)*, Acta Litt. ac Scient. Szeged **2** (1925), 75–86; in [**33**], pp. 128–138.

28. ____, *Einige Sätze, die sich auf das Vorzeichen einer ganzen rationalen Funktion beziehen;...*, Monat. für Math. und Phys. **35** (1928), 305–344; in [**33**], pp. 202–237.

29. ____, *Gestaltliches über die Partialsummen und ihre Mittelwerte bei der Fourierreihe und der Potenzreihe*, Z. Angew. Math. Mech. **13** (1933), 80–88; in [**33**], pp. 479–492.

30. ____, *Neue Eigenschaften der Mittelwerte bei den Fourierreihen*, J. London Math. Soc. **8** (1933), 53–62; in [**33**], pp. 493–501.

31. ____, *Trigonometrische Reihen und Potenzreihen mit mehrfach monotoner Koeffizientenfolge*, Trans. Amer. Math. Soc. **39** (1936), 18–59; in [**33**], pp. 581–620.

32. ____, *Gesammelte Arbeiten*. I, Birkhäuser, Basel, 1970.

33. ____, *Gesammelte Arbeiten*. II, Birkhäuser, Basel, 1970.

34. E. Feldheim, with appendix by G. Szegö, *On the positivity of certain sums of ultraspherical polynomials*, J. Analyse Math. **11** (1963), 275–284; reprinted in G. Szegö, *Collected papers*, Vol. 3, Birkhäuser, Boston, 1982, pp. 821–830.

35. C. H. FitzGerald and C. Pommerenke, *The de Branges theorem on univalent functions*, Trans. Amer. Math. Soc. **290** (1985), 683–690.

36. I. Fuchs, *Potenzreihen mit mehrfach monotonen Koeffizienten*, Arch. Math. **22** (1971), 275–278.

37. ____, *Power series with multiply monotonic coefficients*, Math. Ann. **190** (1971), 289–292.

38. P. R. Garabedian, G. G. Ross, and M. Schiffer, *On the Bieberbach conjecture for even n*, J. Math. Mech. (now Indiana J. Math.) **14** (1965), 975–989.

39. P. R. Garabedian and M. Schiffer, *The local maximum theorem for the coefficients of univalent functions*, Arch. Rational Mech. Anal. **26** (1967), 1–32.

40. G. Gasper, *Nonnegative sums of cosine, ultraspherical and Jacobi polynomials*, J. Math. Anal. Appl. **26** (1969), 60–68.

41. ____, *Banach algebras for Jacobi series and positivity of a kernel*, Ann. of Math. (2) **95** (1972), 261–280.

42. ____, "Positivity and special functions," in *Theory and application of special functions*, edited by R. Askey, Academic Press, New York, 1975, pp. 375–433.

43. ____, *Positive integrals of Bessel functions*, SIAM J. Math. Anal. **6** (1975), 868–881.

44. ____, *Positive sums of the classical orthogonal polynomials*, SIAM J. Math. Anal. **8** (1977), 423–447.

45. ____, *A short proof of an inequality used by de Branges in his proof of the Bieberbach, Robertson and Milin conjectures*, Complex Variables Theory Appl. **7** (1986) (to appear).

46. C. F. Gauss, *Développement des fonctions périodiques en séries*. Werke VII, K. Gesell. Wiss., Göttingen, 1906, pp. 469–472.

47. L. Gegenbauer, *Zur Theorie der Functionen $C_n^\nu(x)$*, Denkschriften Akad. Wissen. Wien, Math. Klasse **48** (1884), 293–316.

48. T. H. Gronwall, *Über die Gibbssche Erscheinung und die trigonometrischen Summen $\sin x + \frac{1}{2}\sin 2x + \cdots + \frac{1}{n}\sin nx$*, Math. Ann. **72** (1912), 228–243.

49. E. Hewitt and R. E. Hewitt, *The Gibbs-Wilbraham phenomenon: an episode in Fourier analysis*, Arch. Hist. Exact Sci. **21** (1979), 129–160.

50. L. K. Hua, *Harmonic analysis of functions of several complex variables in the classical domains*, Transl. Math. Mono. vol. 6, Amer. Math. Soc., Providence, R. I., 1963.

51. D. Jackson, *Über eine trigonometrische Summe*, Rend. Circ. Mat. Palermo **32** (1911), 257–262.

52. E. Kogbetliantz, *Recherches sur la sommabilité des séries ultersphériques par la méthode des moyennes arithmétiques*, J. Math. Pures Appl. (9) **3** (1924), 107–187.

53. T. Koornwinder, "Jacobi functions and analysis on noncompact semisimple Lie groups," in *Special functions: group theoretical aspects and applications*, edited by R. Askey, T. Koornwinder, and W. Schempp, Reidel, Dordrecht, 1984, pp. 1–85.

54. ____, *Squares of Gegenbauer polynomials and Milin type inequalities*, Centrum voor Wisk. en Infor. Report PM-R8412, Amsterdam.

55. J. L. Lewis, *Applications of a convolution theorem to Jacobi polynomials*, SIAM J. Math. Anal. **10** (1979), 1110–1120.

56. M. S. Robertson, *The coefficients of univalent functions*, Bull. Amer. Math. Soc. **51** (1945), 733–738.

57. ____, *Power series with multiply monotonic coefficients*, Michigan Math. J. **16** (1969), 27–37.

58. L. J. Rogers, *Second memoir on the expansion of certain infinite products*, Proc. London Math. Soc. **25** (1894), 318–343.

59. ____, *Third memoir on the expansion of certain infinite products*, Proc. London Math. Soc. **26** (1895), 15–32.

60. W. Rogosinski and G. Szegö, *Über die Abschnitte von Potenzreihen, die in einem Kreise beschränkt bleiben*, Math. Z. **28** (1928), 73–94; also in *Gabor Szegö: Collected papers*. Vol. 2, Birkhäuser, Boston, 1982, pp. 89–110.

61. N. N. Royall, Jr., *Laplace transforms of multiply monotonic functions*, Duke Math. J. **8** (1941), 546–558.

62. St. Ruscheweyh, *On the Kakeya-Eneström theorem and Gegenbauer polynomial sums*, SIAM J. Math. Anal. **9** (1978), 682–686.

63. I. J. Schoenberg, *Positive definite functions on spheres*, Duke Math. J. **9** (1942), 96–108.

64. M. Schweitzer, *The partial sums of second order of the geometric series*, Duke Math. J. **18** (1951), 527–533.

65. T. Sheil-Small, "Applications of the Hadamard product," in *Aspects of contemporary complex analysis*, edited by D. A. Brannan and J. G. Clunie, Academic Press, London, 1980, pp. 515–523.

66. G. Szegö, *Über einen Satz des Herrn Serge Bernstein*, Schriften Königsberger Gel. Ges. (1928), 59–70; also in *Collected papers*. Vol. 2, Birkhäuser, Boston, 1982, pp. 207–218.

67. ____, *Zur Theorie der Legendreschen Polynome*, Jahre. Deutsch. Math.-Verein. **40** (1931), 163–166; also in *Collected papers*. Vol. 2, Birkhäuser, Boston, 1982, pp. 261-264.

68. ____, *Power series with multiply monotonic sequences of coefficients*, Duke Math. J. **8** (1941), 559–564; also in *Collected papers*. Vol. 2, Birkhäuser, Boston, 1982, pp. 797–802.

69. ____, *Orthogonal polynomials*, Amer. Math. Soc. Colloq. Publ., vol. 23, 4th ed., Amer. Math. Soc., Providence, R. I., 1975.

70. P. Turán, *On a trigonometric sum*, Ann. Polon. Math. **25** (1953), 155–163.

71. L. Vietoris, *Über das Vorzeichen gewisser trigonometrischer Summen*, S.-B. Öst. Akad. Wiss. **167** (1958), 125–135; Anzeiger Öst. Akad. Wiss. (1959), 192–193.

72. R. E. Williamson, *Multiply monotonic functions and their Laplace transforms*, Duke Math. J. **23** (1955), 189–207.

73. W. H. Young, *On a certain series of Fourier*, Proc. London Math. Soc. (2) **11** (1912), 357–366.

74. ____, *On the Fourier series of bounded functions*, Proc. London Math. Soc. **12** (1913), 41–70.

On Interpolation, Blaschke Products, and Balayage of Measures

ARNE BEURLING

I. Introduction. During the past decades a considerable interest has been focused on the Pick-Nevanlinna interpolation problem initiated by the two authors in the 1920's. The popularity gained by this field seems to be due to its inherent nature, amenable to a variety of elementary methods. This is the case particularly in its connection with Banach algebras as evidenced in Fisher's book [8], where many of the basic results are obtained by amazingly simple methods. The same can hardly be said about the purely function-theoretic part of the theory where sometimes well-known classical methods have been passed by in favor of procedures not indigenous to the problem at hand. Two cases in point: the principle of argument has not been used to its full potential in connection with uniqueness questions; still wanting is a study of the boundary behavior of infinite Blaschke products and of similar functions occurring in the theory even in the case of simply connected regions.

The following definitions and notations will be used: by Δ we shall denote the open unit disk, and $H^p(\Delta)$, $1 \leq p < \infty$, will stand for the Hardy spaces of functions analytic in Δ and with norms

$$\|f\|_p = \lim_{r=1} \left\{ \frac{1}{2\pi} \int |f(re^{i\theta})|^p \, d\theta \right\}^{1/p},$$

and if $p = \infty$, $\|f\|_\infty = \sup_{z \in \Delta} |f(z)|$.

If α, β, are two points in Δ, the quantity

$$[\alpha; \beta] = \left| \frac{\alpha - \beta}{1 - \alpha\bar{\beta}} \right|$$

is a conformal invariant under selfmappings of Δ and shares this property with the hyperbolic distance $d = d(\alpha, \beta)$ in the geometry in Δ where the linear element is $2|dz|/(1 - |z|^2)$. Since $[\ ;\ ]$ only assumes values $\in [0, 1)$, this distance is sometimes called the pseudo-euclidean in addition to the more familiar term noneuclidean. Its relations to d and to the Green function $g = g(\alpha, \beta)$ are as follows:

$$[\ ;\ ] = \frac{1 - e^{-d}}{1 + e^{-d}} = e^{-g}.$$

We note in passing that the modulus of a Blaschke product with simple zeros at $\{\alpha_i\} \subset \Delta$ can be written

$$|B(z)| = \prod_i [z; \alpha_i].$$

Another thing to remember is that the geodesics in the hyperbolic metric in Δ are formed by circular arcs orthogonal to $\partial \Delta$. Any such arc divides the space into two hyperbolic halfspaces, one of which is open and the other closed in case it incorporates γ itself. The hyperbolic hull of a closed set $K \subset \Delta$ is by definition the intersection of all closed hyperbolic halfspaces containing K.

An infinite set $E = \{\alpha_i\}_1^\infty \subset \Delta$ will be called uniformly separated in the hyperbolic metric if

$$\rho = \rho(E) = \inf_{i \neq j} [\alpha_i; \alpha_j] > 0.$$

If E is finite and consists of distinct points $\{\alpha_i\}_1^n$, then the Pick-Nevanlinna interpolation problem

(1) $$f(\alpha_i) = w_i, \quad 1 \leq i \leq n, \, f \in H_\infty(\Delta),$$

always has a solution, and there is always at least one of them which is minimal in the sense of having minimal norm $\|f\|_\infty$. By $\lambda = \lambda(E)$ we shall denote the interpolation bound for E defined as the least number λ such that (1) has a solution f with $\|f\|_\infty \leq \lambda \max_i |w_i|$. An infinite set E is called an interpolation set if the system

$$f(\alpha_i) = w_i, \quad i = 1, 2, \ldots, \, f \in H^\infty(\Delta),$$

has a solution for any choice of bounded w_i. This is the case if and only if

$$\lambda(E) = \lim_{n=\infty} \lambda(E_n) < \infty, \quad E_n = \{\alpha_i\}_1^n, \, \bigcup_i^\infty E_n = E.$$

The numbers $\rho(E)$ and $\lambda(E)$ are evidently invariant under conformal mapping of Δ onto itself or onto any other simply connected region D. Together with two other invariants ρ' and ρ'' to be defined later, they characterize basic properties of E.

In certain situations it will be convenient to replace Blaschke products by a slightly more general notion to be called Blaschke functions defined as follows.

If D is a simply connected region with a rectifiable Jordan boundary and $f \in H_\infty(\infty)$, then f is a Blaschke function of the first kind if its zeros are distinct and if $|f(z)|$ tends uniformly to $\|f\|_\infty$ as z tends to ∂D. The notation $\mathfrak{b}_n$, $n = 1, 2, \ldots$, will stand for all such f having at most n zeros in D. A function f with an infinity of zeros $\{\alpha_i\}_1^\infty \subset D$ is a Blaschke function of the second kind if these conditions are fulfilled:

(i) $|f(z)|$ tends a.e. to $\|f\|_\infty$ at ∂D (w.r.t. harmonic measure);

(ii) there exist two numbers ρ'' and ρ', $0 < \rho'' < \rho' < \|f\|_\infty$, such that the regions $D_i = \{z : [z; \alpha_i] < \rho'\}$ do not intersect and $|f(z)| \geq \rho''$ in the closed arcwise-connected set $\overline{D} \setminus \bigcup_i^\infty D_i$.

These definitions extend to multiply connected regions but require considerable modification in order to be useful.

We shall next prove an elementary lemma containing the gist of the classic results of Pick and Nevanlinna and therefore deserving a place in this introduction, even though its content is well-known today.

LEMMA. *If $E = \{\alpha_i\}_1^n \subset \Delta$ consists of distinct points, the problem (1) always has as its minimal solution a Blaschke function φ of the first kind with at most $n-1$ zeros. Hence $\varphi \in \mathfrak{b}_{n-1}$. No other minimal solution of (1) exists throughout $H_\infty(\Delta)$.*

The simplest proof is by induction. For $n = 1$ the lemma is true, since (1) has no other solution than $f = w_1$ and $\mathfrak{b}_0$ consists only of constants. We have, therefore, to show that if the first part of the lemma holds for $n - 1$ it remains true for n. To this end, assume that ρ is the minimal norm for solutions of the given system (1). Choose an index j, $1 \leq j \leq n$, and let S be a selfmapping of the disk $|w| < \rho$ taking w_j to the origin. Similarly, let T be a selfmapping of Δ taking α_j to the origin. Consider now the $(n-1)$-dimensional system where we define $\alpha_i' = T\alpha_i$, $w_i' = Sw_i$, $1 \leq i \leq n$,

$$(2) \qquad g(\alpha_i') = w_i'/\alpha_i', \qquad i \neq j, \ 1 \leq i \leq n,$$

and note that the α_i' occurring here are $\neq 0$. Since the first part of the lemma is true for $n-1$, (2) has a minimal solution $g \in \mathfrak{b}_{n-2}$. Consequently $h(z) = zg(z) \in \mathfrak{b}_{n-1}$, $\|h\| = \|g\|$, and since $0 = h(\alpha_j') = w_j'$, $h(z)$ solves the n-dimensional system

$$(3) \qquad h(\alpha_i') = w_i', \qquad 1 \leq i \leq n.$$

In order to obtain the solution of (1), we first form $f(z) = h \circ T^{-1}(z)$ and note that $f \in \mathfrak{b}_{n-1}$ and has norm ρ and solves the equation $f(\alpha_i) = w_i'$, $1 \leq i \leq n$. Finally $g(z) = S^{-1} \circ f(z)$ is the requested minimal solution of (1).

Still to be established is the general uniqueness stated in the lemma. Let f be the solution constructed above and assume, as we may, $\|f\|_\infty = 1$. Let k be a competing minimal solution. In what follows, $N(h; r)$ shall denote the number of zeros of h in the disk $|z| < r$. Since $|f|$ tends uniformly to 1 at $\partial \Delta$ and $|f + k| < 2$ in Δ, there will exist a function $r_\varepsilon \in (0, 1)$ increasing continuously to 1 as $\varepsilon \downarrow 0$ and such that $(1 + \varepsilon)f$ will be a strict majorant of $\frac{1}{2}(f + k)$ in the region $r_\varepsilon \leq |z| < 1$ if ε is sufficiently small. By an application of the theorem of Rouché to the function

$$(4) \qquad h_\varepsilon = (1 + \varepsilon)f - \tfrac{1}{2}(k + f)$$

and the disk $|z| < r_\varepsilon$, we shall have

$$(5) \qquad N(h_\varepsilon; r_\varepsilon) = N(f; r_\varepsilon) \leq n - 1,$$

povided all the zeros of f are contained in the disk. The assumption $f - k \not\equiv 0$ implies that the minimum modulus

$$m(r) = \min_{|z|=r} \tfrac{1}{2}(f - k)$$

is positive for a sequence r_ν tending to 1. If $\varepsilon_\nu > 0$ are chosen $< m(r_\nu)$ and (4) is written in the form $h_\varepsilon = \varepsilon f + \frac{1}{2}(f - k)$, we conclude by a second application of the theorem of Rouché that

(6) $$N(h_{\varepsilon_\nu}; r_\nu) = N(f - k; r_\nu) \geq n$$

if the disk $|z| < r_\nu$ contains all the α_i. When ε tends to 0, h_ε converges uniformly to $\frac{1}{2}(f - k)$ on each relative compact subset of Δ. Relation (5) implies that the limit function has at most $n - 1$ fixed zeros in Δ, while (6) proves that the same function has at least n fixed zeros there. This contradiction proves that our assumption $f - k \neq 0$ is wrong, and this establishes the uniqueness of f.

Rolf Nevanlinna was the first to use an induction proof of the previous kind starting with $n = 1$. It is interesting to note that even if two such schemes have different intermediate steps, they will in the end arrive at the same Blaschke function of the first kind. The general uniqueness result can therefore not be proved by induction from $n = 1$.

II. The interpolation bound, the set E^* associated to E, the Gabriel problem, and the linear interpolation operator.

To begin with we shall give for a finite set $E = \{\alpha_i\}_i^n \subset \Delta$ a new definition of $\lambda(E)$ independent of its role in interpolation. As before E shall always consist of distinct points.

THEOREM I. *Let B_{n-1} be the set of all Blaschke products with exactly $n-1$ zeros. Then the minimum maximorum problem*

(7) $$\tau = \min_{\varphi \in B_{n-1}} \max_{\alpha \in E} |\varphi(\alpha)|$$

has a solution φ with these properties:

(i) *φ is unique up to a unimodular constant and its $n - 1$ zeros are not necessarily distinct. They form by definition the associated set E^* of E. The operator $S: E \to E^*$ commutes with conformal selfmappings T of Δ: $ST(E) = TS(E)$.*

(ii) $|\varphi(\alpha)| = \tau$, $\alpha \in E$.

(iii) $\lambda = \lambda(E) = \tau^{-1}$.

The set E^* as defined in Theorem I will later be proven identical with a set derived from another variational problem. The ensuing proof of Theorem I depends essentially on the results on Gabriel's problem given in seminars at Uppsala University during the 40's and early 50's. In a modified form the problem of Gabriel can be stated this way: If γ is a rectifiable Jordan curve in Δ not meeting $\partial \Delta$, try to estimate

$$\lambda(\gamma) = \sup_{\|f\|_1 = 1} \int_\gamma |f(z)\, dz|, \qquad f \in H_1(\Delta),$$

and find the characteristic properties of maximal solutions f. Since $f(z)$ will remain uniformly bounded on γ, there will always exist maximal solutions always free of zeros in Δ. A more flexible formulation of the problem is obtained on replacing the arc measure $|dz|$ on γ by a more general positive measure μ with support restricted to some disk $|z| \leq r < 1$. In what follows we shall prefer this formulation and write

$$\lambda(\mu) = \sup \int_\Delta |f(z)|\, d\mu, \qquad \|f\|_1 = 1. \tag{8}$$

We note that $\lambda(\mu)$ is subadditive: $\lambda(\mu_1 + \mu_2) \leq \lambda(\mu_1) + \lambda(\mu_2)$.

This kind of Gabriel's problem has one interesting feature: it can easily be solved as a Hilbert space problem and lends itself with equal ease to a potential-theoretic treatment. This duality results in two different sets of necessary and sufficient conditions on maximal solutions. It is this multiple information that creates its usefulness.

In order to adapt the problem sketched in (8) to the proof of our theorem, we shall choose this discrete measure

$$\mu = \sum_{i=1}^n b_i \delta_{\alpha_i}; \qquad b_i = |B'(\alpha_i; E)|^{-1}, \tag{9}$$

where $B(z; E)$ is the Blaschke product vanishing on E and δ_{α_i} denotes the Dirac measure at α_i. Since a maximal solution f of (8) has no zeros in Δ, we can normalize f by the condition $f(0) > 0$ and define a unique function $F(z) \in H_2(\Delta)$ by the relations $f(z) = F^2(z)$, $F(0) > 0$. The two different methods of solving (8) yield these conditions for an extremal $F(z)$:

$$\int_\Delta \frac{F(z)\, d\mu(z)}{1 - \varsigma \bar{z}} = \lambda(\mu) F(\varsigma), \qquad |\varsigma| < 1/r,\ r = \max |\alpha_i|. \tag{10}$$

$$\int_\Delta \frac{1 - |z|^2}{|\varsigma - z|^2} |F(z)|^2\, d\mu(z) = \lambda(\mu) |F(\varsigma)|^2, \qquad |\varsigma| = 1. \tag{11}$$

The first of these relations is obtained by a Hilbert space method and the second by a potential-theoretic approach. We learn by (10) that $F(\varsigma)$ is analytic and single-valued in the disk $|\varsigma| < 1/r$ and can be continued analytically along any path not meeting points $1/\bar\varsigma$, $\varsigma \in \operatorname{supp}\mu$. We know already that $F \neq 0$ in Δ, and by (11) this property is extended to hold on $\partial \Delta$ also.

In order to prove (10) we introduce two scalar products for functions $\in H_2(\Delta)$:

$$(F, G)_\Delta = \int_\Delta F\overline{G}\, d\mu, \qquad (F, G)_{\partial \Delta} = \int_{\partial \Delta} F\overline{G} \frac{|dz|}{2\pi}.$$

It is readily seen that F is maximal if and only if $(F, G)_\Delta = \lambda(\mu)(F, G)_{\partial \Delta}$ for all $G \in H_2(\Delta)$ and thus for $G(z) = 1/(1 - \bar\varsigma z)$, $|\varsigma| < 1/r$. This choice of G makes the scalar equation above identical with (10).

In proving (11) we set $|f(z)| = |F(z)|^2 = e^{u(z)}$, where $u(z)$ is harmonic in the closure $\overline\Delta$ of Δ. Let h be any similar but nonconstant harmonic function in $\overline\Delta$. $F(z)$ is then maximal if and only if

$$\int_\Delta h e^u\, d\mu - \lambda(\mu) \int_{\partial \Delta} h e^u \frac{|d\varsigma|}{2\pi} = 0, \tag{12}$$

$$\int_\Delta h^2 e^u \, d\mu - \lambda(\mu) \int_{\partial\Delta} h^2 e^u \frac{|d\varsigma|}{2\pi} < 0. \qquad (13)$$

It should be pointed out that (13) is automatically satisfied since the harmonic extension to Δ of $h^2(\varsigma)$, $|\varsigma|=1$, is a strict majorant of the subharmonic $h^2(z)$ in Δ. If $h(z)$ in (12) is replaced by its Poisson representation

$$h(z) = \frac{1}{2\pi} \int_{\partial\Delta} \frac{1-|z|^2}{|\varsigma-z|^2} h(\varsigma) |d\varsigma|$$

and the order of integration in (12) is reversed, then

$$\int_\Delta e^{u(z)} \frac{1-|z|^2}{|\varsigma-z|^2} d\mu = \lambda e^{u(\varsigma)}, \qquad |\varsigma|=1, \qquad (14)$$

and this proves (11). Here we multiply both sides by $e^{-u(\varsigma)}$, insert the measure μ defined by (9), and (14) will read:

$$\sum_1^n \frac{|F(\alpha_i)|^2}{|F(\varsigma)|^2} \frac{(1-|\alpha_i|^2)}{|\varsigma-\alpha_i|^2} \frac{1}{|B'(\alpha_i)|} = \lambda. \qquad (14')$$

Similarly, if μ is inserted in relation (10), we obtain

$$\sum_1^n \frac{b_i F(\alpha_i)}{1-\varsigma\overline{\alpha}_i} = \lambda(\mu) F(\varsigma), \qquad |\varsigma| < 1/r. \qquad (15)$$

Since $b_i > 0$ and $F(\alpha_i) \neq 0$, the equation above can be written

$$\frac{P_{n-1}(\varsigma)}{\prod_1^n (1-\varsigma\overline{\alpha}_i)} = \lambda(\mu) F(\varsigma), \qquad (16)$$

where P_{n-1} is a polynomial of degree $n-1$, uniquely derived from F and with all its zeros located in $|\varsigma| > 1$. We shall denote these zeros by $1/\overline{\beta}_i$, $1 \leq i \leq n-1$, and later show that $E^* = \{\beta_i\}_1^{n-1}$ is the associated set of E occurring in Theorem I.

As a result of (15) and (16) we shall prove

$$\frac{F(\varsigma)}{F(0)} = \frac{\prod_1^{n-1}(1-\varsigma\overline{\beta}_i)}{\prod_1^n (1-\varsigma\overline{\alpha}_i)}. \qquad (17)$$

On squaring this formula and multiplying both sides by φ/B, where φ and B are the Blaschke products vanishing on E^* and E respectively, we get

$$\frac{F^2(\varsigma)\varphi(\varsigma)}{F^2(0)B(\varsigma)} = \frac{\prod_1^{n-1}(\varsigma-\beta_i)(1-\varsigma\overline{\beta}_i)}{\prod_1^n(\varsigma-\alpha_i)(1-\varsigma\overline{\alpha}_i)}. \qquad (18)$$

It should be noted that on $|\varsigma|=1$ the right-hand member above equals

$$\frac{\varsigma^{-1} \prod_1^{n-1} |\varsigma-\beta_i|^2}{\prod_1^n |\varsigma-\alpha_i|^2}. \qquad (19)$$

The equations given so far are not only steps in proving Theorem I but are also aimed to exhibit the relations between the two sets E and E^*. As a consequence a combination of (18) and (19) together with the normalization $F(0) > 0$ now yields the equation

$$(20) \qquad 1 = \int_{\partial \Delta} \frac{F^2(\varsigma)\varphi(\varsigma)\,d\varsigma}{B(\varsigma)2\pi i} = \sum_{\alpha \in E} \frac{F^2(\alpha)\varphi(\alpha)}{B'(\alpha)},$$

where the integral reduces to $\int_{\partial \Delta} |F^2(\varsigma)||d\varsigma|/2\pi = 1$ since the modulus of the integrand is $|F^2(\varsigma)d\varsigma|/2\pi$ and its argument by (19) is equal to $d\varsigma/2\pi i = |d\varsigma|/2\pi$. A maximal solution F^2 of the Gabriel problem will by definition satisfy the equation

$$(20') \qquad \sum_{\alpha \in E} \left| \frac{F^2(\alpha)}{B'(\alpha)} \right| = \lambda.$$

If we could show that

$$(21) \qquad |\varphi(\alpha)| \leq \lambda^{-1}, \quad \alpha \in E,$$

then the sign of equality had to be valid above in order not to violate relation (20), and the argument of $\varphi(\alpha)$ would have to satisfy the condition $F^2(\alpha)\varphi(\alpha)/B'(\alpha) > 0$. In other words, if (21) holds, then the values of $\varphi(\alpha)$ on E are uniquely determined. For the proof of (21) we shall use the second alternative solution of Gabriel's problem written in the form (14'):

$$(21') \qquad \sum_1^n \left| \frac{F^2(\alpha_i)}{F^2(\varsigma)} \cdot \frac{1 - |\alpha_i|^2}{|\varsigma - \alpha_i|^2} \cdot \frac{1}{B'(\alpha_i)} \right| = \lambda, \qquad |\varsigma| = 1.$$

We aim to show for each index i the existence of a function $\psi_i(\varsigma) \in H_\infty(\Delta)$ complying with these conditions: $\psi_i(\alpha_j) = 1$ if $i = j$, and $= 0$ if $i \neq j$; secondly, $|\psi_i(\varsigma)|$ should equal the term of that index in the series; thirdly, $\psi_i(\varsigma)$ is allowed to equal a "pseudofactorization" within $|\cdot\cdot|$, meaning that $a = bc$ if $|a| = |bc|$ even though $\arg a \neq \arg bc$. A problem of this kind is in general of diophantine character, but in this case the solution is quite obvious: $F(\varsigma)$ should remain untouched and what remains now is to write

$$|\varsigma - \alpha_i|^2 = (\varsigma - \alpha_i)|1 - \varsigma\bar{\alpha}_i|$$

and

$$\frac{1 - |\alpha_i|^2}{(\varsigma - \alpha_i)^2} \cdot \frac{1}{B'(\alpha_i)} = \frac{1 - |\alpha_i|^2}{|1 - \varsigma\bar{\alpha}_i|} \cdot \frac{B(\varsigma)}{B'(\alpha_i)(\varsigma - \alpha_i)}.$$

Now the Kronecker condition is satisfied, and we can choose

$$\psi_i(\varsigma) = \frac{F^2(\alpha_i)}{F^2(\varsigma)} \cdot \frac{1 - |\alpha_i|^2}{1 - \varsigma\bar{\alpha}_i} \cdot \frac{B(\varsigma)}{B'(\alpha_i)(\varsigma - \alpha_i)}$$

and thus,

$$(22) \qquad \sum_1^n |\psi_i(\varsigma)| = \lambda, \qquad |\varsigma| = 1.$$

This proves $|\varphi(\alpha)| = 1/\lambda$, $\alpha \in E$, and therefore also (21) as well as Theorem I except for the uniqueness stated there, which rests on the assumption that a normalized F is itself unique. If that assumption were untrue, there would exist two different normalized F, say F and G, leading to two different polynomials of degree $n-1$ in equation (16) and finally to two different systems, say $\{\psi_i\}_1^n$ and $\{\tilde{\psi}_i\}_1^n$, in the linear interpolation formula satisfying

$$(22) \qquad \sum_1^n |\psi_i(\varsigma)| = \sum_1^n |\tilde{\psi}_i(\varsigma)| = \lambda, \qquad \varsigma \in \partial\Delta,$$

and consequently

$$(22') \qquad \sum_1^n |\psi_i(\varsigma) + \tilde{\psi}_i(\varsigma)| = \sum_1^n (|\psi_i(\varsigma)| + |\tilde{\psi}_i(\varsigma)|), \qquad \varsigma \in \partial\Delta.$$

Since ψ_i and $\tilde{\psi}_i$ are all analytic and $\neq 0$ in a neighborhood of $\partial\Delta$, it follows by (22) and (22') that the functions $f_i = \psi_i/\tilde{\psi}_i$ have constant arguments on each point of $|\varsigma| = 1$. The equations (15) and (16) are valid independently of whether or not F is unique and yield together with (22) this explicit relation:

$$(23) \qquad \psi_i(\varsigma) = \frac{c_i \prod_{i\neq j=1}^n (\varsigma - \alpha_i)(1 - \varsigma\overline{\alpha}_i)}{\prod_{j=1}^{n-1}(1 - \varsigma\overline{\beta}_j)^2},$$

c_i being constants making $\psi_i(\alpha_i) = 1$. Now if $\{1/\tilde{\gamma}_i\}_1^n$ denote the numbers similarly derived from G we would have

$$\psi_i(\varsigma)/\tilde{\psi}_i(\varsigma) = k_i \prod_{j\neq i}(1 - \varsigma\overline{\tilde{\gamma}}_j)^2/(1 - \varsigma\overline{\beta}_i)^2,$$

with $k_i = c_i/\tilde{c}_i$. Each function $f_i(\varsigma)$ is analytic and zero-free in $\overline{\Delta}$ with constant argument on $\partial\Delta$. Therefore f_i is a constant function and the constant has to be 1 since by the Kronecker condition $f_i(\alpha_i) = 1$. This ends the proof of Theorem I and the uniqueness of F and the associated set E^* of E.

Of course, the minimum maximorum approach in Theorem I cannot be used in case E is an infinite interpolation set, but it applies to its finite subsets E_n. There exists therefore unique functions $\{\psi_i\}_1^n$ derived from the set E_n and such that the linear operator T_n taking any n-tuple $w = \{w_i\}_1^n$ into $H_\infty(\Delta)$ by the formula

$$(24) \qquad T_n(w) = \sum_{i=1}^n w_i \psi_i(\varsigma) \in H_\infty(\Delta).$$

The most interesting fact about T_n is that its operator norm equals the interpolation bound $\lambda(E_n)$.

Due to the equality sign in (22), it is not difficult to establish existence and uniqueness of the operator T even for infinite interpolation sets.

III. The boundary behavior of infinite Blaschke products. If in Gabriel's problem the measure μ does not charge $\partial\Delta$ but its support meets $\partial\Delta$, then $\lambda(\mu)$ may not be finite. The most direct way to approach this question is to consider the restriction of μ to a disk $|z| \leq 1-\varepsilon$ and then try to find whether or not $\lambda(\mu_\varepsilon)$ remains bounded as $\varepsilon \downarrow 0$. This in turn requires good estimates of $\mu(\omega)$ for simply connected regions ω close to $\partial\Delta$ and of small Euclidean diameter, leading to the boundary behavior of infinite Blaschke products. Prior to considering this problem, we want to apply another method based on the linear interpolation operator for finite sets E. The following theorem gives important results on $\lambda(E)$ as related to E and the two new invariants $\tau(E)$ and $\bar{\tau}(E)$ defined below.

THEOREM II. *Let $\{\alpha_i\} \subset \Delta$ be a finite or infinite sequence of distinct points such that $\sum_{\alpha\in E}(1 - |\alpha|^2) < \infty$, and let B be the Blaschke product vanishing on E. Define*

(25)
$$\begin{cases} \tau(\alpha) = |B'(\alpha)|(1 - |\alpha|^2), \\ \tau(E) = \inf_{\alpha \in E} \tau(\alpha), \\ \bar{\tau}(E) = \sup_{\alpha \in E} \tau(\alpha). \end{cases}$$

The condition

(26)
$$\tau(E) > 0$$

implies

(27)
$$1 \le \lambda(E)\tau(E) \le \sum_{\alpha \in E} |F^2(\alpha)|(1 - |\alpha|^2) \le \lambda(E)\bar{\tau}(E),$$

(28)
$$\lambda(E)\tau(E) \le \sum_{\alpha \in E} \frac{|F^2(\alpha)|\tau(\alpha)}{|B'(\alpha)|} \le \lambda(E)\bar{\tau}(E),$$

where F^2, $\|F\|_2 = 1$, $F(0) > 0$, stands for the maximal solution of the Gabriel problem (8) with μ determined by (9).

Consider first the finite case where E_n are finite subsets of E such that $\bigcup_1^\infty E_n = E$. Write $B_i(z) = \prod_{j \ne i}(z - \alpha_j)/(1 - z\bar{\alpha}_j)$, $1 \le i \le n$, and note the identities

(28')
$$\tau(\alpha_i) = |B'(\alpha_i)|(1 - |\alpha_i|^2) = \prod_{j \ne i}[\alpha_j; \alpha_i].$$

Since $\{\min |B_i(z)| : z \in E_n = \tau(E_n)\}$ and $B_i(\alpha_j) = 0$, $j \ne i$, it trivially follows that $\lambda(E_n) \ge 1/\tau(E_n)$. This proves the first part of (27). The remaining inequalities depend on the linear interpolation operator for finite sets E_n. By (20') and (25)

(29)
$$\lambda(E_n) = \sum_{\alpha \in E_n} \frac{|F^2(\alpha)|}{|B'(\alpha)|} \equiv \sum_{\alpha \in E_n} \frac{|F^2(\alpha)|(1 - |\alpha|^2)}{\tau(\alpha)},$$

and the two other relations in (27) are consequences of the definitions of $\tau(E)$ and $\bar{\tau}(E)$. In the infinite case it is sufficient to apply the previous proof to a sequence $\{E_n\}_1^\infty$ of finite subsets of E exhausting E. Furthermore, the limits of $\lambda(E_n)$ and

$\tau(E_n)$ as $n \nearrow \infty$ do always exist, $\tau(E_n)$ being decreasing and $\lambda(E_n)$ increasing for increasing n. This finishes the proof of (27). As for (28) we first note that the charge to Δ given by the measure μ is always $\mu(\Delta) = \sum |B'(\alpha)|^{-1} \leq \lambda(E)$. The rest the proof of (28) follows the previous lines if we write $1 - |\alpha|^2 = \tau(\alpha)/|B'(\alpha)|$.

Theorem II has this obvious corollary: *A set E satisfying the conditions of Theorem II is an interpolation set if and only if*

(30) $$\tau(E) = \inf_{\alpha \in E} |B'(\alpha)|(1 - |\alpha|^2) > 0.$$

Relations (27) and (28) also give a sharper form and a simple proof of the inequality

(31) $$\sum_{i=1}^{\infty} |F^2(\alpha_i)|(1 - |\alpha_i|^2) \leq c\|F\|_2, \qquad F \in H_2(\Delta), \ \|F\|_2 = 1,$$

proved by Carleson with an unspecified constant c under the condition $\tau(E) > 0$, (cf. [2]). Several other proofs of (31) are known. Among these we mention the results by Shapiro and Shields [12], Hörmander [10], and Duren [5].

We shall next take up the study of Blaschke products of the second kind. By the definition in the introduction, $B(z)$ is of the second kind if there exist two constants $0 < \rho'' < \rho' < 1$ such that the open disks $\Delta_i = \{z : [z; \alpha_i] < \rho'\}$ do not intersect and $|B(z)| \geq \rho''$ on the closed arcwise-connected set $\overline{\Delta} \setminus \bigcup_{i=1}^{\infty} \Delta_i$. If the invariant $\rho = \rho(E) = \inf_{j \neq i}[\alpha_j; \alpha_i]$ is given, which we always assume, then the nonintersection condition implies that the range for ρ' has an upper bound ρ_0 in order to be compatible with the value of ρ. More precisely,

(31') $$\rho' \leq \rho_0 \equiv \rho/\left(1 + \sqrt{1 - \rho^2}\right).$$

The sign of equality here does not prevent two different circles $\partial \Delta_i$ and $\partial \Delta_j$ from being tangential, but it does prevent the set $\overline{\Delta} \setminus \bigcup_i^{\infty} \Delta_i$ from being disconnected. For a given set E there may exist an infinity of values for ρ' and ρ'', compatible with E and with each other. It is therefore appropriate to consider methods enabling ρ' and ρ'' to be given determined values. This shall be done later.

If $B(z)$ is a Blaschke product of the second kind, then

$$f_i(z) \equiv B(z)(1 - z\overline{\alpha}_i)/(z - \alpha_i) \neq 0$$

in Δ_i and $|f_i(z)| \geq \rho''/\rho'$, in $\partial \Delta_i$. Hence by the minimum principle

$$|f_i'(\alpha_i)| = |B'(\alpha_i)|(1 - |\alpha_i|^2) \geq \rho''/\rho', \qquad \alpha_i \in E.$$

By Theorem II this implies $\tau(E) \geq \rho''/\rho'$ and $\lambda(E) < \infty$, proving that E is an interpolation set. If, on the other hand, the zeros of $B(z)$ form an interpolation set, then $\lambda(E) < \infty$ and $\tau(E) = \inf_i \prod_{j \neq i}[\alpha_i; \alpha_j] > 0$ follows.

The inequalities previously proved in connection with Theorem II now yield the existence of two constants ρ' and ρ'' such that

$$0 < \rho'' < \rho' < \rho/\left(1 + \sqrt{1 - \rho^2}\right),$$

hence fulfilling the conditions for $B(z)$ to be a product of the second kind. The following result is herewith established.

THEOREM III. *The notion of Blaschke products of the second kind is identical with the notions of infinite Blaschke products with zeros forming an interpolation set.*

In connection with Theorem II we want to pay some attention to a problem regarding the five $\leq$ signs occurring in relations (27) and (28). All of them can be replaced by equalities provided the $\tau(\alpha)$ are constant on E. This case occurs if E is finite and contained in a disk $\Delta_0 = \{z : [z; z_0] \leq r_0 < 1\}$ and if there exists a selfmapping of Δ taking Δ_0 into itself and E into itself. Such a mapping has to have z_0 as a fixpoint. After transforming z_0 to $\varsigma = 0$, the action of T will be a rotation of Δ by the amount $2\pi/p$ with $p = n$ if E consists of the roots of the equation $\varsigma^n - r^n = 0$, $r < 1$. It is easy to see that the minimax function φ in this case has to be ς^{n-1}, and thus

$$\lambda(E) \geq r^{-(n-1)}.$$

The underlying problem is to find as $n \to \infty$ the configuration of E and the value of $r < 1$ making $\lambda(E)$ minimal when $\#E = n$. This question is also of interest due to its close relation to Euler's work on integration and summation of series entering in the problem outlined above.

Our next objective is to give fixed values to ρ' and ρ'' and then to investigate their relation to the boundary behavior of $B(z)$ and to the distribution of the measure μ. For this purpose let E be the given set and B the Blaschke product vanishing on E. Denote by T_i the linear mapping $(z - \alpha_i)/(1 - z\bar{\alpha}_i) \to \xi$.

It takes α_i to the origin $\varsigma = 0$, Δ_i into the disk $|\varsigma| < \rho'$, and Δ onto itself. As proved before we shall have $|B(z)(1 - z\bar{\alpha}_i)/(z - \alpha_i)| \geq \rho''/\rho'$ in $\partial\Delta_i$ and $|B'(\alpha_i)|(1 - |\alpha_i|^2) \geq \rho''/\rho'$.

If $B_i(z) \equiv B(z)(1 - z\bar{\alpha}_i)/(z - \alpha_i)$ and $m_i(r)$ is the minimum of $|B_i(z)|$ on $\{z[z; \alpha_i] = r\}$, then $\log m_i(r)$ is a concave function of $\log r$ for $r \leq \rho'$, and the same is true of the lower envelope $\log m(r)$ of the family $\{\log m_i(r)\}_1^\infty$, according to the three circle theorem of Hadamard. Being interested mainly in the product $r\, m(r)$, we have to consider the graph of $u(t) = t + \log m(e^t)$ for $t = \log r < \log \rho$. As $t \to -\infty$, $u(t)$ has a majorizing asymptote $u(t) = t - a$ with $a = -\log \tau(E)$. As $t \to \log \rho$, $u(t) \to -\infty$ since by assumption there exists either at least 2 indices i such that $m_i(\rho) = 0$ or an infinity of indices i making $m_i(\rho) \to 0$. In either case the line $t = \log \rho$ is a vertical asymptote for the graph, and the concave function $u(t)$ increases to an optimal value $u(t_0)$ when it increases from $-\infty$ to t_0 and then decreases to $-\infty$ as $t \to \log \rho$. The coordinates of the optimal point of $u(t)$, $\rho'(E)$ and $\rho''(E)$, shall be defined as e^{t_0} and e^{u_0} respectively. It should be noted that du/dt is strictly decreasing on each interval I lest $\log m(e^t)$ be linear on I which would contradict the fact that the $B_i(z)$ are zero-free in $\{z : [z; \alpha_i] < r\}$. The graph of $u(t)$ can therefore not have a flat roof, and this makes the optimal point uniquely determined, together with the two new invariants. It should be noted that each point on the graph stays invariant under selfmapping of Δ.

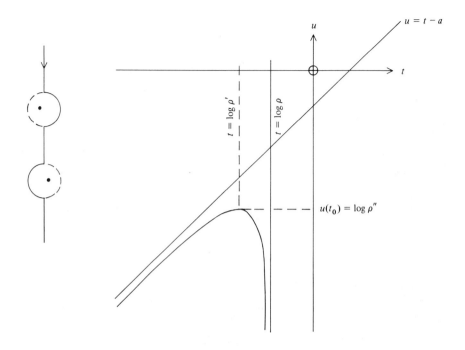

FIGURE 1

In what follows the set $A = \overline{\Delta} \setminus \bigcup_1^\infty \Delta_i$ will be called admissible and $P = \bigcup_1^\infty \Delta_i$ the prohibited set. If γ is a rectifiable Jordan arc or curve in Δ, we shall by $|A \cap \gamma|$ and $|P \cap \gamma|$ denote the total arc length of $A \cap \gamma$ and $P \cap \gamma$ respectively. Each γ to be considered will have its endpoints in A. Moreover, if γ is assumed convex, it will be supplied with a direction chosen so that the directional tangent turns to the left, and γ will intersect any disk Δ_i only at the point of entry and point of exit. The two basic problems to be considered are:

1°: If γ consists of a line segment joining its two endpoints in A, construct a rectifiable Jordan arc $\gamma' \subset A$ with the same endpoints as γ, and find a constant k depending only on E such that

(32) $$|\gamma'| \leq |A \cap \gamma| + k|P \cap \gamma|.$$

2°: If K is a closed convex subset of $\overline{\Delta}$, construct a rectifiable Jordan curve γ' encircling K and such that

(33) $$|\gamma'| \leq k|\partial K|,$$

where the constant k depends only on E.

Problem 1° belongs to elementary plane geometry while 2° is more complicated. In connection with the first case we shall prove:

THEOREM IV. *Let $B(z)$ be a Blaschke product of the second kind with simple zeros at E. Then there exists for each $\varepsilon > 0$ a Jordan curve γ' contained in the* annulus $1 - \varepsilon < |z| \leq 1$ homotopic with $\partial \Delta$, of length $\leq \pi^2$, and such that $|B(z)| \geq \rho''$ on γ'.

If the origin is contained in A, then each point $e^{i\theta} \subset \partial \Delta$ can be joined to $z = 0$ by a Jordan arc of length $\leq \pi/2$ on which $|B(z)| \geq \rho''$.

In order to find a replacement γ' for a linear segment γ with endpoints in A, the following rule shall be used. If γ penetrates a disk Δ_i, we choose as replacement for the chord $\gamma \cap \Delta_i$ the shorter of the two arcs into which $\partial \Delta_i$ is divided, or in case of equal length, any of them (cf. Figure 1). The lengthening factor acting on each such chord is therefore $\leq \pi/2$, and this constant is obviously the best possible. It should be noted that if γ has an endpoint $e^{i\theta}$ on $\partial \Delta$ the Euclidean as well as the hyperbolic distance from $z \in \gamma$ to γ' can be easily estimated as $z \to e^{i\theta}$.

In proving the first part of the theorem, we inscribe in Δ a polygon with vertices $\in \partial \Delta$ and sides short enough so as not to meet the circle $|z| = 1 - \delta$, $\delta < \varepsilon$. On adding up the result for each side, one finds if δ is small enough that $\pi/2$ is still the best lengthening factor for $|P \cap \gamma|$. Hence the perimeter of the polygon is $\leq \pi^2$.

The previous results are included only in order to show some characteristic dissimilarities between an ordinary convergent Blaschke product and products of the second kind or with zeros on an interpolation set.

We still have to consider the case $2°$ of a closed convex subset K of $\overline{\Delta}$. Define for $\varepsilon > 0$, $K_\varepsilon = K \cap \{z : |z| \leq 1 - \varepsilon\}$, let $E = \{\alpha_i\}$ be the zeros of a product $B(z, E)$ of the second kind, and denote by A and A_ε the two subsets of E with α_i contained in K and K_ε respectively. Define

$$[z; K] = \inf_{z' \in K}[z; z'], \quad z \notin K, \qquad [z; K_\varepsilon] = \inf_{z' \in K_\varepsilon}[z; z'], \quad z \in K_\varepsilon,$$

and note that the inclusion $A \subset E$ implies $\lambda(A) \leq \lambda(E)$, $\rho'(A) \geq \rho'(E)$, and $\rho''(A) \geq \rho''(E)$. Keep in mind that K may very well contain all of E, in which case the equality sign will hold above.

In order to study the density problem for the measure μ close to $\partial \Delta$, we choose for K a closed hyperbolic halfspace K^θ limited by a geodesic g with endpoints $e^{\pm i\pi\theta}$, $0 < \theta \leq \frac{1}{2}$. K^θ is convex only if $\theta \leq \frac{1}{2}$, while K_ε^θ is convex and closed both in the euclidean and the hyperbolic space if $\varepsilon > 0$, $\theta \leq \frac{1}{2}$. Here K_ε^θ denotes the truncated K^θ defined as $K_\varepsilon^\theta \cap \{|z| : |z| \leq 1 - \varepsilon\}$ where ε is so small that $K_\varepsilon^\theta \neq \emptyset$. A_ε will be the subset of E contained in K_ε^θ. At this point it is appropriate to introduce the harmonic measure $\omega(z, \beta)$ of the arc β and to consider the level lines $l_t = \{z : \omega(z, \beta) = t\}$, $0 < t < 1$. As $t \searrow 0$, $l_t \nearrow (\partial \Delta \setminus \beta)$, while $l_{1/2} = g$ and $l_t \searrow \beta$ as $t \nearrow 1$. Each l_t consists of a circular arc joining the points $e^{\pm i\pi\theta}$ and forming there with β an interior angle $= \pi(1 - t)$. At the same points the angle between g and l_t is $\varphi = \pi/2 - \pi t$, provided $t \subset (0, \frac{1}{2})$. This implies $\cos \varphi = \sin \pi t$ and makes the radius of $l_t = \sin \pi \theta / \cos \varphi$ and its length

(34) $$|l_t| = \sin \pi \theta (\pi + 2\varphi)/\cos \varphi.$$

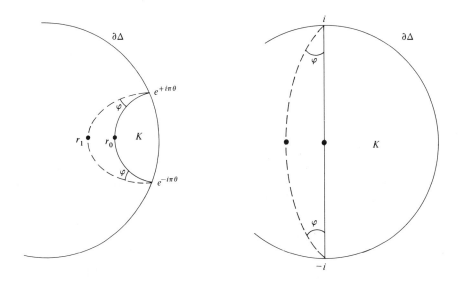

Figure 2

Among the level lines l_t, $0 < t \leq \frac{1}{2}$, the geodesic $g = l_{1/2}$ alone can be transformed by a selfmapping of Δ to a diameter of the disk. The left side of Figure 2 shows $K = K^\theta$, $g = l_{1/2}$, and an l_t with $t \subset (0, \frac{1}{2})$ together with the points r_0 and r_1 where g and l_t meet the radius $(0,1)$. The right side shows the images of K^θ, $g = l_{1/2}$, and the points r_0, r_1 under the mapping T taking $e^{\pm i\pi\theta}$ to $\pm i$.

The transformation T by necessity takes g to the vertical diameter of Δ and, by reason of symmetry, the point r_0 to the origin. Thus $T(z) = (z - r_0)/(1 - zr_0)$.

We apply this to the problem at hand and choose t so that $[r_0; r_1] = |T(r_1)| = \rho'$. By virtue of the definition of K^θ_ε, A_ε, ρ', and ρ'', this choice of t implies for $\varepsilon > 0$

$$(35) \qquad |B(z; A_\varepsilon)| \geq \rho''(A) \geq \rho''(E)$$

in the closed region limited by the arc $\partial\Delta\backslash\beta$ and the dotted level line in the left part of Figure 2, this region being identical with the set $\{z : 0 < \omega(z;\beta) < t\}$.

Our aim is to deduce for $\mu(K^\theta)$ an asymptotic estimate as $\theta \searrow 0$, a problem requiring an explicit relation between ρ' and φ and t. The right part of the figure shows that the radius R' of the dotted arc satisfies $R' \sin\varphi = 1$. The previous choice of t together with an elementary circle theorem yields the equation $\rho'(2R' - \rho') = 1$ with $\rho' = R' - \sqrt{R'^2 - 1}$ as sole acceptable root. Hence

$$(36) \qquad \rho' = (1 - \cos\varphi)/\sin\varphi = \tan(\varphi/2),$$

$$(37) \qquad \varphi = 2\arctan\rho',$$

$$\cos\varphi = \frac{1-\tan^2(\varphi/2)}{1+\tan(\varphi/2)} = \frac{1-\rho'^2}{1+\rho'^2}. \tag{38}$$

In conjunction with (34) we shall have for the length of l_t, since $\rho' < 1$,

$$|l_t| = \sin\pi\theta(\pi + 4\arctan\rho')\frac{1+\rho'^2}{1-\rho'^2} < 2\pi\sin\pi\theta\frac{1+\rho'^2}{1-\rho'^2}. \tag{39}$$

Together with (35) this yields as $\varepsilon \searrow 0$

$$\frac{1}{2\pi\theta}\int_{l_t}\left|\frac{dz}{B(z;A_\varepsilon)}\right| + \frac{1}{|\beta|}\int_\beta || \leq \frac{\sin\pi\theta}{\theta\rho''}\left(\frac{1+\rho'^2}{1-\rho'^2}\right) + \frac{\pi}{\rho''} \tag{40}$$

$$< \frac{\sin\pi\theta}{\theta}\frac{2}{\rho''(1-\rho'^2)},$$

where the majorant π/ρ'' of the second integral is obtained by replacing β by a chain of short chords γ_m tending to β and of total length $\leq 2\pi^2|\beta|$ according to the proof of Theorem IV. Now $f(z)$ be the minimal solution of this interpolation problem on the sequence E:

$$f(\alpha_i) = \begin{cases} \arg B'(\alpha_i;E), & \alpha_i \in A_\varepsilon, \\ 0, & \alpha_i \in E\setminus A_\varepsilon, \end{cases} \tag{41}$$

and note that $\|f\|_\infty \leq \lambda(E)$. If K^θ denotes the region limited by l_t and the arc β, then by (40)

$$\lim_{\varepsilon=0}\frac{1}{2\pi}\int_{\partial K^\theta}\left|\frac{dz}{B(z;A_\varepsilon)}\right| < \frac{2\sin\pi\theta}{\rho''(1-\rho'^2)}(1+\rho'^2),$$

and consequently,

$$\lim_{\varepsilon=0}\mu(K_\varepsilon^\theta) = \sum_{\alpha_i\in A_\varepsilon}\frac{1}{|B'(\alpha_i;E)|} = \left|\frac{1}{2\pi i}\int_{\partial K^\theta}\frac{f(z)\,dz}{B(z;E)}\right| \tag{42}$$

$$< \frac{2\lambda\sin\pi\theta}{\rho''(1-\rho'^2)}.$$

This yields the following explicit estimate of the "boundary density" for the measure μ

$$\lim_{\theta=0}\lim_{\varepsilon=0}\frac{\mu(K_\varepsilon^\theta)}{\theta} < \frac{2\pi\lambda}{\rho''(1-\rho'^2)}, \tag{43}$$

where λ, ρ'', and ρ' are referred to the set E. The constant 2π here can only be diminished by a fraction without violating the result.

It should be noticed that (42) as $\theta \searrow 0$ can be written in this more interesting form

$$\mu(K^\theta) < C_1(E)\sin\pi\theta, \qquad C_1(E) = 2\lambda/\rho''(1-\rho'^2), \tag{44}$$

where $C_1(E)$ is invariant under selfmappings of Δ, whereas the second factor depends only on K^θ and is invariant only under the subgroup leaving θ invariant.

The inequality (44) has a counterpart in the more exacting problem of majorizing the measure

$$\nu(K) = \int_K |F^2|\, d\mu,$$

where $F^2 = f$ is the normalized maximal solution of problem (8) and K^θ an open or closed hyperbolic halfspace limited by a geodesic g and an arc β of $\partial\Delta$ of length $2\pi\theta$, $0 < \theta \leq \frac{1}{2}$. The resolution of this problem depends essentially on a sophisticated lemma due to K. T. Smith [13], leading to a constant $C_0(E)$ such that in Δ

(45) $$|F^2(z)| \leq C_0(E).$$

Readers interested in the method will find an instructive account of this problem by Hörmander included in his paper [11].

For hyperbolic halfspaces a combination of (42) and (45) yields this result:

(46) $$\nu(K^\theta) < \frac{2\lambda}{\rho''(1-\rho'^2)} \sin \pi\theta \equiv \frac{2\lambda C_0 C_1}{\rho''(1-\rho'^2)} \sin \pi\theta,$$

where the compounded factor again is invariant under selfmappings of Δ.

In conclusion we want to focus attention on another method akin to the balayage of measures introduced in potential theory by Poincaré a century ago. In Poincaré's theory the vehicle for the balayage (sweeping) of measures was based on minimizing a certain energy integral. This does not imply that the process is limited to Hilbert spaces or to any other space of uniform convexity. It works well even in a space like $H_\infty(\Delta)$, and its rather singular norm $\|\cdot\|_\infty$ does not prevent the minimax problem, now replacing the energy integral, from having a unique solution. In some situations the balayage process is more efficient than the geometric method used earlier in this paper. Let $\{\psi_i\}_1^n$ be the Kronecker functions associated to a finite E_n in the linear interpolation formula, and let e be a subset of indices i. Let the sweeping procedure be determined by the linear interpolation formula, taking the Dirac measure at α_i to the measure on $\partial\Delta$ with density $|\psi_i(\varsigma)\, d\varsigma|/2\pi$. If c_i are complex numbers, then the measure $\sum c_i \delta_{\alpha_i}$ will be swept to a positive measure on $\partial\Delta$ with density less than or equal to

$$\max_{i \in e} |c_i| \sum_{i \in e} |\psi_i(\varsigma)|\, |d\varsigma|/2\pi,$$

independently of the geometry of the region containing the α_i and excluding all other points $\in E_n$. This leads to a considerably simpler proof of the corona theorem in Δ.

Finally, a problem about infinite Blaschke products. Let $B(z)$ be of the second kind with its zeros satisfying the condition

$$\sum_1^\infty (1 - |\alpha_i|^2) \log\left(\frac{1}{1-|\alpha_i|^2}\right) < \infty.$$

Does this imply that the Taylor series of $B(z)$ can diverge on $\partial\Delta$ only on sets "smaller" than null sets?

The first part of this paper, focusing on the minimax problem and the linear interpolation formula, was done during 1959; the following academic year these were the subjects for a couple of seminars at the Institute for Advanced Study in Princeton. The remaining part of the paper is of recent vintage (1985).

Bibliography

1. L. Ahlfors, *Conformal invariants: Topics in geometric function theory*, McGraw-Hill, 1973.
2. L. Carleson, *An interpolation problem for bounded analytic functions*, Amer. J. Math. **80** (1958), 921–930.
3. ___, *Interpolation by bounded analytic functions and the corona problem*, Ann. of Math. (2) **76** (1962), 547–559.
4. ___, *Interpolation by bounded analytic functions and the corona problem*, Proc. Internat. Congr. Math. (Stockholm, 1962), Inst. Mittag-Leffler, Djursholm, 1963, pp. 314–316.
5. P. Duren, *Theory of H^p spaces*, Academic Press, 1970.
6. J. P. Earl, *On the interpolation of bounded sequences by bounded functions*, J. London Math. Soc. (2) **2** (1970), 544–548.
7. ___, *A note on bounded interpolation in the unit disc*, J. London Math. Soc. (2) **13** (1976), 419–423.
8. S. Fisher, *Function theory on planar domains*, Wiley, New York, 1983.
9. J. Garnett, *Bounded analytic functions*, Academic Press, 1981.
10. L. Hörmander, *Generators for some rings of analytic functions*, Bull. Amer. Math. Soc. **73** (1967), 943–949.
11. ___, *L^p estimates for (pluri)-subharmonic functions*, Math. Scand. **20** (1967), 65–78.
12. H. S. Shapiro and A. L. Shields, *On some interpolation problems for analytic functions*, Amer. J. Math. **83** (1961), 513–532.
13. K. T. Smith, *A generalization of an inequality of Hardy and Littlewood*, Canad. J. Math. **8** (1956), 157–170.

A more complete bibliography can be found in S. Fisher [8] and J. Garnett [9].

Powers of Riemann Mapping Functions

LOUIS DE BRANGES

A Riemann mapping function is a power series which represents a function with distinct values at distinct points of the unit disk. The Riemann mapping functions which now appear are normalized with constant coefficient zero and coefficient of z positive. An estimate of the coefficients of the νth power of such a function is obtained for any number $\nu > -\frac{1}{2}$. When $\nu = 0$, this result reduces to the author's proof [6] of a conjecture of Milin [11] which implies the Robertson conjecture and the Bieberbach conjecture. The Robertson conjecture [12] is itself an estimate of the coefficients of square roots of Riemann mapping functions. So it would be natural to expect that the estimates obtained in the case $\nu = \frac{1}{2}$ would imply the Robertson conjecture. But this turns out not to be the case. A new proof of the Bieberbach conjecture is not given by the present methods. The interest of the results seems to be principally in the new information obtained when ν is negative. The proofs are an application of the Löwner theory [10]. A survey of relevant information about the Löwner differential equation [5] will first be given.

A power series $f(z)$ is said to be subordinate to a power series $g(z)$ if $f(z) = g(B(z))$ for a power series $B(z)$ with constant coefficient zero which represents a function which is bounded by one in the unit disk. If $f(z)$ and $g(z)$ are normalized Riemann mapping functions, then $f(z)$ is subordinate to $g(z)$ if, and only if, the region onto which $f(z)$ maps the unit disk is contained in the region onto which $g(z)$ maps the unit disk. A Löwner family is a family of Riemann mapping functions, $F(t, z)$, indexed by a positive parameter t, such that each series $F(t, z)$ has constant coefficient zero and coefficient of z equal to t and such that $F(a, z)$ is subordinate to $F(b, z)$ when $a < b$.

Assume that $f(z)$ and $g(z)$ are Riemann mapping functions with constant coefficient zero and coefficient of z positive. If $f(z)$ is subordinate to $g(z)$, then $f(z)$ and $g(z)$ are members of a Löwner family of Riemann mapping functions $F(t, z)$.

Assume as given a Löwner family of Riemann mapping functions $F(t, z)$. Then the coefficients of $F(t, z)$ are absolutely continuous functions of t which satisfy

---

Research supported by the National Science Foundation.

the Löwner differential equation

$$t\frac{\partial}{\partial t}F(t,z) = \varphi(t,z)z\frac{\partial}{\partial z}F(t,z),$$

where $\varphi(t,z)$ is a power series with constant coefficient one which represents a function with positive real part in the unit disk for every index t, and the coefficients of $\varphi(t,z)$ are measurable functions of t.

Assume as given a family of power series $\varphi(t,z)$ with constant coefficient one, t positive, which represent functions with positive real part in the unit disk. If the coefficients of $\varphi(t,z)$ are measurable functions of t, then a unique Löwner family of Riemann mapping functions $F(t,z)$ exists which satisfies the Löwner differential equation with the given coefficient function $\varphi(t,z)$.

Coefficient estimates for Riemann mapping functions are derived from contractive properties of substitution transformations in generalizations of the space of square summable power series [7]. Let ν be a given number, $\nu > -\frac{1}{2}$, and let σ_n be a nonnegative function of positive integers n. Define $\mathfrak{G}_\sigma^\nu$ to be the Hilbert space of equivalence classes of generalized power series

$$f(z) = \sum_{n=1}^{\infty} a_n z^{\nu+n}$$

such that

$$\|f(z)\|_{\mathfrak{G}_\sigma^\nu}^2 = \sum_{n=1}^{\infty}(\nu+n)\sigma_n|a_n|^2$$

is finite. Equivalence of such series $f(z)$ and $g(z)$ means that the coefficient of $z^{\nu+n}$ in $f(z)$ is equal to the coefficient of $z^{\nu+n}$ in $g(z)$ whenever σ_n is positive.

The estimates obtained by this method are initially applicable only to the Riemann mapping functions for subregions of the unit disk.

An element $f(z)$ of a space $\mathfrak{G}_\sigma^\nu$ does not define a function analytic in the unit disk when ν is not an integer. But the ambiguities in the definition of function values for $f(z)$ are not any worse than they are for z^ν. The quotient $f(z)/z^\nu$ is meaningful as a function analytic in the unit disk.

Assume that $B(z)$ is the Riemann mapping function for a subregion of the unit disk, that $B(z)$ has constant coefficient zero, and that the coefficient of z in $B(z)$ is positive. Then $(B(z)/z)^\nu$ is meaningful as a power series with positive constant coefficient. If $f(z)$ belongs to $\mathfrak{G}_\sigma^\nu$, then $f(B(z))$ is meaningful as the product of z^ν and a power series. It is sufficient to give the interpretation when $f(z) = z^{\nu+n}$ for a positive integer n. The desired factorization is

$$f(B(z)) = z^\nu(B(z)/z)^\nu B(z)^n.$$

The expression $B(z)^\nu$ has a similar interpretation, but it cannot belong to $\mathfrak{G}_\sigma^\nu$ because the coefficient of z^ν is nonzero. An element of the space is, however, obtained when the first term of the series is deleted. The difference-quotient

$$\frac{B(z)^\nu - B'(0)^\nu z^\nu}{\nu}$$

can be estimated as an element of the space. When $\nu = 0$, the difference-quotient is interpreted as
$$\log \frac{B(z)}{zB'(0)}.$$
A family of spaces $\mathfrak{G}^\nu_{\sigma(t)}$, $t \geq 1$, is said to be ν-admissible if $t^{-2\nu}\sigma_n(t)$ is a nonincreasing and absolutely continuous function of t and if the differential equation
$$\frac{n}{2\nu + n}\sigma_n(t) + \frac{t\sigma'_n(t)}{2\nu + n} = \frac{2\nu + n + 1}{n+1}\sigma_{n+1}(t) - \frac{t\sigma'_{n+1}(t)}{n+1}$$
holds for every positive integer n.

The estimates which were previously obtained [6] in the case $\nu = 0$ are now extended to the general case $\nu > -\frac{1}{2}$.

THEOREM 1. *Assume as given a ν-admissible family of spaces $\mathfrak{G}^\nu_{\sigma(t)}$ such that $\sigma_1(t)$ is not identically zero, but such that $\sigma_n(t)$ is eventually identically zero. Then the inequality*
$$\left\| \frac{B(z)^\nu - B'(0)^\nu z^\nu}{\nu} + f(B(z)) \right\|^2_{\mathfrak{G}^\nu_{\sigma(a)}}$$
$$\leq \left(\frac{a}{b}\right)^{2\nu} \|f(z)\|^2_{\mathfrak{G}^\nu_{\sigma(b)}}$$
$$+ 4\left(\frac{a}{b}\right)^{2\nu} \sum_{n=1}^\infty \frac{\Gamma(2\nu + n)^2}{\Gamma(2\nu + 1)^2 \Gamma(n+1)^2}(\nu + n)[\sigma_n(a) - \sigma_n(b)]$$
holds for every element $f(z)$ of $\mathfrak{G}^\nu_{\sigma(b)}$ whenever $B(z)$ is the Riemann mapping function for a subregion of the unit disk which has constant coefficient zero and coefficient of z positive, $1 \leq a = bB'(0)$. Equality holds if, and only if, a complex number ω of absolute value one exists such that
$$\frac{B(z)}{(1 + \omega B(z))^2} = \frac{B'(0)z}{(1+\omega z)^2}$$
and such that the coefficient of $z^{\nu+n}$ in $f(z)$ is equal to the coefficient of $z^{\nu+n}$ in
$$\frac{z^\nu}{\nu(1+\omega z)^{2\nu}} - \frac{z^\nu}{\nu}$$
whenever $\sigma_n(t)$ is not identically zero.

When $\nu = 0$, this last expression is interpreted as
$$\log \frac{1}{(1+\omega z)^2}.$$

PROOF OF THEOREM 1. The differential equation satisfied by the coefficients is first used in integral form,
$$\frac{n}{2\nu + n}\int_a^b \frac{\sigma_n(t)}{t}dt - \frac{2\nu + n + 1}{n+1}\int_a^b \frac{\sigma_{n+1}(t)}{t}dt$$
$$= \frac{\sigma_n(a) - \sigma_n(b)}{2\nu + n} + \frac{\sigma_{n+1}(a) - \sigma_{n+1}(b)}{n+1}.$$

The identity
$$\int_a^b \frac{(2\nu+1)\sigma_1(t) - t\sigma_1'(t)}{t} dt$$
$$= 2 \sum_{n=1}^{\infty} \frac{\Gamma(2\nu+n)^2}{\Gamma(2\nu+1)^2 \Gamma(n+1)^2} (\nu+n)[\sigma_n(a) - \sigma_n(b)]$$

follows because $\sigma_n(t)$ is eventually identically zero.

This information is used to reformulate the inequality in the statement of the theorem. Use is now made of the Löwner theory.

A Löwner family of Riemann mapping functions $F(t, z)$ exists such that the identity
$$F(a, z) = F(b, B(z))$$
holds for some numbers a and b such that $1 \leq a = bB'(0)$. The family can for example be chosen so that $F(b, z)$ is a constant multiple of the identity transformation.

For any numbers a and b such that $0 < a \leq b < \infty$, a unique Riemann mapping function $B(b, a, z)$ exists, which maps the origin into the origin and the unit disk into the unit disk, such that the identity
$$F(a, z) = F(b, B(b, a, z))$$
is satisfied. The coefficients of $B(b, a, z)$ are absolutely continuous functions of a which satisfy the Löwner equation
$$a(\partial/\partial a)B(b, a, z) = \varphi(a, z)z(\partial/\partial z)B(b, a, z).$$

Hold b fixed. The desired inequality is verified by showing that the expression
$$a^{-2\nu}\|h(a,z)\|_{\mathfrak{G}^\nu_{\sigma(a)}}^2 - 2b^{-2\nu}\int_a^b \frac{(2\nu+1)\sigma_1(t) - t\sigma_1'(t)}{t} dt$$
is a nondecreasing function of a, $a \leq b$, where
$$h(a, z) = \frac{B(b, a, z)^\nu - (a/b)^\nu z^\nu}{\nu} + f(B(b, a, z)).$$
This is done by showing that the expression is an absolutely continuous function of a whose derivative is nonnegative almost everywhere.

For a computation of the derivative, observe that the differential equation
$$a(\partial/\partial z)h(a, z) = \varphi(a, z)a(\partial/\partial z)h(a, z) + (a/b)^\nu [\varphi(a, z) - 1]z^\nu$$
is satisfied. Write
$$h(a, z) = \sum_{n=1}^{\infty} a^\nu h_n(a) z^{\nu+n}.$$
The function which is to be shown nondecreasing is
$$\sum_{n=1}^{\infty} h_n(a)\bar{h}_n(a)(\nu+n)\sigma_n(a) - 2b^{-2\nu}\int_a^b \frac{(2\nu+1)\sigma_1(t) - t\sigma_1'(t)}{t} dt.$$

The indefinite integral and each term in the sum is absolutely continuous. The desired conclusion is obtained by showing that the result of formally applying the operator $a(\partial/\partial a)$ is nonnegative.

The expression to be shown nonnegative is

$$\sum_{n=1}^{\infty} ah'_n(a)\bar{h}_n(a)(\nu+n)\sigma_n(a) + \sum_{n=1}^{\infty} h_n(a)a\bar{h}'_n(a)(\nu+n)\sigma_n(a)$$

$$+ \sum_{n=1}^{\infty} h_n(a)\bar{h}_n(a)(\nu+n)a\sigma'_n(a) + 2b^{-2\nu}((2\nu+1)\sigma_1(a) - a\sigma'_1(a)).$$

Apply the Riesz representation of a function which is analytic and has positive real part in the unit disk. For each index a, a unique nonnegative measure $\mu(a,\cdot)$ exists on the Borel subsets of the unit disk such that the identity

$$\varphi(a,z) = \int \frac{1+wz}{1-wz} d\mu(a,w)$$

is satisfied in the sense of formal power series. (The coefficient of z^n on the left is equal to the coefficient of z^n on the right for every nonnegative integer n.)

It is convenient to introduce the notation

$$s_n(a,w) = (\nu+n)h_n(a) + w(\nu+n-1)h_{n-1}(a) + \cdots + w^{n-1}(\nu+1)h_1(a)$$

with the interpretation $s_0(a,w) = 0$. Then the identities

$$a\frac{\partial}{\partial a}h_n(a) = \int \left[2b^{-\nu}w^n + \frac{n}{\nu+n}s_n(a,w) + \frac{2\nu+n}{\nu+n}ws_{n-1}(a,w) \right] d\mu(a,w)$$

and

$$h_n(a) = \frac{s_n(a,w) - ws_{n-1}(a,w)}{\nu+n}$$

are satisfied.

The expression to be shown nonnegative is the integral with respect to $\mu(a,\cdot)$ of the sum

$$\sum_{n=1}^{\infty} \left[2b^{-\nu}w^n + \frac{n}{\nu+n}s_n(a,w) + \frac{2\nu+n}{\nu+n}ws_{n-1}(a,w) \right]$$

$$\times [s_n(a,w) - ws_{n-1}(a,w)]^- \sigma_n(a)$$

$$+ \sum_{n=1}^{\infty} [s_n(a,w) - ws_{n-1}(a,w)]$$

$$\times \left[2b^{-\nu}w^n + \frac{n}{\nu+n}s_n(a,w) + \frac{2\nu+n}{\nu+n}ws_{n-1}(a,w) \right]^- \sigma_n(a)$$

$$+ \sum_{n=1}^{\infty} |s_n(a,w) - ws_{n-1}(a,w)|^2 \frac{a\sigma'_n(a)}{\nu+n}$$

$$+ 2b^{-2\nu}[(2\nu+1)\sigma_1(a) - a\sigma'_1(a)]$$

(continued)

(*continued*)

$$= \sum_{n=1}^{\infty} [2b^{-\nu}\omega^n + s_n(a,\omega) + \omega s_{n-1}(a,\omega)][s_n(a,\omega) - \omega s_{n-1}(a,\omega)]^{-}\sigma_n(a)$$

$$+ \sum_{n=1}^{\infty} [s_n(a,\omega) - \omega s_{n-1}(a,\omega)]$$
$$\times [2b^{-\nu}\omega^n + s_n(a,\omega) + \omega s_{n-1}(a,\omega)]^{-}\sigma_n(a)$$

$$+ \sum_{n=1}^{\infty} |s_n(a,\omega) - \omega s_{n-1}(a,\omega)|^2 \frac{a\sigma_n'(a) - 2\nu\sigma_n(a)}{\nu + n}$$

$$+ 2b^{-2\nu}[(2\nu + 1)\sigma_1(a) - a\sigma_1'(a)]$$

$$= \sum_{n=1}^{\infty} 2b^{-\nu}[\omega^n \bar{s}_n(a,\omega) - \omega^{n-1}\bar{s}_{n-1}(a,\omega)]\sigma_n(a)$$

$$+ \sum_{n=1}^{\infty} 2b^{-\nu}[\omega^{-n} s_n(a,\omega) - \omega^{1-n} s_{n-1}(a,\omega)]\sigma_n(a)$$

$$+ \sum_{n=1}^{\infty} 2[s_n(a,\omega)\bar{s}_n(a,\omega) - s_{n-1}(a,\omega)\bar{s}_{n-1}(a,\omega)]\sigma_n(a)$$

$$+ \sum_{n=1}^{\infty} |\omega^{-n} s_n(a,\omega) - \omega^{1-n} s_{n-1}(a,\omega)|^2 \frac{a\sigma_n'(a) - 2\nu\sigma_n(a)}{\nu + n}$$

$$+ 2b^{-2\nu}[(2\nu + 1)\sigma_1(a) - a\sigma_1'(a)]$$

$$= \sum_{n=1}^{\infty} 2b^{-\nu}\omega^n \bar{s}_n(a,\omega)[\sigma_n(a) - \sigma_{n+1}(a)]$$

$$+ \sum_{n=1}^{\infty} 2b^{-\nu}\omega^{-n} s_n(a,\omega)[\sigma_n(a) - \sigma_{n+1}(a)]$$

$$+ \sum_{n=1}^{\infty} 2s_n(a,\omega)\bar{s}_n(a,\omega)[\sigma_n(a) - \sigma_{n+1}(a)]$$

$$+ \sum_{n=1}^{\infty} |\omega^{-n} s_n(a,\omega) - \omega^{1-n} s_{n-1}(a,\omega)|^2 \frac{a\sigma_n'(a) - 2\nu\sigma_n(a)}{\nu + n}$$

$$+ 2b^{-2\nu}[(2\nu + 1)\sigma_1(a) - a\sigma_1'(a)]$$

$$= \sum_{n=1}^{\infty} 2|b^{-\nu} + \omega^{-n} s_n(a,\omega)|^2 [\sigma_n(a) - \sigma_{n+1}(a)]$$

$$+ \sum_{n=1}^{\infty} |\omega^{-n} s_n(a,\omega) - \omega^{1-n} s_{n-1}(a,\omega)|^2 \frac{a\sigma_n'(a) - 2\nu\sigma_n(a)}{\nu + n}$$

$$- 2b^{-2\nu}[a\sigma_1'(a) - 2\nu\sigma_1(a)]$$

$$= -\sum_{n=1}^{\infty} 2|b^{-\nu} + \omega^{-n} s_n(a,\omega)|^2 \frac{a\sigma_n'(a) - 2\nu\sigma_n(a)}{2\nu + n}$$

$$-\sum_{n=1}^{\infty} 2|b^{-\nu} + \omega^{-n} s_n(a,\omega)|^2 \frac{a\sigma_{n+1}'(a) - 2\nu\sigma_{n+1}(a)}{n+1}$$

$$+\sum_{n=1}^{\infty} |\omega^{-n} s_n(a,\omega) - \omega^{1-n} s_{n-1}(a,\omega)|^2 \frac{a\sigma_n'(a) - 2\nu\sigma_n(a)}{\nu + n}$$

$$- 2b^{-2\nu}[a\sigma_1'(a) - 2\nu\sigma_1(a)]$$

$$= -\sum_{n=1}^{\infty} 2|b^{-\nu} + \omega^{-n} s_n(a,\omega)|^2 \frac{a\sigma_n'(a) - 2\nu\sigma_n(a)}{2\nu + n}$$

$$-\sum_{n=1}^{\infty} 2|b^{-\nu} + \omega^{1-n} s_{n-1}(a,\omega)|^2 \frac{a\sigma_n'(a) - 2\nu\sigma_n(a)}{n}$$

$$+\sum_{n=1}^{\infty} |\omega^{-n} s_n(a,\omega) - \omega^{1-n} s_{n-1}(a,\omega)|^2 \frac{a\sigma_n'(a) - 2\nu\sigma_n(a)}{\nu + n}$$

$$= -\sum_{n=1}^{\infty} \left| 2b^{-\nu} + \frac{n}{\nu+n} \omega^{-n} s_n(a,\omega) + \frac{2\nu+n}{\nu+n} \omega^{1-n} s_{n-1}(a,\omega) \right|^2$$

$$\times (\nu+n) \frac{a\sigma_n'(a) - 2\nu\sigma_n(a)}{n(2\nu+n)}$$

≥ 0.

The required interchange of summation and differentiation is justified by the hypothesis that $\sigma_n(t)$ is eventually identically zero. Use will be made of the fact, which is established after the proof of Theorem 2, that $\sigma_n(t)$ is a polynomial in $1/t$ which has a zero of order at least n at the origin. When $\sigma_n(t)$ is not identically zero, $\sigma_n'(t)$ is, with isolated exceptions, nonzero.

Equality holds if, and only if, the expression

$$\int \left| 2b^{-\nu} + \frac{n}{\nu+n} \omega^{-n} s_n(a,\omega) + \frac{2\nu+n}{\nu+n} \omega^{1-n} s_{n-1}(a,\omega) \right|^2 d\mu(a,\omega)$$

vanishes whenever $2\nu\sigma_n(a) - \sigma_n'(a)$ is nonzero. Since $s_0(a,\omega) = 0$ and since

$$s_1(a,\omega) = (\nu+1)h_1(a),$$

the expression

$$\int |2b^{-\nu}\omega + h_1(a)|^2 d\mu(a,\omega)$$

vanishes whenever $2\nu\sigma_1(a) - a\sigma_1'(a)$ is nonzero. The measure $\mu(a,\cdot)$ is therefore the measure with mass one concentrated at some point $\omega = \omega(a)$ and

$$h_1(a) = -2b^{-\nu}\omega(a).$$

Since

$$ah_1'(a) = 2b^{-\nu}\omega(a) + h_1(a) = 0,$$

both $h_1(a)$ and $\omega(a)$ are independent of a. These conclusions apply because of the hypothesis that $\sigma_1(t)$ does not vanish identically.

Write $\omega = \omega(a)$. The desired form of $B(b, a, z)$ follows since
$$\varphi(a, z) = \frac{1 + \omega z}{1 - \omega z}.$$

The identity
$$n s_n(a, \omega) + (2\nu + n)\omega s_{n-1}(a, \omega) = -2(\nu + n)b^{-\nu}\omega^n$$
holds when $\sigma_n(t)$ is not identically zero. An inductive argument shows that
$$s_n(a, \omega) = b^{-\nu} \frac{\Gamma(2\nu + n + 1)}{\Gamma(2\nu + 1)\Gamma(n + 1)}(-1)^n \omega^n - b^{-\nu}\omega^n$$
and that
$$h_n(a, \omega) = 2b^{-\nu} \frac{\Gamma(2\nu + n)}{\Gamma(2\nu + 1)\Gamma(n + 1)}(-1)^n \omega^n.$$

Thus the coefficient of $z^{\nu+n}$ in $h(a, z)$ is equal to the coefficient of $z^{\nu+n}$ in
$$\frac{(a/b)^\nu z^\nu}{\nu(1 + \omega z)^{2\nu}} - \frac{(a/b)^\nu z^\nu}{\nu}.$$

The theorem follows.

An estimate of coefficients of powers of unbounded Riemann mapping functions is obtained in the limit of large b.

THEOREM 2. *Assume that a ν-admissible family of spaces $\mathfrak{G}^\nu_{\sigma(t)}$ is given such that $\sigma_1(t)$ is not identically zero, but such that $\sigma_n(t)$ is eventually identically zero. Then the inequality*
$$\left\| \frac{F(z)^\nu - F'(0)^\nu z^\nu}{\nu} \right\|^2_{\mathfrak{G}^\nu_{\sigma(a)}}$$
$$\leq 4F'(0)^{2\nu} \sum_{n=1}^{\infty} \frac{\Gamma(2\nu + n)^2}{\Gamma(2\nu + 1)^2 \Gamma(n + 1)^2}(\nu + n)\sigma_n(a)$$

holds for every Riemann mapping function $F(z)$ with constant coefficient zero and coefficient of z positive, $a \geq 1$. Equality holds if, and only if, a complex number ω of absolute value one exists such that
$$F(z) = \frac{F'(0)z}{(1 + \omega z)^2}.$$

PROOF OF THEOREM 2. Since $F(z)$ can be replaced by $F(\lambda z)$ for a complex number λ of absolute value one, it can be assumed that the coefficient of z in $F(z)$ is positive. Then a Löwner family of power series $F(t, z)$ exists which contains $F(z)$. Define the bounded Riemann mapping functions $B(b, a, z)$ as in the proof of Theorem 1. Apply the estimate of Theorem 1 with
$$f(z) = \frac{F(b, z)^\nu - b^\nu z^\nu}{\nu b^\nu}$$

and $B(z) = B(b, a, z)$. The inequality can be written

$$\left\| \frac{F(a,z)^\nu - a^\nu z^\nu}{\nu} \right\|^2_{\mathfrak{G}^\nu_{\sigma(a)}} \leq (a/b)^{2\nu} \left\| \frac{F(b,z)^\nu - b^\nu z^\nu}{\nu} \right\|^2_{\mathfrak{G}^\nu_{\sigma(b)}}$$

$$+ 4a^{2\nu} \sum_{n=1}^{\infty} \frac{\Gamma(2\nu+n)^2}{\Gamma(2\nu+1)^2 \Gamma(n+1)^2}$$

$$\times (\nu+n)[\sigma_n(a) - \sigma_n(b)].$$

Since $F(z)$ can be multiplied by a positive constant, it can be assumed that $F'(0) \geq 1$. If $F(z)$ represents a function which is bounded in the unit disk, then the choice of $F(b, z)$ can be made equal to a constant multiple of z and with $F(z) = F(a, z)$. The inequality then reads

$$\left\| \frac{F(a,z)^\nu - a^\nu z^\nu}{\nu} \right\|^2_{\mathfrak{G}^\nu_{\sigma(a)}}$$

$$\leq 4a^{2\nu} \sum_{n=1}^{\infty} \frac{\Gamma(2\nu+n)^2}{\Gamma(2\nu+1)^2 \Gamma(n+1)^2} (\nu+n)[\sigma_n(a) - \sigma_n(b)].$$

Since $t^{-2\nu} \sigma_1(t)$ is a nonnegative, nonincreasing function of t, the expression

$$(2\nu+1)\sigma_1(t) - t\sigma'_1(t)$$

is nonnegative. An identity at the start of the proof of Theorem 1 now shows that the right side of the inequality is a nondecreasing function of b. It will be shown after the proof of Theorem 2 that $\sigma_n(t)$ is a polynomial in $1/t$ with constant coefficient zero. Since $\sigma_n(t)$ is eventually identically zero, the inequality

$$\left\| \frac{F(a,z)^\nu - a^\nu z^\nu}{\nu} \right\|^2_{\mathfrak{G}^\nu_{\sigma(a)}} \leq 4a^{2\nu} \sum_{n=1}^{\infty} \frac{\Gamma(2\nu+n)^2}{\Gamma(2\nu+1)^2 \Gamma(n+1)^2} (\nu+n)\sigma_n(a)$$

is obtained in the limit of large b.

This obtains the desired inequality when $F(z)$ is a bounded Riemann mapping function. The inequality follows for an unbounded Riemann mapping function because any such function $F(z)$ is a limit

$$F(z) = \lim_{t \uparrow 1} F(tz)$$

of bounded Riemann mapping functions.

To determine the cases of equality, return to a general Löwner family of functions $F(t, z)$. The identity

$$\lim_{b \uparrow \infty} b^{-2\nu} \left\| \frac{F(b,z)^\nu - b^\nu z^\nu}{\nu} \right\|^2_{\mathfrak{G}^\nu_{\sigma(b)}} = 0$$

now follows because $\sigma_n(b)$ is zero in the limit of large b for every index n. Another derivation of the last inequality is obtained from the inequality at the start of the proof of the limit of large b. If equality holds in the last inequality, it holds in the inequality at the start of the proof for every b. By Theorem 1, a complex

number ω of absolute value one exists such that the identity
$$\frac{B(b,a,z)}{(1+\omega B(b,a,z))^2} = \frac{za/b}{(1+\omega z)^2}$$
is satisfied. The condition implies that the coefficient $\varphi(t,z)$ in the Löwner equation is equal to
$$\frac{1+\omega z}{1-\omega z}$$
for $a \le t \le b$. The number ω is independent of b.

By Theorem 1, it also follows that the coefficient of $z^{\nu+n}$ in
$$\frac{F(b,z)^\nu - b^\nu z^\nu}{\nu b^\nu}$$
is equal to the coefficient of $z^{\nu+n}$ in
$$\frac{z^\nu}{\nu(1+\omega z)^{2\nu}} - \frac{z^\nu}{\nu}$$
when $\sigma_n(t)$ is not identically zero.

Replace z by $B(b,a,z)$. The coefficient of $z^{\nu+n}$ in
$$\frac{F(a,z)^\nu - b^\nu B(b,a,z)^\nu}{\nu b^\nu}$$
is equal to the coefficient of $z^{\nu+n}$ in
$$\frac{B(b,a,z)^\nu}{\nu(1+\omega B(b,a,z))^{2\nu}} - \frac{B(b,a,z)^\nu}{\nu}.$$
It follows that the coefficient of $z^{\nu+n}$ in
$$\frac{F(a,z)^\nu - a^\nu z^\nu}{\nu a^\nu}$$
is equal to the coefficient of $z^{\nu+n}$ in
$$\frac{z^\nu}{\nu(1+\omega z)^{2\nu}} - \frac{z^\nu}{\nu}.$$

Since $\sigma_1(t)$ is not identically zero by hypothesis, these considerations apply when $n = 1$. It has been shown that the coefficient of z^2 in $F(a,z)$ is equal to $-2a\omega$. Since equality holds in the Bieberbach conjecture for the second coefficient, a theorem of Bieberbach [4] gives the desired form of $F(z)$. This completes the proof of the theorem.

The notation
$$F(a,b;c;z) = 1 + \frac{ab}{1c}z + \frac{a(a+1)b(b+1)}{12c(c+1)}z^2 + \cdots$$
is used for the hypergeometric series. An identity due to Gauss states that
$$F(a,b;c;1) = \frac{\Gamma(c)\Gamma(c-a-b)}{\Gamma(c-a)\Gamma(c-b)},$$
when c and $c-a-b$ have positive real part.

A generalization of the hypergeometric series is
$$F(a,b,c;d,e;z) = 1 + \frac{abc}{1de}z + \frac{a(a+1)b(b+1)c(c+1)}{12d(d+1)e(e+1)}z^2 + \cdots.$$
A theorem of Clausen [3] states in this notation that the identity
$$F(a,b;c;z)^2 = F(2a,2b,a+b;c,2c-1;z)$$
holds when $a + b + \frac{1}{2} = c$.

A solution of the system of differential equations for the coefficient functions is obtained in the form
$$\sigma_n(t) = \Delta_n(t) + \frac{(2\nu+n)(2\nu+2n+2)}{(-1)(n+1)}\Delta_{n+1}(t)$$
$$+ \frac{(2\nu+n)(2\nu+n+1)(2\nu+2n+3)(2\nu+2n+4)}{(-1)(-2)(n+1)(n+2)}\Delta_{n+2}(t) + \cdots,$$
where $\Delta_n(t)$ satisfies the elementary differential equation
$$t\Delta'_n(t) = -n\Delta_n(t)$$
with solution
$$\Delta_n(t) = \Delta_n(1)t^{-n}.$$

The expansion is applied in cases when $\Delta_n(t)$, and hence $\sigma_n(t)$, is eventually identically zero. It is possible to solve for the functions $\Delta_n(t)$ in terms of the functions $\sigma_n(t)$ because a square matrix with zeros below the diagonal and ones on the diagonal has an inverse of the same form. Each function $\Delta_n(t)$ is expressible as a linear combination of the functions $\sigma_k(t)$ with $k \geq n$, and the coefficient of $\sigma_n(t)$ is one. The coefficient of $\sigma_k(t)$ when $k > n$ is determined inductively in the unique way such that the differential equation for $\Delta_n(t)$ is satisfied. The resulting identity is
$$(\nu+n)\Delta_n(t) = (\nu+n)\sigma_n(t) + \frac{(2\nu+n)(2\nu+2n)}{(1)(n+1)}(\nu+n+1)\sigma_{n+1}(t)$$
$$+ \frac{(2\nu+n)(2\nu+n+1)(2\nu+2n)(2\nu+2n+1)}{(1)(2)(n+1)(n+2)}$$
$$\times (\nu+n+2)\sigma_{n+2}(t) + \cdots.$$

Clausen's identity can now be used to construct a ν-admissible family of spaces $\mathfrak{G}^\nu_{\sigma(t)}$ such that $\sigma_n(t) = 0$ when $n > r$, where r is a given positive integer. A nonnegative parameter λ is used in the construction. Choose
$$\Delta_n(t) = \frac{\Gamma(n+1)\Gamma(2\nu+2\lambda+r+n+1)(4t)^{-n}}{\Gamma(\nu+n+1)\Gamma(\nu+\lambda+n+1)\Gamma(2\nu+n+1)\Gamma(r+1-n)}$$
for $n = 1, \ldots, r$. Then the identity
$$t^{-2\nu}\sigma_n(t) = \frac{\Gamma(n+1)\Gamma(2\nu+2\lambda+r+n+1)}{\Gamma(\nu+n+1)\Gamma(\nu+\lambda+n+1)\Gamma(2\nu+n)\Gamma(r+1-n)}$$
$$\times 4^{-n}\int_t^\infty F(n-r, 2\nu+2\lambda+r+n+1, \nu+n+\tfrac{1}{2}; \nu+\lambda+n+1,$$
$$2\nu+2n+1; s^{-1})s^{-2\nu-n-1}\,ds$$
holds for $n = 1, \ldots, r$.

A theorem of Askey and Gasper [**2**] states that the integrand is nonnegative. The desired conclusion is immediate from Clausen's formula when $\lambda = 0$. When λ is positive, the nonnegativity of the integrand is obtained from an expansion which can be obtained from the uniqueness of solutions of differential equations with given initial conditions. Unique real numbers $c_1, \ldots, c_r$ exist such that the identities

$$\frac{\Gamma(2\nu + 2\lambda + r + n + 1)}{\Gamma(\nu + \lambda + n + 1)\Gamma(r + 1 - n)}$$
$$\times F(n - r, 2\nu + 2\lambda + r + n + 1, \nu + n + \tfrac{1}{2}; \nu + \lambda + n + 1, 2\nu + 2n + 1; z)$$
$$= \sum_{k=n}^{r} \frac{\Gamma(2\nu + k + n + 1)}{\Gamma(\nu + n + 1)\Gamma(k + 1 - n)} c_k$$
$$\times F(n - k, 2\nu + k + n + 1, \nu + n + \tfrac{1}{2}; \nu + n + 1; 2\nu + 2n + 1; z)$$

are satisfied for $n = 1, \ldots, r$. A direct computation shows that the numbers $c_1, \ldots, c_r$ are nonnegative.

Indeed the numbers are already determined by the identities in the case when $z = 0$, which then read

$$\frac{\Gamma(2\nu + 2\lambda + r + n + 1)}{\Gamma(\nu + \lambda + n + 1)\Gamma(r + 1 - n)} = \sum_{k=n}^{r} \frac{\Gamma(2\nu + k + n + 1)}{\Gamma(\nu + n + 1)\Gamma(k + 1 - n)} c_k$$

for $n = 1, \ldots, r$. The Gegenbauer expansion

$$\frac{\Gamma(2\nu + 2\lambda + r + 2)}{\Gamma(\nu + \lambda + 2)\Gamma(r)} F(1 - r, 2\nu + 2\lambda + r + 2; \nu + \lambda + 2; z)$$
$$= \sum_{n=1}^{r} \frac{\Gamma(2\nu + n + 2)}{\Gamma(\nu + 2)\Gamma(n)} c_n F(1 - n, 2\nu + n + 2; \nu + 2; z)$$

is now verified by comparing the coefficient of z^n on each side.

To make the Gegenbauer computation of c_n, observe that the expression $F(-n, 2\nu + n + 3; \nu + 2; z)$ is a Jacobi polynomial of degree n which satisfies the identity

$$F(-n, 2\nu + n + 3; \nu + 2; z) = (-1)^n F(-n, 2\nu + n + 3; \nu + 2; 1 - z).$$

Since $F(-r, 2\nu + 2\lambda + r + 3; \nu + \lambda + 2; z)$ satisfies a similar identity, it follows that $c_n = 0$ when n does not have the same parity as r.

The orthogonality properties of Jacobi polynomials give the identity

$$\frac{\Gamma(2\nu + 2\lambda + r + 2)}{\Gamma(\nu + \lambda + 2)\Gamma(r)} \int_0^1 F(1 - r, 2\nu + 2\lambda + r + 2; \nu + \lambda + 2; t)$$
$$\times F(1 - n, 2\nu + n + 2; \nu + 2; t) t^{\nu+1}(1 - t)^{\nu+1} dt$$
$$= \frac{\Gamma(2\nu + n + 2)}{\Gamma(\nu + 2)\Gamma(n)} c_n \int_0^1 F(1 - n, 2\nu + n + 2; \nu + 2; t)^2 t^{\nu+1}(1 - t)^{\nu+1} dt.$$

Use is made of the Rodrigues formula

$$(\nu + 2) \cdots (\nu + n) z^{\nu+1}(1-z)^{\nu+1} F(1-n, 2\nu + n + 2; \nu + 2; z)$$
$$= (d/dz)^{n-1}[z^{\nu+n}(1-z)^{\nu+n}]$$

to obtain, on integration by parts, the identity

$$\frac{\Gamma(2\nu + 2\lambda + r + n + 1)}{\Gamma(\nu + \lambda + n + 1)\Gamma(r + 1 - n)}$$
$$\times \int_0^1 F(n - r, 2\nu + 2\lambda + r + n + 1; \nu + \lambda + n + 1; t) t^{\nu+n}(1-t)^{\nu+n} dt$$
$$= \frac{\Gamma(2\nu + 2n + 1)}{\Gamma(\nu + n + 1)} c_n \int_0^1 t^{\nu+n}(1-t)^{\nu+n} dt,$$

where

$$\int_0^1 t^{\nu+n}(1-t)^{\nu+n} dt = \frac{\Gamma(\nu+n+1)\Gamma(\nu+n+1)}{\Gamma(2\nu + 2n + 2)}.$$

A theorem of Gauss [8] states that the formal power series identity

$$F(2a, 2b; c; z) = F(a, b; c; 4z(1-z))$$

holds when $a + b + \frac{1}{2} = c$. The identity

$$\int_0^1 F(n - r, 2\nu + 2\lambda + r + n + 1; \nu + \lambda + n + 1; t) t^{\nu+n}(1-t)^{\nu+n} dt$$
$$= 2^{-2\nu-2n-1} \int_0^1 F(\tfrac{1}{2}n - \tfrac{1}{2}r, \nu + \lambda + \tfrac{1}{2}r + \tfrac{1}{2}n + \tfrac{1}{2}; \nu + \lambda + n + 1; t)$$
$$\times t^{\nu+n}(1-t)^{-1/2} dt$$

holds when n has the same parity as r, where

$$\int_0^1 F(\tfrac{1}{2}n - \tfrac{1}{2}r, \nu + \lambda + \tfrac{1}{2}r + \tfrac{1}{2}n + \tfrac{1}{2}; \nu + \lambda + n + 1; t) t^{\nu+n}(1-t)^{-1/2} dt$$
$$= \frac{\Gamma(\nu + n + 1)\Gamma(\tfrac{1}{2})}{\Gamma(\nu + n + \tfrac{3}{2})}$$
$$\times F(\tfrac{1}{2}n - \tfrac{1}{2}r, \nu + \lambda + \tfrac{1}{2}r + \tfrac{1}{2}n + \tfrac{1}{2}, \nu + n + 1; \nu + \lambda + n + 1, \nu + n + \tfrac{3}{2}; 1).$$

A theorem of Saalschütz [3] states that

$$F(a, b, c; d, e; 1) = \frac{\Gamma(d)\Gamma(1 + a - e)\Gamma(1 + b - e)\Gamma(1 + c - e)}{\Gamma(1 - e)\Gamma(d - a)\Gamma(d - b)\Gamma(d - c)},$$

when $a, b,$ or c is a negative integer and $a + b + c + 1 = d + e$. The identity

$$F(\tfrac{1}{2}n - \tfrac{1}{2}r, \nu + \lambda + \tfrac{1}{2}r + \tfrac{1}{2}n + \tfrac{1}{2}, \nu + n + 1; \nu + \lambda + n + 1, \nu + n + \tfrac{3}{2}; 1)$$
$$= \frac{\Gamma(\nu + \lambda + n + 1)\Gamma(-\nu - \tfrac{1}{2}r - \tfrac{1}{2}n - \tfrac{1}{2})\Gamma(\lambda + \tfrac{1}{2}r - \tfrac{1}{2}n)\Gamma(\tfrac{1}{2})}{\Gamma(\nu + \lambda + \tfrac{1}{2}r + \tfrac{1}{2}n + 1)\Gamma(-\nu - n - \tfrac{1}{2})\Gamma(\lambda)\Gamma(\tfrac{1}{2} - \tfrac{1}{2}r + \tfrac{1}{2}n)}$$

holds when n has the same parity as r.

From these calculations it now follows that
$$c_n = 2^{2\lambda}\frac{\Gamma(\nu+\lambda+\tfrac{1}{2}r+\tfrac{1}{2}n+\tfrac{1}{2})\Gamma(\nu+n+\tfrac{3}{2})}{\Gamma(\nu+\tfrac{1}{2}r+\tfrac{1}{2}n+\tfrac{3}{2})}\frac{\Gamma(\lambda+\tfrac{1}{2}r-\tfrac{1}{2}n)}{\Gamma(\lambda)\Gamma(1+\tfrac{1}{2}r-\tfrac{1}{2}n)}$$
is positive when n has the same parity as r. This completes the proof of the Askey-Gasper inequality.

The differential equations for the coefficient functions imply the identity
$$\sigma_n(1) - \sigma_{n+1}(1) = \frac{2\nu\sigma_n(1) - \sigma'_n(1)}{2\nu+n} + \frac{2\nu\sigma_{n+1}(1) - \sigma'_{n+1}(1)}{n+1},$$
where
$$2\nu\sigma_n(1) - \sigma'_n(1)$$
$$= \frac{\Gamma(n+1)\Gamma(2\nu+2\lambda+r+n+1)}{\Gamma(\nu+n+1)\Gamma(\nu+\lambda+n+1)\Gamma(2\nu+n)\Gamma(r+1-n)}$$
$$\times 4^{-n} F(n-r, 2\nu+2\lambda+r+n+1, \nu+n+\tfrac{1}{2}; \nu+\lambda+n+1, 2\nu+2n+1; 1)$$
for $n = 1, \ldots, r$. A theorem of G. N. Watson [9] states that the identity
$$F(a,b,c; \tfrac{1}{2}a+\tfrac{1}{2}b+\tfrac{1}{2}, 2c; 1) = \frac{\Gamma(\tfrac{1}{2})\Gamma(\tfrac{1}{2}+c)\Gamma(\tfrac{1}{2}+\tfrac{1}{2}a+\tfrac{1}{2}b)\Gamma(\tfrac{1}{2}-\tfrac{1}{2}a-\tfrac{1}{2}b+c)}{\Gamma(\tfrac{1}{2}+\tfrac{1}{2}a)\Gamma(\tfrac{1}{2}+\tfrac{1}{2}b)\Gamma(\tfrac{1}{2}-\tfrac{1}{2}a+c)\Gamma(\tfrac{1}{2}-\tfrac{1}{2}b+c)}$$
holds when $a, b,$ or c is a negative integer.

The identity
$$F(n-r, 2\nu+2\lambda+r+n+1, \nu+n+\tfrac{1}{2}; \nu+\lambda+n+1, 2\nu+2n+1; 1)$$
$$= \frac{\Gamma(\nu+n+1)\Gamma(\nu+\lambda+n+1)\Gamma(\tfrac{1}{2}r-\tfrac{1}{2}n+\tfrac{1}{2})\Gamma(\tfrac{1}{2}r-\tfrac{1}{2}n+\lambda+\tfrac{1}{2})}{\Gamma(\nu+\tfrac{1}{2}r+\tfrac{1}{2}n+1)\Gamma(\nu+\lambda+\tfrac{1}{2}r+\tfrac{1}{2}n+1)\Gamma(\tfrac{1}{2})\Gamma(\lambda+\tfrac{1}{2})}$$
holds when n has the same parity as r. The left side vanishes when n has opposite parity.

The identity
$$2\nu\sigma_n(1) - \sigma'_n(1)$$
$$= 2^{2\nu+2\lambda}\frac{\Gamma(n+1)\Gamma(\nu+\lambda+\tfrac{1}{2}r+\tfrac{1}{2}n+\tfrac{1}{2})\Gamma(\tfrac{1}{2}r-\tfrac{1}{2}n+\lambda+\tfrac{1}{2})}{\Gamma(2\nu+n)\Gamma(\nu+\tfrac{1}{2}r+\tfrac{1}{2}n+1)\Gamma(\lambda+\tfrac{1}{2})\Gamma(\tfrac{1}{2}r-\tfrac{1}{2}n+1)\Gamma(\tfrac{1}{2})}$$
holds when n has the same parity as r. The left side vanishes when n has opposite parity.

The identities
$$\sigma_n(1) - \sigma_{n+1}(1)$$
$$= 2^{2\nu+2\lambda}\frac{\Gamma(n+1)\Gamma(\nu+\lambda+\tfrac{1}{2}r+\tfrac{1}{2}n+\tfrac{1}{2})\Gamma(\tfrac{1}{2}r-\tfrac{1}{2}n+\lambda+\tfrac{1}{2})}{\Gamma(2\nu+n+1)\Gamma(\nu+\tfrac{1}{2}r+\tfrac{1}{2}n+1)\Gamma(\lambda+\tfrac{1}{2})\Gamma(\tfrac{1}{2}r-\tfrac{1}{2}n+1)\Gamma(\tfrac{1}{2})}$$
and
$$\sigma_{n-1}(1) - \sigma_n(1)$$
$$= 2^{2\nu+2\lambda}\frac{\Gamma(n)\Gamma(\nu+\lambda+\tfrac{1}{2}r+\tfrac{1}{2}n+\tfrac{1}{2})\Gamma(\tfrac{1}{2}r-\tfrac{1}{2}n+\lambda+\tfrac{1}{2})}{\Gamma(2\nu+n)\Gamma(\nu+\tfrac{1}{2}r+\tfrac{1}{2}n+1)\Gamma(\lambda+\tfrac{1}{2})\Gamma(\tfrac{1}{2}r-\tfrac{1}{2}n+1)\Gamma(\tfrac{1}{2})}$$
hold when n has the same parity as r.

These calculations supply explicit numbers in the estimates of Theorem 2.

THEOREM 3. *Let r be a given positive integer and let λ be a given nonnegative number. If $F(z)$ is a Riemann mapping function with constant coefficient zero and coefficient of z positive, then*

$$\frac{F(z)^\nu - F'(0)^\nu z^\nu}{\nu} = \sum_{n=1}^\infty a_n z^{\nu+n},$$

where the inequality

$$\sum_{n=1}^r \frac{\Gamma(n+1)\Gamma(\nu+\lambda+\tfrac{1}{2}r+\tfrac{1}{2}n'+\tfrac{1}{2})\Gamma(\tfrac{1}{2}r-\tfrac{1}{2}n'+\lambda+\tfrac{1}{2})}{\Gamma(2\nu+n+1)\Gamma(\nu+\tfrac{1}{2}r+\tfrac{1}{2}n'+1)\Gamma(\tfrac{1}{2}r-\tfrac{1}{2}n'+1)} \sum_{k=1}^n (\nu+k)|a_k|^2$$

$$\leq 4F'(0)^{2\nu} \sum_{n=1}^r \frac{\Gamma(n+1)\Gamma(\nu+\lambda+\tfrac{1}{2}r+\tfrac{1}{2}n'+\tfrac{1}{2})\Gamma(\tfrac{1}{2}r-\tfrac{1}{2}n'+\lambda+\tfrac{1}{2})}{\Gamma(2\nu+n+1)\Gamma(\nu+\tfrac{1}{2}r+\tfrac{1}{2}n'+1)\Gamma(\tfrac{1}{2}r-\tfrac{1}{2}n'+1)}$$

$$\times \sum_{k=1}^n (\nu+k)\frac{\Gamma(2\nu+k)^2}{\Gamma(2\nu+1)^2\Gamma(k+1)^2}$$

is satisfied. Equality holds if, and only if, a complex number ω of absolute value one exists such that

$$F(z) = \frac{F'(0)z}{(1+\omega z)^2}.$$

The notation n' is used for $n+1$ when n does not have the same parity as r, and for n when n has the same partity as r.

Theorem 3 is an immediate consequence of Theorem 2 and the remarks which precede the statement of Theorem 3. The strongest inequality is obtained when $\lambda = 0$, but the inequality is interesting when $\lambda = \tfrac{1}{2}$ because of simplifications which then occur.

THEOREM 4. *Assume that σ_n is a nonincreasing function of positive integers n such that the inequality*

$$(n+1)[\sigma_n - \sigma_{n+1}] \geq (2\nu+n+1)[\sigma_{n+1} - \sigma_{n+2}]$$

holds for every positive integer n and such that

$$\lim_{n\to\infty} \sigma_n = 0;$$

then the inequality

$$\sum_{n=1}^\infty (\nu+n)\sigma_n |a_n|^2 \leq 4F'(0)^{2\nu} \sum_{n=1}^\infty \frac{\Gamma(2\nu+n)^2}{\Gamma(2\nu+1)^2\Gamma(n+1)^2}(\nu+n)\sigma_n$$

holds whenever $F(z)$ is a Riemann mapping function with constant coefficient zero and coefficient of z positive,

$$\frac{F(z)^\nu - F'(0)^\nu z^\nu}{\nu} = \sum_{n=1}^\infty a_n z^{\nu+n}.$$

Equality holds with σ_n not identically zero if, and only if, a complex number ω of absolute value one exists such that

$$F(z) = \frac{F'(0)z}{(1+\omega z)^2}.$$

The hypotheses of Theorem 4 imply the representation

$$\sigma_n = \sum_{r=n}^{\infty} \frac{\Gamma(r+1)}{\Gamma(2\nu+r+1)} \rho_r$$

for a nonnegative and nonincreasing function ρ_n of positive integers n. By convexity, it is sufficient to give a proof of the theorem in the case that a positive integer r exists such that $\rho_n = 1$ for $n = 1, \ldots, r$ and $\rho_n = 0$ for $n > r$. The desired conclusion is then given by Theorem 3.

Theorem 4 contains the Milin conjecture when $\nu = 0$, but it does not contain the Robertson conjecture when $\nu = \frac{1}{2}$. This raises the question whether the Robertson conjecture can be obtained using Theorem 2 in the case $\nu = \frac{1}{2}$. A negative answer has been given by George Gasper (private communication).

Let r be a given positive integer, and consider any admissible family of spaces $\mathfrak{G}_{\sigma(t)}^{\nu}$ such that $\sigma_n(t)$ vanishes identically when $n > r$. Then the solution of the differential equations for the coefficient functions is obtained with

$$\sigma_r(t) = \sigma_r(1) t^{-r}$$

and

$$\sigma_{r-1}(t) = \sigma_{r-1}(1) t^{1-r} + 2 \frac{(\nu+r)(2\nu+r-1)}{r} \sigma_r(1)(t^{1-r} - t^{-r}).$$

The condition that $t^{-2\nu} \sigma_{r-1}(t)$ is nonincreasing is the inequality

$$\sigma_{r-1}(1) \geq 2 \frac{(\nu+r)}{r} \sigma_r(1).$$

Gasper observes that the inequality is not satisfied when $(\nu+n)\sigma_n(1) = 1$ for $n = r$ and $n = r - 1$, at least if $\nu = \frac{1}{2}$. And the inequality is not satisfied for any other choice of ν.

The proof of the Askey-Gasper inequality here given is substantially the original argument of Askey and Gasper [2]. The reader is referred to Askey [1] for information about the Gegenbauer expansion which appears in the proof. See also the article by Askey and Gasper in these proceedings on "Inequalities for polynomials." The idea of reviewing the proof in the light of its present application was suggested by T. H. Koornwinder in "Squares of Gegenbauer polynomials and Milin-type inequalities," CWI Report PM-R8412, Amsterdam, October 1984. The case $\nu = 0$ and $\lambda = 0$ of the present Theorem 3 is due to Koornwinder, who applies a method which was discovered independently and slightly earlier by the author, as recorded in a letter to Gasper. See the comments made by Askey and Gasper in their contribution to these proceedings.

The author thanks Richard Askey for locating the reference to Gauss [8], which the author had failed to record in his notes on hypergeometric series.

References

1. R. Askey, *Orthogonal polynomials and special functions*, SIAM, Philadelphia, Pa., 1975.
2. R. Askey and G. Gasper, *Positive Jacobi polynomial sums.* II, Amer. J. Math. **98** (1976), 709–737.
3. W. Bailey, *Generalized hypergeometric series*, Cambridge University Press, London, 1935.
4. L. Bieberbach, *Über die Koeffizienten derjenigen Potenzreihen, welche eine schlichte Abbildung des Einheitskreises vermitteln*, Sitzungsberichte Preussische Akademie der Wissenschaften, 1916, pp. 940–955.
5. L. de Branges, *Coefficient estimates*, J. Math. Anal. Appl. **82** (1981), 420–450.
6. ____, *A proof of the Bieberbach conjecture*, Acta Math. **154** (1985), 137–152.
7. ____, *Square summable power series*, Springer-Verlag (to appear).
8. C. F. Gauss, "Determinatio seriei nostrae per aequationem differentialem secundi ordinis," *Werke*. Vol. III, Königlichen Gesellschaft der Wissenschaften zu Göttingen, 1976.
9. G. H. Hardy, *A chapter from Ramanujan's note-book*, Proc. Camb. Philos. Soc. **21** (1923), 492–503.
10. K. Löwner, *Untersuchungen über schlichte konforme Abbildungen des Einheitskreises*, I. Math. Ann. **80** (1923), 102–121.
11. I. M. Milin, *Univalent functions and orthonormal systems*, Nauka, Moscow, 1971; English transl, Transl. Math. Mono., vol. 49, Amer. Math. Soc., Providence, R. I., 1977.
12. M. S. Robertson, *A remark on the odd schlicht functions*, Bull. Amer. Math. Soc. **42** (1936), 366–370.

300 Years of Analyticity

J. DIEUDONNÉ

The idea of "geometric progression" goes back to early Greek mathematics, and the formula
$$a + aq + aq^2 + \cdots + aq^n = a(q^{n+1} - 1)/(q - 1)$$
was probably known, at least for small values of n. Zeno's paradox of Achilles and the tortoise is proof that the "passage to the limit" giving the sum of the series
$$1 + a + a^2 + \cdots + a^n + \cdots = 1/(1 - a)$$
for $a < 1$ was a step Greek mathematicians at first were reluctant to take; however, one of the methods used by Archimedes to compute the area of a segment of parabola can be interpreted as a geometric expression of that sum for $a = 1/4$.

After the Renaissance, mathematicians had no more scruples to use such infinite sums, and the convergence of the geometric series not only was commonplace, but was used to prove the convergence of less familiar series by comparison with a geometric series, as Lord Brouncker did in 1668 for a series giving $\log 2$ as its sum.

In 1664–1665, Newton, then a student at Cambridge, was reading Wallis's *Algebra* in which "infinite products" were introduced; in order to understand them, he started to manipulate power series, which he wrote essentially as we do now; he very rapidly discovered the formation of the coefficients for the series
$$(1 + x)^{p/q}, \quad \log(1 + x), \quad e^x, \quad \sin x, \quad \text{arc} \sin x$$
(without at that time bothering about questions of convergence). But until very recently, these results remained in manuscript form, although later Newton communicated some of them to friends and colleagues. The first power series to appear in print was the series for $\log(1 + x)$ (or more precisely for the primitive $\int \log(1+x)\, dx$), obtained independently by N. Mercator in 1668; this immediately gave rise to similar investigations by James Gregory and by Leibniz after 1673.

At the beginning of the eighteenth century, with the discovery of Taylor's formula, it was implicitly agreed that *any* function was equal to the sum of its Taylor series; and the fact that all operations in algebra and analysis (even the "inversion" of a function), applied to power series, again gave power series as their result fascinated mathematicians of that time and even much later.

It is common knowledge that Cauchy, the same man who was most influential in founding analysis on a rigorous basis, is also the mathematician who, by his bold use of integration along curves in the complex plane, created the modern theory of analytic functions; in spite of many competitors, that theory has retained a central position in mathematics and has shown its vitality in the tremendous variety of its applications to other parts of mathematics and to mathematical physics.

I will not here bring coals to Newcastle and reel off all the wonderful features of the theory of analytic functions, with which you are more familiar than myself. I shall instead focus my remarks on some connections of the concept of analyticity with parts of mathematics which may be unfamiliar and even seem outlandish to some of you.

I. Analyticity and the theory of distributions. Since the beginning of the eighteenth century, De Moivre's formula $e^{ix} = \cos x + i \sin x$ has naturally linked power series and trigonometric series; and after 1800, analysts have been aware of (and have made use of) the kinship between holomorphic functions, Fourier series and integrals, and the Laplace transform. But that kinship has been obscured by the difficulties of the theory of Fourier series, too strictly bound to the idea of ordinary convergence. The theory of tempered distributions has brought these questions in a much clearer light. To simplify exposition, I will restrict myself most of the time to functions of one variable, but most results I shall mention are special cases of theorems valid in spaces of arbitrary finite dimension.

Since Cauchy and Weierstrass, the central fact in complex analysis has been the one-to-one correspondence

$$(1) \qquad (a_n)_{n \geq 0} \rightleftarrows \sum_{n=0}^{\infty} a_n z^n$$

between sequences of complex numbers which do not increase too fast and functions holomorphic in a neighborhood of 0. Innumerable problems (including the theme of this Symposium!) have been concerned with what may be said of one of these objects when something is known of the corresponding one; this, as you know, constitutes an enormous and fascinating "dictionary."

When you turn to Fourier series, you immediately meet the same kind of correspondence

$$(2) \qquad (c_n)_{n \in \mathbb{Z}} \underset{\overline{F}}{\overset{F}{\rightleftarrows}} \sum_{n=-\infty}^{+\infty} c_n e^{ni\theta}$$

between families of coefficients and sums of trigonometric series, which has been one of the most unsatisfactory and thickest jungles of classical analysis. A situation as satisfactory as the correspondence (1) has only been achieved by substituting distributions in place of functions. More precisely, there is a one-to-one correspondence (2), when, on the left-hand side, only families of *polynomial growth* are considered, that is, families such that

(3) $$|c_n| \leq (1+|n|)^k \quad \text{for some } k > 0 \text{ and } n \geq n_0,$$

and the right-hand side is replaced by *any periodic distribution* T on $\mathbf{R}$, of period 1. The beauty of this correspondence is that it is *stable* for derivative, primitive, and convolution; so Euler was perfectly justified in taking derivatives of Fourier series and considering them again as Fourier series!

The right-hand side of (2) is a distribution which can be split into a sum of two periodic distributions $T_1 + T_2$, corresponding to families (c_n) having $c_n = 0$ for all $n < 0$ (resp. $n \geq 0$), and which have holomorphic extensions

(4) $$f_1 \colon \quad z \mapsto \sum_{n=0}^{+\infty} c_n z^n \quad \text{for } |z| < 1,$$

(5) $$f_2 \colon \quad z \mapsto \sum_{n=-\infty}^{-1} c_n z^n \quad \text{for } |z| > 1,$$

and $f_1(re^{i\theta})$ (resp. $f_2(r^{-1}e^{i\theta})$), defined for $0 < r < 1$, *tends* indeed to T_1 (resp. T_2) when $r \to 1$ for the *weak topology* of distributions. This of course includes as special cases Laurent series and Hardy spaces.

Both sides of (2) are special cases of *tempered distributions* on $\mathbf{R}$, those with support in $\mathbf{Z}$ on the left-hand side and the periodic ones on the right-hand side. The Fourier transformation $\mathbf{F}$ and its inverse $\overline{\mathbf{F}}$ can be defined for *any* tempered distribution T, by the relation

(6) $$\langle \mathbf{F}T, f \rangle = \langle T, \mathbf{F}f \rangle$$

for every *declining function* f (a C^∞ function such that the product of any derivative of f by any polynomial is bounded). The relation of this operator $\mathbf{F}$ with holomorphic functions constitutes the modern aspect of the Laplace transform.

Let us start with the simplest tempered distributions, those with *compact support*. For such a distribution T, $\mathbf{F}T$ can be identified with a C^∞ function

(7) $$\mathbf{F}T(\xi) = \langle T, \exp(-2\pi i x \xi) \rangle,$$

which (i) has all derivatives of polynomial growth, (ii) can be extended to an *entire function*, the *Fourier-Laplace transform* $\mathrm{FL}(T)$ obtained by replacing in (7) the real number ξ by a complex number $\xi + i\eta$, and (iii) satisfies the inequality

(8) $$|\mathrm{FL}(T)(\xi + i\eta)| \leq C(1 + |\xi + i\eta|)^N \exp(2\pi A|\eta|).$$

The converse is the famous *Paley-Wiener theorem*, extended to distributions by L. Schwartz.

What can be said of *arbitrary* tempered distributions? Suppose first that T has its support in the interval $[0, +\infty[$. Then $T \cdot \exp(2\pi\eta x)$ is tempered for all $\eta < 0$, and
$$FL(T)(\xi + i\eta) = \langle T, \exp(-2\pi i x(\xi + i\eta))\rangle$$
is defined and holomorphic for $\eta < 0$. Furthermore, one has an inequality
$$(9) \qquad |FL(T)(\xi + i\eta)| \leq A(\xi)|\eta|^{-N}$$
for some $N \geq 0$, $A(\xi)$ being bounded in every compact subset of $\mathbf{R}$. Similarly, if T has its support in $]-\infty, 0]$, $FL(T)(\xi + i\eta)$ is defined and holomorphic for $\eta > 0$ and satisfies again an inequality (9). Finally, any tempered distribution T can be written $T_1 + T_2 + T_3$, where T_2 has compact support, T_1 has its support in $[0, +\infty[$, and T_3 has its support in $]-\infty, 0]$. The traditional Laplace transforms for T_1 and T_3 are obtained by turning the axes $90°$.

Now you may well ask what has all this to do with classical analysis? Very much indeed; the use of tempered distributions and their Fourier transforms has brought about more results during the last 30 years in the theory of partial differential equations than had been obtained in the previous 200 years. I shall only give an example (Malgrange-Ehrenpreis): let $P(D)$ be a differential operator with *constant* coefficients, of arbitrary order; then, if Ω is an open convex set in $\mathbf{R}^n$ and v a C^∞ function in Ω, there exists at least a C^∞ function u in Ω such that $P(D) \cdot u = v$; try to prove this without distributions!

II. p-adic analysis. The origin of p-adic numbers is very remote from analysis; it lies in the theory of *diophantine equations*
$$(10) \qquad P(x_1, x_2, \ldots, x_n) = 0,$$
where P is a polynomial with *integral* coefficients, and solutions are sought in *integers*; these problems go back to the Greeks and are of considerable difficulty. A time-honored device is to *reduce modulo p* for a prime number p, that is, look for solutions of the *congruence*
$$(11) \qquad P(x_1, x_2, \ldots, x_n) \equiv 0 \pmod{p}.$$
(If there are none, (10) is impossible.) The next step is to see if there are also solutions of the congruence $\bmod p^2, p^3$, etc. Here, if $(x_1, x_2, \ldots, x_n)$ is a solution of (11), one may look for the congruence $\bmod p^2$ to solutions of the form
$$x'_j = x_j + py_j \qquad (1 \leq j \leq n).$$
As
$$P(x_1 + py_1, x_2 + py_2, \ldots, x_n + py_n) \equiv P(x_1, x_2, \ldots, x_n) + p\sum_{j=1}^n \frac{\partial P}{\partial x_j} y_j \pmod{p^2}$$
and by (11) there is an integer b_0 such that $P(x_1, \ldots, x_n) = b_0 p$, one is reduced to solving
$$\sum_{j=1}^n \frac{\partial P}{\partial x_j} y_j + b_0 \equiv 0 \pmod{p}.$$

But if the $\partial P/\partial x_j$ are not all multiples of p, that congruence *always* has solutions. Then the process may be indefinitely repeated, and it yields a sequence of solutions $(x_1^{(k)}, \ldots, x_n^{(k)})$ of the congruence mod p^{k+1}, such that

(13) $\qquad x_j^{(k)} \equiv a_0 + a_1 p + \cdots + a_k p^k \pmod{p^{k+1}}$ with $0 \leq a_j \leq p-1$.

For arithmeticians, these are "better and better" approximations for the solutions of (10) when they exist. Instead of writing the infinite system of solutions (13), Hensel took the bold step of introducing the symbols

(14) $\qquad x = a_0 + a_1 p + \cdots + a_n p^n + \cdots$

with infinitely many terms and computing with them as if they were convergent power series, writing for instance, for $p = 3$,

$$-5 = 1 + 2 \cdot 3 + 2 \cdot 3^2 + \cdots + 2 \cdot 3^n + \cdots.$$

People thought he was crazy, although by this method he was able to derive new results and simpler proofs in the theory of algebraic numbers.

In fact, the theory of these "p-adic numbers," as he called them, exactly reproduced what had happened for complex numbers, which for a long time had been meaningless symbols with which one could prove theorems concerning accepted notions; but here, the progress of algebra and topology at the beginning of the twentieth century soon could give to p-adic numbers the status of genuine mathematical objects.

This came about by a generalization of the classical definition of real numbers given by Cantor, based on a *completion* of the rational field $\mathbf{Q}$: it is given the structure of a metric space by the distance $d(x,y) = |x-y|$, and the general process of completion of a metric space described by Hausdorff yields the real line as the completed metric space. But the Hausdorff process applies to any distance, and in particular to a distance of the form $d(x,y) = q(x-y)$, where $q: \mathbf{Q} \to \mathbf{R}$ is a function not identically 0 and such that

(15) $\qquad q(x) \geq 0, \qquad q(xy) = q(x)q(y), \qquad q(x+y) \leq q(x) + q(y).$

It turns out that these functions were *all* determined by Ostrowski in 1918: if one leaves aside the trivial case in which $q(x) = 1$ for all $x \neq 0$, they are

(i) the functions $x \mapsto |x|^\alpha$ for an exponent $0 < \alpha \leq 1$, $|x|$ being the "usual" absolute value, and

(ii) a *new* series of "p-adic absolute values" $x \mapsto |x|_p$, each attached to a prime number p and defined as follows: if

(16) $\qquad x = \pm p^m p_1^{m_1} p_2^{m_2} \cdots p_k^{m_k}$

is the unique decomposition of any $x \neq 0$ in $\mathbf{Q}$ (with exponents $m_j \in \mathbf{Z}$), then

(17) $\qquad |x|_p = a^{-m}$ for a fixed $a > 1$.

(In number theory, it is convenient to take $a = p$.) The completion $\mathbf{Q}_p$ for the absolute value $|x|_p$ is called the *field of p-adic numbers*, and the function

$x \mapsto |x|_p$ extends by continuity to Q_p with the same properties (15), all its values still being p^{-k} with $k \in Z$. The p-adic numbers such that $|x|_p \leq 1$ are called *p-adic integers*; they form a ring Z_p, the closure of Z in Q_p, and every $x \in Q_p$ can be written uniquely $x = p^m z$ with $|z|_p = 1$, $m \in Z$, and $|x|_p = p^{-m}$.

There are, however, plenty of differences between R and Q_p; whereas R has only *one* algebraic extension $C \neq R$ of degree 2, Q_p has algebraic extensions of arbitrarily high degree; on its algebraic closure $\tilde{Q}_p$ there is a unique absolute value, again written $|x|_p$, which extends the p-adic absolute value, but its values are now of the form p^r with r a *rational* number. For that absolute value $\tilde{Q}_p$ is *not* complete, but its completion C_p is also algebraically closed; it is that field which can be considered as corresponding to C in the classical case, and in which it is possible to do *p-adic analysis*.

This starts, as in the classical case, with power series; it has many common features with complex analysis, but also some pecularities which at first seem very strange. They all stem from the *ultrametric inequality*

$$(18) \qquad |x+y|_p \leq \text{Max}(|x|_p, |y|_p).$$

A first consequence is that a series $\sum_{n=1}^{\infty} a_n$ in C_p (or in any ultrametric complete field) converges *if and only if* $\lim_{n \to \infty} a_n = 0$. The radius of convergence R of a power series $\sum_{n=0}^{\infty} a_n z^n$ in C_p is then defined as usual and given by the usual formula of Cauchy (with of course the p-adic absolute value); $f(z) = \sum_{n=0}^{\infty} a_n z^n$ is equal to its Taylor series for $|z|_p < R$, and the series may or may not converge for $|z|_p = R$. The maximum principle and the Cauchy inequalities are valid.

But the geometry of C_p is very different from the geometry of C: C_p is a totally discontinuous space, a disk $|z|_p < r$ or $|z|_p \leq r$ has *no boundary*, and for $|z_0|_p < r$ the relations $|z - z_0|_p < r$ and $|z|_p < r$ are *equivalent*. This implies that if R is the radius of convergence of a series $\sum_{n=0}^{\infty} a_n z^n$ in C_p, then R is *also* the radius of convergence of the series $\sum_{n=0}^{\infty} a_n (z_0 + z)^n$ for $|z_0|_p < R$; in other words, it is not possible to define analytic continuation by the method of Weierstrass.

This seemingly insuperable obstacle has nevertheless been overcome by M. Krasner, taking his cue from another approximation theorem in complex analysis, namely the Runge theorem. If A is a subset of C_p, an *analytic element* is by definition a uniform limit in A of *rational* functions which have no pole in A. Provided suitable properties are assumed for A, the principle of analytic continuation holds for analytic elements in A, that is, if such a function is 0 in a disk contained in A, it is 0 everywhere in A. Once this is obtained, a sizable part of the classical theory of entire and meromorphic functions can be developed along the same pattern in p-adic analysis.

The p-adic numbers now occupy a central position in number theory, algebraic geometry, and group theory; a simple enumeration of the various results in which they are a fundamental tool would already be very long. I shall again limit myself to illustrate my remarks by a single result, a beautiful theorem of Dwork on *usual* power series $f(z) = \sum_{n=0}^{\infty} a_n z^n$ with *integral* coefficients a_n. An old theorem of E. Borel states that if $f(z)$ is meromorphic in a disk of radius $R > 1$, then f is a *rational* function. This was much improved by Dwork in the following way: consider a prime number p; as $\mathbb{Z} \subset \mathbb{Z}_p$ the power series can be considered for $z \in \mathbb{C}_p$; let R_p be the radius of a disk in which the analytic continuation of the series exists and is meromorphic; then, if the product $RR_p > 1$, $f(z)$ is again *rational*. The method of proof is similar to Borel's, namely, a majoration of the Hankel determinants which, due to the fact that these determinants are *integers*, can only be satisfied when they are 0; but whereas E. Borel only used the classical Cauchy inequalities to obtain that majoration, Dwork could apply the Cauchy inequalities *both* in $\mathbb{C}$ and in $\mathbb{C}_p$. His theorem created quite a stir when it was proved, for at that time it provided the first proof of one of the "Weil conjectures," namely, the rationality of the "zeta function" of a hypersurface over a finite field.

III. Formal power series. The notion of absolute value can clearly be defined in *any commutative ring*, and not only in $\mathbb{Q}$. One of the simplest and most useful is defined in the ring $A[T]$ of polynomials over any domain of integrity A: for any polynomial $P \neq 0$, let $\omega(P)$ be the smallest degree of monomials in P with coefficient not 0; then $|P| = a^{-\omega(P)}$ (for any real number $a > 1$) is an absolute value satisfying the ultrametric inequality (18). Here the process of completion yields something very classical, namely the *formal power series*, which usually are written $\sum_{n=0}^{\infty} a_n T^n$, with *arbitrary* elements $a_n \in A$; if one objects to such an apparently meaningless symbol, it can simply be replaced by the sequence (a_n). In other words, the set of formal power series is just the infinite product $A^{\mathbb{N}}$, with the usual addition, but the multiplication is the so-called "Cauchy multiplication"; one thus obtains a domain of integrity $A[[T]]$, in which T is written for the sequence $(0,1,0,0,\ldots)$; the topology defined by the absolute value is the product of the discrete topologies on the factors A, and *then* the series $\sum_{n=0}^{\infty} a_n T^n$ does effectively converge for that topology to the element (a_n).

Of course, if A itself is a topological ring (for instance $\mathbb{R}$ or $\mathbb{C}$), it is also possible to take on $A^{\mathbb{N}}$ the product of the topologies of the factors. It is perhaps not inappropriate to point out that we then have come full circle to the conception of power series which was more or less clear in the minds of mathematicians from Newton to Euler and Lagrange: think of Newton claiming (in an anagram!) that he "could integrate all differential equations," or of Euler manipulating the series $\sum_{n=0}^{\infty} (-1)^n n! x^n$ to give it a meaning. Many arguments of that period which are usually dismissed as unrigorous are perfectly rigorous for the product topology on $\mathbb{C}^{\mathbb{N}}$, such as

$$e^x = \lim_{n \to \infty} \left(1 + \frac{x}{n}\right)^n \qquad \frac{\sin x}{x} = \lim_{n \to \infty} \prod_{\nu=1}^{n/2} \left(1 - \frac{x^2}{\nu^2} \frac{1 + \cos \frac{2\nu\pi}{n}}{1 - \cos \frac{2\nu\pi}{n}}\right).$$

The introduction of formal power series in algebraic geometry is linked to the concept of *local ring*; this is the purely algebraic concept which substitutes for the intuitive idea of "neighborhood" of a point when there is no topology on the field of scalars. In classical algebraic geometry, Riemann had emphasized the paramount importance of the *field of rational functions* K on an algebraic variety V; for a point $x \in V$, the *local ring* O_x at that point is the subring of K consisting of functions *defined* at the point x. In the simplest case $V = C$, O_x is the ring of functions f/g, where f and g are two polynomials in $C[T]$ such that $g(x) \neq 0$; it can of course be imbedded as a subring of the ring of power series in $z - x$, convergent in some neighborhood of x. This observation had been one of the ideas which guided Hensel in his conception of p-adic numbers: the analogies between the ring of polynomials $C[T]$ and the ring of integers Z are classical, and O_x can be considered as the ring of fractions f/g such that the polynomial g does not belong to the *maximal ideal* $\mathfrak{m} = (T - x)$ of $C[T]$; as the maximal ideals of Z are the ideals generated by prime numbers, what corresponds to O_x is the subring of Q consisting of fractions a/b with $b \notin (p)$, and Hensel wanted to find a bigger ring which would correspond to the ring of convergent power series in $z - x$.

Modern algebraic geometry generalizes this to algebraic varieties defined over an arbitrary algebraically closed "field of scalars" k. If such a variety is irreducible, it is possible to define "rational functions" with values in k, which in general are not defined everywhere in V but nevertheless are the elements of a field K, a transcendental extension of k. The definition of the local ring O_x at a point $x \in V$ is then the same as above; but the word "local" is slightly misleading, because even when $k = C$, the structure of O_x involves much more than an ordinary neighborhood of x: for instance, the local rings at two points of a smooth curve of genus 2 in the plane are not isomorphic, in general. To get something which corresponds to our intuition of a "local" behavior, a method consists in applying to O_x a process of "completion" which generalizes the one based on absolute values; then it can indeed be shown that for all *simple* points $x \in V$, the completion $\hat{O}_x$ is always isomorphic to a ring of formal power series $k[[T_1, T_2, \ldots, T_n]]$ (defined by an obvious generalization of $A[[T]]$ for an arbitrary number of indeterminates).

This embedding of a ring into a ring of much simpler structure can be a very useful tool to prove properties of O_x by "descent" from the same properties of $\hat{O}_x$ which are much easier to prove for that ring. This device can also be applied to the theory of ordinary differential equations, where the existence of "formal" solutions is usually much simpler to prove and sometimes implies the existence of genuine solutions, as in recent work of Malgrange.

I should also mention that already in 1949 Zariski had been able to define for an algebraic variety over an arbitrary field a concept of "holomorphic functions" by a process akin to the later Krasner analytic continuation in p-adic analysis; this enabled him to prove deep theorems in algebraic geometry.

Finally, the classical theory of Lie groups, which is based on analytic laws of composition, can be generalized both for p-adic power series and for formal power series; they provide an efficient method of attack in the difficult theory of algebraic and arithmetic groups.

Of course, it is perfectly legitimate to ask what all these "abstract" theories have to do with problems of classical mathematics, and if they are not mere empty generalizations. The answer is that they have been imagined precisely to provide more powerful tools to deal with hard questions in classical algebra and number theory; let me only mention that in the very recent past, it is by using such tools that formidable problems like the Weil conjectures and the Mordell conjecture have finally been solved.

Problems in Mathematical Physics Connected with the Bieberbach Conjecture

P. R. GARABEDIAN

1. Variational methods. Attempts to settle the Bieberbach conjecture have motivated both a variety of special techniques in function theory and more systematic methods in the calculus of variations. The recent proof of de Branges may be viewed as a happy combination of these two approaches [3]. The goal of the present paper will be to describe a few successes of the variational method, not only in solving problems for univalent functions, but also in contributing to the solution of questions from fluid mechanics and magnetohydrodynamics that arise in the applications.

Our point of departure is the Hadamard variational formula

$$\delta G(z,\varsigma) = \frac{1}{2\pi} \int \frac{\partial G(w,z)}{\partial \nu} \frac{\partial G(w,\varsigma)}{\partial \nu} \delta\nu \, ds$$

for the Green's function G of a domain that has been perturbed by a normal shift $\delta\nu$ of its boundary [11]. Loewner's representation of univalent functions, which is used in the de Branges proof, can be derived by specialization of this formula. Similarly one can arrive in a formal way at Schiffer's ordinary differential equation

$$P(1/f) \, df^2 = Q(z) \, dz^2$$

for any extremal function maximizing $|a_n|$ among normalized univalent functions

$$f(z) = z + a_2 z^2 + a_3 z^3 + \cdots$$

in the unit circle [4, 14]. In particular, it becomes apparent why this differential equation involves quadratic differentials.

A major achievement of the variational theory is the proof of regularity of the arcs bounding the domain onto which the unit circle is mapped by an extremal function. This is made possible by Schiffer's method of interior variation, or by his technique of varying the boundary. We note that a novel version of the latter technique enables one to deduce the Riemann mapping theorem by solving an extremal problem for univalent functions in the exterior of an arbitrary continuum [8].

Work supported by NSF grant DMS-8320430.

Both the Bieberbach conjecture and the Riemann mapping theorem have served as excellent testing grounds for new methods of function theory. Another test of new ideas is their usefulness in the study of more general problems of mathematical physics. Here we shall mention a few examples of that kind to which concepts from function theory apply.

2. Fluid mechanics. Consider a normalized conformal mapping

$$\varsigma = z + b_0 + \frac{b_1}{z} + \cdots$$

of the exterior of some curve C in the z-plane onto the region outside a horizontal slit in the ς-plane. The analytic function ς defines the complex potential of an irrotational flow of incompressible fluid past the profile C. The virtual mass

$$M = \iint \left|\frac{d\varsigma}{dz} - 1\right|^2 dx\,dy$$

of the flow is related to the power series coefficient b_1 by the formula

$$2\pi \operatorname{Re}\{b_1\} = M + A,$$

where A is the area enclosed by C. This result of G. I. Taylor in fluid mechanics is equivalent to the area theorem in function theory [2], and it shows that $\operatorname{Re}\{b_1\} \geq 0$. It is the problem of maximizing $\operatorname{Re}\{b_1\}$ that we referred to above as an approach to the Riemann mapping theorem. Since b_1 is additive under composition of mappings, these are related issues.

The variational formula

$$2\pi\delta \operatorname{Re}\{b_1\} = \int \left|\frac{d\varsigma}{dz}\right|^2 \delta\nu\,ds$$

for the virtual mass follows from Hadamard's analysis. From it we conclude that a curve C enclosing specified sets and minimizing $\operatorname{Re}\{b_1\}$ may include arcs across which the speed $|d\varsigma/dz|$ of the flow must remain continuous. Such arcs are referred to in fluid mechanics as vortex sheets. Thus we arrive at a variational principle for the construction of vortex sheets [9]. This concept generalizes to space of three dimensions. Moreover, it appears that the analytic arcs bounding extremal domains for the coefficient problem of univalent function theory have an interpretation as vortex sheets of appropriate flows.

For a prescribed choice of the parameter λ let us consider the extremal problem

$$2\pi \operatorname{Re}\{b_1\} - \lambda A = \text{minimum},$$

where the curve C is once more constrained to enclosed specified bodies [10]. The extremal domain is seen to be bounded by free streamlines satisfying the constant pressure condition

$$\left|\frac{d\varsigma}{dz}\right|^2 = \lambda$$

for cavitational flow. Existence and regularity of the free streamline flow past a given profile can be established by the method of interior variation. This is true for axially symmetric as well as two-dimensional models, and there is a related treatment of nonlinear surface waves. However, similar arguments seem to imply nonexistence for corresponding three-dimensional flows without symmetry. The variational principle also furnishes a numerical algorithm for the computation of flows with free boundaries that turns out to be successful in practice.

In the case of axially symmetric cavitational flow, the proof of regularity of the free streamlines lies deeper. It has led to a formula for the solution of the inverse problem of determining a flow that has a prescribed analytic curve as its free boundary. Questions of analytic continuation that arise in this connection have suggested a more general procedure for the solution of inverse problems in two-dimensional fluid mechanics that is based on the theory of functions of two complex variables. This is called the method of complex characteristics. It has become widely accepted as a means of designing shockless airfoils in transonic flow.

3. Magnetohydrodynamics. Hadamard's formula for the second variation of domain functionals is also an important tool of function theory and partial differential equations. Long ago we applied it to a variety of problems in mathematical physics [9], and later it was used to investigate the Bieberbach conjecture [5]. The most significant progress along these lines seems to have been made in the field of plasma physics. Fundamental results about the stability of a plasma confined in toroidal equilibrium by a strong magnetic field appeared in the scientific literature when research on controlled nuclear fusion was declassified in 1958.

The variational principle of ideal magnetohydrodynamics asserts that toroidal equilibrium and stability of a plasma are characterized by minimizing the potential energy

$$E = \iiint \left[\frac{B^2}{2} + \frac{p}{\gamma - 1}\right] dx\, dy\, dz,$$

subject to appropriate constraints, where B stands for the confining magnetic field, p is the fluid pressure, and γ is a gas constant [1]. For plasma at constant pressure there is a sharp boundary version of this principle that is quite similar to the one for virtual mass described in the preceding section. A special example of the magnetohydrodynamic problem that is relevant to univalent functions will now be formulated in the language of conformal mapping.

Let D stand for the doubly connected domain lying between two curves C_1 and C_2 in the complex plane. Let the outer curve, say C_1, be fixed, and consider the problem of maximizing the area A enclosed by the inner curve C_2, subject to the constraint that the conformal modulus of D be fixed. The variational theory shows that under suitable hypotheses there is an extremal domain bounded by an analytic curve C_2 along which the modulus of the logarithmic derivative of the conformal mapping of D onto a circular ring remains constant. This is the form that the sharp boundary condition

$$B^2 = \text{const}.$$

for magnetohydrodynamic equilibrium takes in the present case. The area inside C_2 represents the plasma, and the conformal modulus of D plays the role of the potential energy E.

The most important problems of magnetohydrodynamics occur, of course, in space of three dimensions. Serious issues arise concerning the existence of equilibria when the geometry has no two-dimensional symmetry. The Kolmogoroff-Arnold-Moser theory of dynamical systems predicts that smooth equilibria of this kind cannot be found. Nevertheless, the variational principle has been used to develop a computational model of equilibrium, stability, and transport that seems to be in remarkably satisfactory agreement with the experimental information that is available [1]. This is an exciting topic of current interest that is, however, somewhat removed from our central theorem of the Bieberbach conjecture for univalent functions.

4. Odd coefficients. Both in the proof of de Branges and in earlier work, success in attacking the Bieberbach conjecture has come through the discovery of intermediate inequalities that imply $|a_n| \leq n$ but are easier to analyze. It is remarkable how long it took to realize that the Grunsky inequalities have such a property for the even coefficients, more specifically in the case $|a_4| \leq 4$. This was because one had to skip over a_3 and go on to estimate a_4 using $f(z^2)^{1/2}$. The theory of the second variation has played a role in deciding which inequalities provide the most promising approach. In that direction it was used both to study the Bieberbach conjecture for functions close to the Koebe function and to test the credibility of Robertson's conjecture [6].

For a while it appeared that the odd coefficients were less accessible than the even ones. At that time the variational method produced new inequalities that did for the odd coefficients what Grunsky's inequalities had accomplished for the even ones [13]. In particular, sharp estimates of a_5 were obtained by looking for circumstances in which Schiffer's ordinary differential equation could be integrated in closed form [7]. Computer experiments served as another way to find out whether these estimates were effective, but it turned out, of course, that the Milin inequality was better [12]. The contribution of Gautschi to the work of de Branges has also exploited numerical computations as a guide. Thus the interaction between methods of mathematical physics and function theory seems to be fruitful from many points of view.

REFERENCES

1. F. Bauer, O. Betancourt, and P. Garabedian, *Magnetohydrodynamic equilibrium and stability of stellarators*, Springer-Verlag, New York, 1984.

2. S. Bergman and M. Schiffer, *Kernel functions and elliptic differential equations in mathematical physics*, Academic Press, New York, 1953.

3. L. de Branges, *A proof of the Bieberbach conjecture*, Steklov Math. Inst. Preprint E-5-84 (1984), 1–21.

4. R. Courant, *Dirichlet's principle, conformal mapping, and minimal surfaces*, Interscience, New York, 1950.

5. P. Duren and M. Schiffer, *The theory of the second variation in extremum problems for univalent functions*, J. Analyse Math. **10** (1962-63), 193–252.

6. S. Friedland, *On a conjecture of Robertson*, Arch. Rational Mech. Anal. **37** (1970), 255–261.

7. P. Garabedian, *Inequalities for the fifth coefficient*, Comm. Pure Appl. Math. **19** (1966), 199–214.

8. ____, *Univalent functions and the Riemann mapping theorem*, Proc. Amer. Math. Soc. **61** (1976), 242–244.

9. P. Garabedian and M. Schiffer, *Convexity of domain functionals*, J. Analyse Math. **2** (1953), 281–368.

10. P. Garabedian and D. Spencer, *Extremal methods in cavitational flow*, J. Rational Mech. Anal. **1** (1952), 359–409.

11. J. Hadamard, *Leçons sur le calcul des variations*, Hermann, Paris, 1910.

12. I. Milin, *Univalent functions and orthonormal systems*, Transl. Math. Mono., vol. 49, Amer. Math. Soc., Providence, R. I., 1977.

13. C. Pommerenke, *Univalent functions*, Vandenhoeck and Ruprecht, Göttingen, 1975.

14. M. Schiffer, *Un calcul de variation pour une famille de fonctions univalentes*, C. R. Acad. Sci. Paris **205** (1937), 709–711.

Extremal Methods

D. H. HAMILTON

1. Prehistory. The idea that a function, extremal for a certain problem over some class, should have special properties is basic to analysis and goes back to the eighteenth century "calculus of variations." However it was Riemann's inspired but naive enunciation of the Dirichlet problem that pointed the way to the extremal theory of conformal mapping. By the turn of this century Hilbert, Koebe, and Poincaré had overcome the initial difficulties and were solving basic mapping problems such as the mapping of a multiply connected domain onto the complement of horizontal slits by such methods as maximizing $\operatorname{Re} f'(\infty)$ over an appropriate class. Similarly Fejér and Riesz's solution of the Riemann mapping of a simply connected domain G inside the unit disk D onto D takes the function f with maximum $|f'(0)|$ and then uses the square root of a bilinear transformation to show that we must have $f(G) = D$.

These first results depended mostly on early notions of the compactness of a family of functions together with some fairly elementary variations. It is interesting to note that many of the tools of modern analysis—inner product spaces, compact families, etc.—in fact evolved out of the process of solving such problems.

In his proof of the uniformization theorem, Koebe was led to study the now well-known class S of holomorphic functions $f(z) = z + a_2 z^2 + \cdots$ which are conformal on the disk D. The normalization $a_0 = 0$, $a_1 = 1$, which is just a scaling to make S compact, also implied that the boundary of $f(D)$ cannot get too close to 0. Just how close was discovered in 1916 by Bieberbach, who found the famous Koebe 1/4-theorem. The extremal function

$$k(z) = \frac{z}{(1-z)^2} = z + 2z^2 + 3z^3 + \cdots$$

also maximized $|a_2|$ in S, which was enough for Bieberbach to conjecture that the so-called Koebe function was extremal for the nth coefficient too. Actually the conjecture that $|a_n| \leq n$ was a rather tentative footnote to a much more ambitious proposal, namely, to characterize the collection of all sequences $(a_2, a_3, \ldots)$ arising as Taylor coefficients of functions in S, in particular, to find the coefficient body V_n. No doubt what Bieberbach had in mind was a sort of universal dictionary of conformal mappings, analogous to the module space of Riemann surfaces.

2. Some functional analysis. I think it is fair to say that we thought we had a good idea of the appropriate functional-analytic context for extremal problems. Let $\mathcal{H}$ be the space of holomorphic functions on D, endowed with the usual topology of convergence on compact subsets of D. Montel immediately tells us that S is a compact subfamily, which means that any continuous functional Φ will achieve its maximum at some extremal f_0.

Actually mere continuity of Φ implies little about the extremal, and we really need Gâteaux derivatives or better still that our functional be in fact linear. Toeplitz [65] described the dual space $\mathcal{H}^*$ of $\mathcal{H}$ as

$$\Phi(f) = \lim_{r \to 1} \int_0^{2\pi} f(re^{i\theta})\overline{g(e^{i\theta})}\, d\theta,$$

for g analytic on the closed disk D. Right from the start (1938), Schiffer [59] considered more general linear problems than $|a_n| \leq n$, although he was content to let g be a rational function.

In fact, $\mathcal{H}$ is not necessarily the most useful embedding. Consider a related class Σ consisting of functions $f = z + \sum_{n=1}^{\infty} a_n z^{-n}$ univalent on $\{|z| > 1\}$. The area theorem of Bieberbach [5], $\sum_{n=1}^{\infty} n|a_n|^2 \leq 1$, may be used to embed Σ in the Dirichlet space $\mathcal{D}$ which consists of those f holomorphic in $\{|z| > 1\}$ for which $\|f\| < \infty$, where $\|f\|^2 = \iint_{|z|>1}(|f' - 1|^2/2\pi)\, dx\, dy$. Furthermore f is on the boundary of the unit ball of $\mathcal{D}$ if and only if $\text{Area}(\mathbb{C} - f(|z| > 1)) = 0$.

3. Variational techniques. The class S is a very nonlinear set, so one cannot merely add small perturbations to obtain a valid variation. The main problem is to ensure that the perhaps bizarre boundary of a simply connected domain is perturbed into a boundary of another simple connected domain. Historically the first nontrivial variation is due to Hadamard [30]. The idea is to move a portion of the boundary in the normal direction δn and compute corresponding changes in the Green's function $G(z, w)$ by means of Green's formula; assuming $\delta n = O(\varepsilon)$,

$$\delta G(z, w) = -\frac{1}{2\pi} \int_{\partial D} \frac{\partial G}{\partial n_t}(z, t) \frac{\partial G}{\partial n_t}(t, w) \delta n |dz_t| + o(\varepsilon);$$

of course this only works if the boundary is smooth. This is quite useful for special problems in mathematical physics (cf. the article by Garabedian in this volume), or even for deformation problems for compact Riemann surfaces (which have no boundaries); see Ahlfors [2]. However, as a general extremal method for S, this is hopeless as there is no a priori reason for assuming that the boundary of the extremal domain has any regularity at all.

The way out of this was obtained by Schiffer. Instead of the usual method (see Duren [22]), let us derive the Schiffer variation the way Ahlfors did in 1955 by means of a quasiconformal variation. This approach is considerably more general and has served as a step in my own work.

We need something of the theory of quasiconformal mappings: ϕ is a quasiconformal homeomorphism if "small disks" are mapped to "small ellipses" of bounded eccentricity. It is found that ϕ satisfies the Beltrami equation

$$\frac{\partial \phi}{\partial \bar{z}} = \mu \frac{\partial \phi}{\partial z},$$

where μ, the complex dilatation, has L^∞ norm $\|\mu\|_\infty = k < 1$. The general existence theorem was done by Morrey [46], although Gauss had done the real analytic case a century earlier.

However the main reason for the importance of the Beltrami equation was the possibility of an explicit solution. Boyarskiĭ [7] in 1957 defined

$$\phi(z) = z + C\mu + C\mu S\mu + C\mu S\mu S\mu + \cdots,$$

where

$$C\mu = -\frac{1}{\pi} \iint \frac{\mu(x+iy)}{(x+iy-z)} \, dx \, dy$$

is the Cauchy transform, and

$$S\rho = (\partial/\partial z)C\rho$$

is the Hilbert transform. The convergence of the series is ensured by the theory of singular integral equations, namely, that S has L^2 norm 1 (Beurling) and is in fact a bounded operator on L^p for $1 < p < \infty$ (Calderón, Zygmund); see [12]. It is then not hard to see that ϕ satisfies the Beltrami equation and is a homeomorphism of $\mathbb{C}$. We may use ϕ to produce a variation of $f \in S$. For let μ have compact support U in $f(D)$ and ϕ be the corresponding solution to the Beltrami equation. Now $\phi \circ f$ is a homeomorphism of D to $\phi(\mathbb{C})$, which is not analytic on $f^{-1}(U)$. To produce an analytic function, we introduce a certain quasiconformal self-map ψ of the disk with Beltrami coefficient λ on $f^{-1}(U)$. By suitable choice of λ as an explicit function of μ, we find that

$$f^* = \phi \circ f \circ \psi^{-1}$$

is an *analytic* homeomorphism. Finally we obtain a variation for S by renormalizing to ensure that $f^*(0) = 0$, $\partial f^*(0)/\partial z = 1$.

The Schiffer variation is produced by concentrating μ on a small disk near $w = f(\xi)$; to first order we have

$$\delta f = \varepsilon \left\{ \frac{f^2}{f-w} \right\} + \frac{\varepsilon}{2} \left\{ \frac{\xi+z}{\xi-z} zf' - f \right\} \left\{ \frac{f(\xi)}{\xi f'(\xi)} \right\}^2$$
$$+ \frac{\bar{\varepsilon}}{2} \left\{ \frac{1+\bar{\xi}z}{1-\bar{\xi}z} zf' - f \right\} \left\{ \frac{f(\xi)}{\xi f'(\xi)} \right\}^2$$

on expanding on compact subsets of D. Thus if f is extremal for Φ, i.e., $\operatorname{Re} \Phi(\delta f) \leq 0$, the Schiffer differential equation is obtained:

$$P(w)\{\xi f'/f\}^2 = Q(\xi).$$

Now $P(w)$ is analytic on $\mathbb{C} - f(D)$, and $Q(\xi)$ is positive and rational on $\{|\xi| = 1\}$. Thus the boundary of $f(D)$ is a horizontal trajectory of a quadratic differential, in particular, it is piecewise analytic.

Unfortunately the solution to this differential equation depends on unspecified parameters (such as earlier coefficients if $\Phi f = a_n$), so explicit solutions are only possible in special cases. This is Teichmüller's Principle [64], which was generalized by Jenkins [39] into the General Coefficient Theorem.

In the 40's Schaeffer and Spencer [57] described the coefficient region V_n exactly but so implicitly as not to give $|a_4| \leq 4$. Later work of Garabedian, Charzyński, Pederson, Ozawa, and Schiffer showed that $|a_n| \leq n$ for $n < 7$; see [25, 14, 50, 51, 48]. In 1962 Ahlfors [3] used second-order terms for Φ to study curvature properties of the module space of Riemann surfaces. At the same time Duren and Schiffer [19] began a systematic development of the second variation; see also Chang, Schiffer, and Schober [14]. Eventually Garabedian and Schiffer and Bombieri (see [26, 27, 6]) were able to prove the local Bieberbach conjecture. It proved very difficult to squeeze information out of the quadratic differential description of V_n.

For the particular case that Φ is a nonconstant linear functional L, the extremals are called *support points*. Then an elementary variation shows $\operatorname{Re} P(w) < 0$ on the complementary arc $\Gamma = \mathbb{C} - f(D)$. Thus Γ is a monotone analytic slit with the "$\pi/4$-property": $|\operatorname{Arg} dw/w| \leq \pi/4$, $w \in \Gamma$. Originally this was proved for functionals L with $g = z^{-n}$ by Golusin [29] and extended to general $L \in H^*$ by Pfluger [52] (see also Brickman and Wilken [10]). An astute example of Pearce [49] showed that $\pi/4$ can be attained.

In any case it seemed that there were few support functions, and these had nice properties. It was natural to try to characterize the set of support functions, $\operatorname{supp}(S)$, and solve general linear problems. This may be viewed as the essential intent of the "Schiffer program." Furthermore this would lead to an extremal theory on general domains, as Schober and Hengartner [36] showed.

4. Extreme points. The Riesz-Herglotz formula (see [55]) is the inspiration for much convexity theory in analysis. It says that every function $p(z) = 1 + p_1 z + \cdots$ analytic with positive real part on D has representation

$$p(z) = \int_T \frac{1 + zw}{1 - zw} d\mu(w), \qquad p \in P,$$

for some probability measure on the circle $T = \{|w| = 1\}$. Actually it is an immediate deduction from the Poisson formula (once Riesz had given a clear idea of what a measure is).

The modern concept lives in a locally convex topological vector space X (see [18, Chapter V]) and says that if $A \subset X$, then $\alpha \in A$ is an extreme point of A if

$$\alpha \neq t\beta + (1-t)\gamma, \qquad 0 < t < 1,$$

for any distinct β and γ in A. For extremal problems, what is important is that if A is compact and X separable, then the closed convex hull of A, $\overline{\operatorname{co}} A$, is

compact (Masur) and if $\text{Ext}(A)$ is the set of extreme points of A, then
$$\max_{A} \text{Re}\, Lf = \max_{\text{Ext}(A)} \text{Re}\, Lf$$
for any $L \in X^*$. Thus, for instance, Carathéodory applied his theory of convexity in R^n to the coefficient body of P, using
$$\text{Ext}\, P = \left\{ \frac{1+zw}{1-zw} \colon w \in \partial D \right\}.$$
By 1960 the correct generalizations to general X had been worked out: if A is compact and convex, then Krein-Milman tells us that
$$A = \overline{\text{co}}\, \text{Ext}(A),$$
while Choquet theory even gives an integral representation. Results followed in the converse direction. de Branges proved the Weierstrass approximation theorem from Krein-Milman, and Holland [37] even directly derived the Riesz-Herglotz formula from Choquet theory.

For the classical subclasses of S, extreme point theory proved useful. It began with contributions from the very young Löwner [45], Nevanlinna [47], and Dieudonné [17]. In 1936 Robertson [55] discovered how to transform the Riesz-Herglotz formula into an integral representation of a subclass. Afterwards this was systematically used by Golusin [28]. Typical is the class $S_t = \{f \in S \colon f(D)$ is starlike with respect to $0\}$ for which Nevanlinna proved the Bieberbach conjecture. Brickman, MacGregor, and Wilken [11] found that
$$\text{Ext}(S_t) = \{z/(1-zw)^2 \colon w \in T\},$$
i.e., any linear problem on S_t is a problem in calculus of one variable w (see also Hummel [38]).

Somewhat deeper is the description of classes of bounded boundary rotation due to Brannan, Clunie, and Kirwan [8], who actually solved the coefficient problem by means of the extreme points. For such classes A several parameters may be needed to characterize $\text{Ext}(\overline{\text{co}}\, A)$. Perhaps deepest of all in this direction is the solution of the Pólya-Schoenberg conjecture by Ruscheweyh and Sheil-Small [56] (see also Suffridge [63]).

For the full class S there is still no characterization of $\text{Ext}(S)$, while for Σ things are complicated. Recall that $\Sigma = \{f = z + \sum_{n=1}^{\infty} a_n z^{-n}$ univalent in $\overline{D}^c\}$. Springer [61] in 1955 used the area theorem to show that
$$\text{Area}(\mathsf{C} - f(\overline{D}^c)) = 0 \Rightarrow f \in \text{Ext}(\Sigma),$$
while a few years ago I [31] found a simple argument to show the converse. Thus there are too many extreme points of Σ to solve coefficient problems!

Let us consider the class S. In 1970 Brickman [9] gave an elegant argument that if $f \in \text{Ext}(S)$, then $\mathsf{C} - f(D)$ is a *monotone arc* Γ. For $f \in \text{Ext}(\overline{\text{co}}\, S)$ we can say a bit more. Kirwan and Pell [41] use the $\pi/4$-result plus a general theorem from functional analysis,
$$\overline{\text{Supp}(S)} \supset \text{Ext}(\overline{\text{co}}\, S),$$

to prove that every $f \in \text{Ext}(\overline{co}\,S)$ satisfies the "$\pi/4$-result (a.e.)." Recently I observed a somewhat stronger result, namely, that Γ is (monotone) quasismooth with constant $\sqrt{2}$. The definition of quasismooth is due to Warschawski (see Pommerenke [54]) and means that for any $z, w \in \Gamma$

$$\text{Length}\,\Gamma(z, w) \leq \sqrt{2}|w - z|,$$

where $\Gamma(z, w)$ is the subarc of Γ which joins z and w.

Any hope of solving general linear problems relies on $\text{Supp}(S)$ being nice. For example, is $\text{Supp}(S)$

(i) parametrized as a smooth variety?,

or weaker

(ii) closed?,

or weaker still do we have

(iii) $\text{Ext}(S) \subset \text{Supp}(S)$

as had been conjectured by Duren [22, 23]. These were recently disproved by me [32].

5. Modern theory. Modern geometric function theory begins in 1955 with Ahlfors' discovery of an explicit analytic solution to the Beltrami equation. Very soon Ahlfors [2] and Bers were applying these quasiconformal variations to resurrect Kleinian groups. In the last decade they have also been used by Weitsman and Drasin for Nevanlinna theory and most spectacularly by Sullivan to investigate Julia sets.

Recently I used this as a step for *boundary extremal problems*. The original impetus was the need to construct a variation for the Hayman class H, which has an angular derivative normalization not amenable to Schiffer's method.

There are other such boundary problems; one which has proved fiendishly difficult is

$$\max_{S} \text{Area}\{f(D) \cap D\}.$$

J. Lewis[43] needed deep results from partial differential equations just to show $\partial f(D)$ is smooth for the extremal functions.

However, let us fix our attention on "boundary linear functionals" which arise in the following way. For $f \in S$ define

$$L(f) = \lim_{r \to 1} \int_0^{2\pi} f(re^{i\theta})\overline{g(e^{i\theta})}\,d\theta.$$

Of course if $g \sim \sum b_n e^{in\theta}$ the coefficient estimate shows that it is sufficient that $\sum n|b_n| < \infty$. Actually this is not enough to develop a reasonable variational theory.

It is classical that $S \subset H^p$, $p < 1/2$, and according to the duality result of Duren, Romberg, and Shields [20] L is bounded on S provided g' is Hölder continuous. In fact one can do better than that (g' need only be Dini continuous), but the exact characterization of S^* is an open question. Certainly $(S^*)' \subset H^\infty$.

In any case for technical reasons we embed S in the Hilbert space

$$\mathfrak{H} = \left\{ \sum a_n z^n : \sum \frac{|a_n|^2}{n^6} < \infty \right\}.$$

Now $\mathfrak{H}$ is a "nice" space, and as $\mathfrak{H}^* \supset \mathcal{H}^*$ many more support points may be generated.

The next problem is to give variations valid for S in the $\mathfrak{H}$ topology. We find that quasiconformal variations (but not Schiffer variations) may be used for such generalized linear functionals. One then attempts to derive quadratic differential equations similar to those in the Schiffer theory.

It is found that $P(w)$ is the jump of a function $P_0(w)$ analytic on $f(D)$. However $P(w)$ is only defined on $\mathbb{C} - f(D)$, and is merely continuous. In fact $P(w)$ has regularity properties dependent on the function g which determines L. Finally one proves that $\Gamma = \mathbb{C} - f(D)$ is a system of arcs with regularity controlled by g.

All this would be vacuous if Γ were an analytic slit. In fact if g is symmetric one easily sees that the Koebe function is extremal. Nevertheless it is possible to contrive a choice of g such that

(i) L nonconstant on S, and

(ii) for any extremal f of L, Γ is *nowhere* analytic on some subarc (not even C^4).

Now one of these extremals is an extreme point of S, so we derive a global result from a local extremal result and find a nonsupporting extreme point of S (recall that Γ is analytic for support points).

6. What is extremal theory good for?

The essentially geometric method of quasiconformal variation proved itself for extremal geometric problems but has been difficult to apply to analytic problems, mainly because of the unexpectedly complicated structure of S.

The variational method has always been useful for counterexamples to coefficient problems. For instance consider powers

$$\left(\frac{f(z)}{z} \right)^\lambda = \sum_{n=0}^{\infty} a_n(\lambda) z^n$$

of functions $f \in S$ and the corresponding "Koebe function"

$$\left(\frac{k(z)}{z} \right)^\lambda = \sum_{n=0}^{\infty} A_n(\lambda) z^n.$$

Schaeffer and Spencer [57] found that, for $n \geq 3$, there is an $f \in S$ such that $|a_n(\frac{1}{2})| > |A_n(\frac{1}{2})|$. (The case $n = 3$ is due to Fekete and Szegö [24].) In fact the "Bieberbach conjecture" fails for $0 < \lambda < \frac{1}{2}$, even in order of magnitude for λ sufficiently small (see Pommerenke [53]). For $\frac{1}{4} \leq \lambda < \frac{1}{2}$ the order of magnitude is correct, but nevertheless the conjecture fails in the following strong way: there

is a positive δ such that for $n \geq 3$, $\frac{1}{4} < \lambda < \frac{1}{2}$, there is an $f \in S$ such that
$$|a_n(\lambda)| \geq (1+\delta)|A_n(\lambda)|.$$
This result of Hayman and Hummel [35] for some values of λ and of Hamilton (unpublished) for $\frac{1}{4} < \lambda < \frac{1}{2}$ makes use of variations in the Hayman class H (see [34]). Hamilton uses higher-order quasiconformal variations.

Now if $\lambda = -1$, i.e., we are essentially in Σ, not even the right order of magnitude is known (see Clunie [15], Clunie and Pommerenke [16]). We do know $|a_n| = O(n^{-\alpha})$, for some α, $\frac{1}{2} < \alpha < 1$. In fact the "natural" conjecture that $(z^n - z^{-n})^{1/n}$ is extremal fails for all n (see Chang, Schiffer, and Schober [13], as well as Tsao [66]).

We conclude by using extreme point theory to derive a new result on integral means. It is a classical result of Littlewood [44] that there is an absolute constant A such that
$$\int_0^{2\pi} |f(re^{i\theta})|\, d\theta \leq A \int_0^{2\pi} |k(re^{i\theta})|\, d\theta.$$
It is a deep and beautiful result of Baernstein [4] that $A = 1$. This may be compared with an older result of Jenkins [39] that for $f \in S$
$$|f(r)| + |f(-\rho)| \leq |k(r)| + |k(-\rho)|$$
for positive r, ρ.

Now it is possible (unpublished) to use Brickman's results on the extreme points of S together with a new quantitative version of the Ahlfors distortion theorem to prove that there is an absolute constant A such that for any positive measure μ on D
$$\iint |f(z)|\, d\mu \leq A \max_\theta \iint |k(ze^{i\theta})|\, d\mu$$
for any $f \in S$. The constant A satisfies $1 \lneq A \leq 16$. Similar results hold for powers $p > 1$, but for $p < 1$ the question remains open.

References

1. L. V. Ahlfors, *Conformality with respect to Riemannian metrics*, Ann. Acad. Sci. Fenn. Ser. A I, no. 206 (1955).

2. ____, "The complex analytic structure of the space of closed Riemann surfaces," in *Analytic functions*, Princeton Univ. Press, Princeton, N. J., 1960, pp. 45–66.

3. ____, *Curvature properties of Teichmüller's space*, J. Analyse Math. **9** (1961/62), 161–176.

4. A. Baernstein, *Integral means, univalent functions and circular symmetrization*, Acta Math. **133** (1974), 139–169.

5. L. Bieberbach, *Über die Koeffizienten derjenigen Potenzreihen, welche eine schlichte Abbildung des Einheitskreises vermitteln*, S. B. Preuss. Adad. Wiss. (1916), 940–955.

6. E. Bombieri, *On the local maximum property of the Koebe function*, Invent. Math. **4** (1967), 26–67.

7. B. V. Boyarskiĭ, *Generalized solutions of systems of differential equations of first order and elliptic type with discontinuous coefficients*, Mat. Sb. **43** (1957), 451–503.

8. D. A. Brannan, J. G. Clunie, and W. E. Kirwan, *On the coefficient problem for functions of bounded boundary rotation*, Ann. Acad. Sci. Fenn. Ser. A I, no. 523 (1973).

9. L. Brickman, *Extreme points of the set of univalent functions*, Bull. Amer. Math. Soc. **76** (1970), 372–374.

10. L. Brickman and D. Wilken, *Support points of the set of univalent functions*, Proc. Amer. Math. Soc. **42** (1974), 523–528.

11. L. Brickman, T. H. MacGregor, and D. R. Wilken, *Convex hulls of some classical families of univalent functions*, Trans. Amer. Math. Soc. **156** (1971), 91–107.

12. A. P. Calderón and A. Zygmund, *On the existence of certain singular integrals*, Acta. Math. **88** (1952), 85–139.

13. A. Chang, M. M. Schiffer, and G. Schober, *On the second variation for schlict functions*, J. Analyse Math. **40** (1981), 203–238.

14. Z. Charzyński and M. M. Schiffer, *A new proof of the Bieberbach conjecture for the fourth coefficient*, Arch. Rational Mech. Anal. **5** (1960), 187–193.

15. J. Clunie, *On schlicht functions*, Ann. of Math. (2) **69** (1959), 511–519.

16. J. Clunie and Ch. Pommerenke, *On the coefficients of univalent functions*, Michigan Math. J. **14** (1967), 71–78.

17. J. Dieudonné, *Sur les fonctions univalentes*, C. R. Acad. Sci. Paris **192** (1931), 1148–1150.

18. N. Dunford and J. T. Schwartz, *Linear operators*. Part I, Interscience, New York, 1958.

19. P. L. Duren and M. Schiffer, *The theory of the second variation in extremum problems for univalent functions*, J. Analyse Math. **10** (1962-63), 193–252.

20. P. L. Duren, B. W. Romberg, and A. L. Shields, *Linear functionals on H^p spaces with $0 < p < 1$*, J. Reine Angew. Math. **238** (1969), 32–60.

21. P. L. Duren, "Extremal problems for univalent functions," in *Aspects of contemporary complex analysis*, edited by D. A. Brannan and J. G. Clunie, Academic Press, London, 1980, pp. 181–208.

22. ___, *Univalent functions*, Springer-Verlag, New York, 1983.

23. ___, "Support points of univalent functions," in *Linear and complex analysis problem book*, Lecture Notes in Math., vol. 1043, Springer-Verlag, Berlin and New York, 1984, pp. 636–637.

24. M. Fekete and G. Szegö, *Eine Bemerkung über ungerade schlichte Funktionen*, J. London Math. Soc. **8** (1933), 85–89.

25. P. R. Garabedian and M. Schiffer, *A proof of the Bieberbach conjecture for the fourth coefficient*, J. Rational Mech. Anal. **4** (1955), 427–465.

26. ___, *The local maximum theorem for the coefficients of univalent functions*, Arch. Rational Mech. Anal. **26** (1967), 1–32.

27. P. R. Garabedian, M. Schiffer, and G. G. Ross, *On the Bieberbach conjecture for even n*, J. Math. Mech. **14** (1965), 975–989.

28. G. M. Golusin, *On typically real functions*, Mat. Sb. **27** (1950), 201–218.

29. ___, *Geometric theory of functions of a complex variable*, Gosudarstv. Izdat. Tehn.-Teor. Lit., Moscow-Leningrad, 1952.

30. J. Hadamard, *Mémoire sur le problème d'analyse relatif à l'équilibre des plaques élastiques encastrées*, Mémoires présentés par divers savants à l'Académie des Sciences, Vol. 33 (1908), 1–128.

31. D. H. Hamilton, *The extreme points of Σ*, Proc. Amer. Math. Soc. **85** (1982), 393–396.

32. ___, *Extremal boundary problems for schlicht functions*, Proc. London Math. Soc. (to appear).

34. W. K. Hayman, *Bounds for the large coefficients of univalent functions*, Ann. Acad. Sci. Fenn. Ser. A I. no. 250 (1958).

35. W. K. Hayman and J. A. Hummel, *Coefficients of powers of univalent functions*, Complex Variables Theory Appl. (to appear).

36. W. Hengartner and G. Schober, *Extreme points for some classes of univalent functions*, Trans. Amer. Math. Soc. **185** (1973), 265–270.

37. F. Holland, *The extreme points of a class of functions with positive real part*, Math. Ann. **202** (1973), 85–87.

38. J. A. Hummel, *The coefficient regions of starlike functions*, Pacific J. Math. **7** (1957), 1381–1389.

39. J. A. Jenkins, *Symmetrization results for some conformal invariants*, Amer. J. Math. **75** (1953), 510–522.

40. ____, *A general coefficient theorem*, Trans. Amer. Math. Soc. **77** (1954), 262–280.

41. W. E. Kirwan and R. Pell, *Extremal properties of a class of slit conformal mappings*, Michigan Math. J. **25** (1978), 223–232.

42. P. Kufarev, *On a method of investigation of extremal problems in the theory of univalent functions*, Dokl. Akad. Nauk SSSR **107** (1956), 633–635.

43. J. Lewis, *On the minimum area problem*, Indiana Univ. Math. J. **34** (1985), 631–661.

44. J. E. Littlewood, *On inequalities in the theory of functions*, Proc. London Math. Soc. **23** (1925), 481–519.

45. K. Löwner, *Untersuchungen über die Verzerrung bei konformen Abbildungen des Einheitskreises $|z| < 1$, die durch Funktionen mit nicht verschwindender Ableitung geliefert werden*, Ber. Verh. Sachs. Ges. Wiss. Leipzig **69** (1917), 89–106.

46. C. B. Morrey, *On the solution of quasilinear elliptic partial differential equations*, Trans. Amer. Math. Soc. **43** (1938), 126–166.

47. R. Nevanlinna, *Über die konforme Abbildung von Sterngebieten*, Overs. Finska Veten.-Soc. Forh. **63(A)**, no. 6 (1920-21), 1–21.

48. M. Ozawa, *On the Bieberbach conjecture for the sixth coefficient*, Kodai Math. Sem. Rep. **21** (1969), 97–128.

49. K. Pearce, *New support points of S and extreme points of HS*, Proc. Amer. Math. Soc. **81** (1981), 425–428.

50. R. N. Pederson, *A proof of the Bieberbach conjecture for the sixth coefficient*, Arch. Rational Mech. Anal. **31** (1968), 331–351.

51. R. N. Pederson and M. Schiffer, *A proof of the Bieberbach conjecture for the fifth coefficient*, Arch. Rational Mech. Anal. **45** (1972), 161–193.

52. A. Pfluger, *Lineare Extremalprobleme bei schlichten Funktionen*, Ann. Acad. Sci. Fenn. Ser. A I., no. 489 (1971).

53. Ch. Pommerenke, *Über die Mittelwerte und Koeffizienten multivalenter Funktionen*, Math. Ann. **145** (1961/62), 285–296.

54. ____, *Univalent functions*, Vandenhoeck und Ruprecht, Göttingen, 1975.

55. M. S. Robertson, *On the theory of univalent functions*, Ann. of Math. **37** (1936), 374–408.

56. St. Ruscheweyh and T. Sheil-Small, *Hadamard products of schlicht functions and the Pólya-Schoenberg conjecture*, Comment. Math. Helv. **48** (1973), 119–135.

57. A. C. Schaeffer and D. C. Spencer, *The coefficients of schlicht functions*, Duke Math. J. **10** (1943), 611–635.

58. A. C. Schaeffer, M. Schiffer, and D. C. Spencer, *The coefficient regions of schlicht functions*, Duke Math. J. **16** (1949), 493–527.

59. M. Schiffer, *On the coefficients of simple functions*, Proc. London Math. Soc. **44** (1938), 450–452.

60. ____, *Extremum problems and variational methods in conformal mapping*, Proc. Internat. Cong. Math. (Edinburgh, 1958), Cambridge Univ. Press, New York, 1960, pp. 211–231.

61. G. Springer, *Extreme Punkte der konvexen Hülle schlichter Funktionen*, Math. Ann. **129** (1955), 230–232.

62. G. Schober, *Univalent functions—Selected topics*, Lecture Notes in Math., vol. 478, Springer-Verlag, New York and Berlin, 1975.

63. T. J. Suffridge, *Extreme points in a class of polynomials having univalent sequential limits*, Trans. Amer. Math. Soc. **163** (1972), 225–237.

64. O. Teichmüller, *Ungleichungen zwischen den Koeffizienten schlichter Funktionen*, S.-B. Preuss. Akad. Wiss. Phys.-Math. Kl. (1938), 363–375.

65. O. Toeplitz, *Die linearen vollkommen Räume der Funktionentheorie*, Comment. Math. Helv. **23** (1949), 222–242.

66. A. Tsao, *Disproof of a coefficient conjecture for meromorphic univalent functions*, Trans. Amer. Math. Soc. **274** (1982), 783–796.

The Method of the Extremal Metric

JAMES A. JENKINS

The method of the extremal metric has its origin in the obvious statement that, for a conformal mapping, area distortion is the square of length distortion. In its most primitive form it appears as the length-area method: this applies to a family of curves which sweep out all or part of a domain which is subjected to a conformal mapping such that there is a lower bound for the length of the images of the curves and an upper bound for the area of the image of the domain. An application of some form of the Schwarz inequality leads to an appropriate conclusion.

Isolated examples of the method go back to the early part of this century. The first I have located occur in a paper of H. Bohr [6] and one by W. Gross [9]. Two noteworthy examples appear in applications to boundary correspondence in a paper by G. Faber [7] and in the book of Hurwitz-Courant [11] (presumably due to Courant).

The method was first developed in a consistent manner in the late 1920's and early 1930's independently in two distinct directions by Herbert Grötzsch and Lars Ahlfors. Remarkably each credits his inspiration to the above-mentioned results respectively due to Faber and Courant.

Grötzsch used what is called the method of strips. The simplest entities which display conformal invariants are the quadrangle and the doubly connected domain. A quadrangle is a simply connected domain of hyperbolic type with four assigned boundary elements. This can be mapped conformally onto a rectangle so that the latter correspond to the vertices. The ratio of the lengths of the sides of the rectangle is then a characteristic conformal invariant. A doubly connected domain with nondegenerate boundary continua can be mapped conformally onto a circular ring $r_1 < |z| < r_2$ $(0 < r_1 < r_2)$, and the ratio r_2/r_1 is then a characteristic conformal invariant. Grötzsch's basic techniques rested on two lemmas which can be stated as follows.

LEMMA I. *Let Q_i be quadrangles (finite or countable in number) lying in the circular ring $r_1 < |z| < r_2$ $(0 < r_1 < r_2)$, each with a pair of opposite sides on the two bounding circles of that ring and such that the ratio of the length of the sides of the corresponding rectangle for them to the length of the other sides of* the rectangle is a_i/b_i. Then

$$\sum_i (a_i/b_i) \leq 2\pi/\log(r_2/r_1),$$

with equality occurring if and only if the quadrangles are obtained from the ring by radial decomposition so that the sum of the areas of the quadrangles is equal to the area of the ring.

LEMMA II. *Let D_i be doubly connected domains (finite or countable in number) lying in the circular ring $r_1 < |z| < r_2$ $(0 < r_1 < r_2)$ with the same topological situation and such that their corresponding circular rings have ratios of radii $s_2^{(i)}/s_1^{(i)}$. Then*

$$\sum_i \frac{1}{2\pi} \log(s_2^{(i)}/s_1^{(i)}) \leq \frac{1}{2\pi} \log(r_2/r_1),$$

with equality occurring if and only if the D_i are obtained from the ring by concentric circumferential decomposition so that the sum of the areas of the D_i is equal to the area of the ring.

It is instructive to go through Grötzsch's method in one of the simplest cases. Consider the circular ring $1 < |z| < R$ and conformal mappings of it by functions f which relate $|z| = 1$ to $|w| = 1$, are regular, and satisfy $|f(z)| > 1$. The problem is to find those functions for which a boundary point in $|w| > 1$ of the image is closest to $|w| = 1$. The answer is that this occurs for a mapping onto a domain D^* bounded by $|w| = 1$ and a slit on a radial ray Y from a point P at distance d (> 1) from the origin to the point at infinity. Suppose that for a competing mapping there were a boundary point at distance $d' \leq d$. Using magnification centered at the origin by ratio d/d' and a rotation, we can bring this point into coincidence with P, the resultant image domain of $1 < |z| < R$ being denoted by Δ. Let there be a maximal decomposition of D^* by strips corresponding to a radial decomposition of $1 < |z| < R$ so that the strips are symmetric under reflection in Y. Forming the union of each symmetric pair of strips and a segment of Y, we then decompose this by intersection with Δ into two quadrangles for which the sum of the invariants is at least equal to that of the previous pair. This is equivalent to saying that if we take a rectangle and divide it into two quadrangles by possibly cutting off the ends and joining the other pair of sides by a continuum, the sum of the invariants of the quadrangles so obtained is minimal when the division is produced by cutting on the midline. This can be proved in various manners. (Grötzsch's proof used a somewhat esoteric decomposition into approximately square subdomains which was evidently carried over from the paper of Faber mentioned above.) The conclusion then is evident.

Grötzsch's method was very advantageous in dealing with domains of infinite connectivity, and most of his many applications are valid in the most general case. Over the years his approach became more and more sophisticated and displayed a strong differential-geometric tendency.

Ahlfors' basic contribution to the method of the extremal metric was given in his thesis [1]. This consists of three essential parts. In the first he gave two results for the conformal mapping of certain general strip domains which he called the First and Second Fundamental Inequalities. For an interval of values of x the strip domain $\mathfrak{S}$ is to have a certain distinguished cross cut σ_x of length $\theta(x)$ on the ordinate at value x. For a conformal mapping of $\mathfrak{S}$ on a horizontal parallel strip S of width 1 on the image γ_x of σ_x the maximal value of the real part is to be $\xi_2(x)$, the minimal value $\xi_1(x)$. The First Fundamental Inequality (now habitually called the Ahlfors distortion theorem) gives a bound of the form $(x_1 < x_2)$

$$\int_{x_1}^{x_2} \frac{dx}{\theta(x)} \leq (\xi_1(x_2) - \xi_2(x_1))^+ + C$$

for an absolute constant C (where $t^+ = \max(0,t)$). The Second Fundamental Inequality gives a bound of the form

$$\xi_2(x_2) - \xi_1(x_1) \leq \int_{x_1}^{x_2} \frac{dx}{\theta(x)} + V,$$

which applies only when the strip domain $\mathfrak{S}$ satisfies certain subsidiary conditions and where V depends on geometric quantities associated with $\mathfrak{S}$. Ahlfors' proofs were rather involved applications of the length-area method.

The second part of Ahlfors' thesis dealt with the problem of the angular derivative which he formulated in the context of strip domains in a manner which has been standard in all subsequent work. Using the previous section he gave some conditions (necessary or sufficient) for the existence of angular derivatives.

The third part of his thesis gave the first published proof of the Denjoy conjecture which, in its most direct formulation, says that an integral function of finite order ρ can have at most 2ρ distinct asymptotic values (in an appropriate sense). The proof was primarily an application of the Ahlfors distortion theorem. He also stated an extension to the case where the asymptotic paths satisfy certain conditions of increase of argument (about which more later).

Ahlfors used the same techniques in several other papers which contained a number of results of considerable interest, but none of them had the same impact as his thesis.

Chronologically the next development in the method of the extremal metric appears in the work of Teichmüller. The apex of his work undoubtedly occurs with the big paper [41] in the *Abhandlungen der preussischen Akademie*. In it he formulates the relationship between quadratic differentials and extremal problems in function theory. A (meromorphic) quadratic differential $Q(z)\,dz^2$ on a Riemann suface is given by assigning to each local uniformizing parameter z a meromorphic function $Q(z)$, where these transform by

$$Q^*(z^*) = Q(z)\left(\frac{dz}{dz^*}\right)^2.$$

It is thus meaningful to speak of a quadratic differential having a zero or pole of given order at a point of the Riemann surface. These points are called the critical points of the quadratic differential. Of extreme importance in the study of a quadratic differential are its trajectories, i.e., maximal curves on which $Q(z)\,dz^2 > 0$. Teichmüller indicated almost without proof the local structure of the trajectories at various points, ordinary and critical, and identified certain types of domains occurring in the global trajectory structure: end domains in which the trajectories are like parallel lines in a half-plane; strip domains in which the trajectories are like parallel lines in a parallel strip; ring domains in which the trajectories are like the concentric circles in a circular ring; circle domains in which the trajectories are like the concentric circles about the center of a disc. He did not mention the question of whether every trajectory had a point set closure which was either an arc or a Jordan curve, i.e., whether conversely there could be recurrent trajectories. Teichmüller enunciated the principle that the solution of a certain type of extremal problem in geometric function theory is in general associated with a quadratic differential. If in the problem a point is assumed to be fixed without further requirement, the quadratic differential will have a simple pole there. If in addition the functions treated in the problem are required to have at the point, in terms of suitably assigned local uniformizing parameters, fixed values for their first n derivatives, the quadratic differential will have a pole of order $n + 1$ there. More generally, the highest derivative occurring may not be required to be fixed, but some condition on its region of variation may be desired. Teichmüller was led to this by his considerations on quasiconformal mappings. However he never gave anything in the nature of an explicit general result embodying this principle.

Actually it is two other papers by Teichmüller which relate most directly to the method of the extremal metric. In the first [**39**] he uses the traditional length-area method albeit with a strong differential-geometric slant. The most novel results in this paper lie, however, in the area of quasiconformal mappings. The other paper [**40**] gives his coefficient theorem according to which, for a simply connected domain on the sphere, containing the point at infinity and bounded by a continuum consisting of trajectory arcs of a quadratic differential $P(z)\,dz^2$ with $P(z)$ a polynomial of degree n, $P(z) = \alpha z^n + \cdots$, if $f \in \Sigma(D)$, $f(z) = z + \sum_{j=0}^{\infty} C_j z^{-j}$, and $C_j = 0$, $j = 0, \ldots, n$, then $\mathcal{R}(\alpha C_{n+1}) \leq 0$. Unlike the situation in the first-mentioned paper the proofs in these papers are worked out in completely precise detail. However Teichmüller did not make any explicit connection between the length-area method and quadratic differentials other than applying the former in his uniqueness proof for extremal quasiconformal mappings.

A most decisive step in the development of the method of the extremal metric occurred at the beginning of the 1940's with the abstraction of the length-area method formulated by Arne Beurling and developed initially by him and Ahlfors. This deals with a curve family Γ, consisting of locally rectifiable curves, in a

domain D and poses the following extremal problem: consider all functions ρ nonnegative and measurable in D such that $\int_\gamma \rho |dz|$ exists, in an appropriate sense, for all $\gamma \in \Gamma$; let $L_\rho(\Gamma) = \text{g.l.b.}_{\gamma \in \Gamma} \int_\gamma \rho |dz|$; then, for P the family of such ρ for which $L_\rho(\Gamma)$ and $A_\rho(D) = \iint_D \rho^2 \, dA_z$ are not simultaneously 0 or ∞, we consider
$$\text{g.l.b.}_{\rho \in P} \frac{A_\rho(D)}{(L_\rho(\Gamma))^2}.$$
This is called the module of the family Γ. The definition is readily extended to Riemann surfaces, and one may also consider multiple curve families. The above quantity is seen at once to be a conformal invariant in the sense that subjecting the configuration to a conformal mapping does not change its value. Ahlfors and Beurling usually worked with the reciprocal of the module, which they called the extremal length of Γ. They gave dual lectures at the Scandinavian Mathematical Congress in 1946, Beurling's consisting of the theoretical background and Ahlfors' dealing with the application of the method to some multiple curve families associated with certain simple geometrical configurations [3]. Unfortunately Beurling's lecture was never published. They later gave a brief account of the method in several publications [4, 5].

The characteristic conformal invariants of a quadrangle and a doubly connected domain are readily obtained in this context by taking the respective curve families as that joining a pair of opposite sides of the quadrangle and that consisting of Jordan curves separating the boundary components of the doubly connected domain. A simple application of Fubini's theorem gives the respective values b/a and $(1/2\pi)\log(r_2/r_1)$ in the notation previously employed. Since quantities like these play the chief role in most applications, the question may be asked whether there is a great advantage in the abstract formulation. The answer is: indeed there is. As an example, let us give a proof of the Ahlfors distortion theorem which appeared first in a paper by Kôtaro Oikawa and myself [26].

Using the notation employed above, the module of the family of cross-cuts σ_x, $x_1 < x < x_2$, is $\int_{x_1}^{x_2} dx/\theta(x)$. The module of the quadrangle bounded by the cross-cuts γ_{x_1}, γ_{x_2} and sides on the horizontal sides of S for the family of curves joining the latter pair of sides of the quadrangle is at least equal to the above quantity and in turn is at most $(\xi_1(x_2) - \xi_2(x_1))^+ + 2$. Thus
$$\int_{x_1}^{x_2} \frac{dx}{\theta(x)} \leq (\xi_1(x_2) - \xi_2(x_1))^+ + 2.$$
One should compare this with other proofs including that of Teichmüller [39] and that in the recent book by Ahlfors [2].

In this paper we gave also an improved version of the Second Fundamental Inequality. The latter part of the paper consists of an examination of a result proved by Hayman using the length-area method in terms of the method of the extremal metric. It provided an explicit quantitative version of Hayman's qualitative result and has had a number of important applications.

Not long after the paper [26] appeared, by using its concepts the angular derivative problem received what must be regarded as its definitive solution, simultaneously by Rodin and Warschawski [38] on the one hand and by Kôtaro Oikawa and myself [27] on the other. It should be remarked that J. Wolff [44] had used the length-area method to treat one aspect of this problem.

A number of variants of the definition of module have been given which in certain special circumstances may provide some advantage but for most purposes that given above is effective.

When I went in the fall of 1949 to spend a year at the Institute for Advanced Study, by a fortuitous circumstance both Schiffer and Spencer were spending the academic year at Princeton University. Schaefer and Spencer had just completed their book on the coefficient regions for univalent functions. In it they gave a derivation of the local structure of the trajectories of a quadratic differential, verifying the conclusions of Teichmüller. They also studied some special cases of the global structure of trajectories of meromorphic quadratic differentials on the sphere, showing in particular that there could be no recurrent trajectory in the case of a differential with one or two poles and obtaining the same result for a particular type of meromorphic quadratic differential with three poles. They expected and were trying to prove that this was the general situation. Quite early I realized that recurrent trajectories were possible. Indeed in the particular case of the differential $C[(z-z_1)(z-z_2)(z-z_3)(z-z_4)]^{-1} dz^2$ (all z_j distinct), for all but a countable number of values of the argument of the constant C all trajectories will be recurrent. Later I analyzed the global structure, obtaining an essentially definitive description for positive quadratic differentials on finite Riemann surfaces [16, 19]. In particular it can be seen that the only general circumstances in which one can affirm the absence of recurrent trajectories for such a quadratic differential are in the case of schlichtartig domains and when the total number of poles and boundary components is at most three (Three Pole Theorem). Sometime later I proved that this result is essentially of topological character [23].

Beginning at the same time I showed the relationship between quadratic differentials and an important class of modules for multiple curve families [15]. On a finite Riemann surface S, let there be given a finite set of disjoint Jordan curves $j_1, \ldots, j_n$, no two homotopic, and let there be given nonnegative quantities $a_1, \ldots, a_n$, not all zero. We consider two problems. First consider the module problem requiring that for rectifiable Jordan curves γ_k respectively homotopic to j_k, $k = 1, \ldots, n$, $\int_{\gamma_k} \rho |dz| \geq a_k$, and take the module $M(a_1, \ldots, a_n)$ defined as g.l.b. $\iint_S \rho^2 \, dA$. Second consider disjoint doubly connected domains $D_1, \ldots, D_n$, lying in S with the same topological situation as $j_1, \ldots, j_n$ and with modules $M_1, \ldots, M_n$. We ask for the maximum of $\sum_{k=1}^n a_k^2 M_k$. These two values are the same, and the unique extremal configuration corresponds to a positive quadratic differential on S whose trajectory structure consists entirely of ring domains which are candidates for the D_j (some possibly being degenerate). The first definition is in the Ahlfors-Beurling format. The second may be regarded as the generalization of Grötzsch's approach. Their equivalence provides a powerful technique for treating extremal problems for univalent functions and is particularly effective when used in conjunction with symmetrization methods. In the particular case of plane domains of finite connectivity, the above considerations were developed in the academic year 1949-50.

Subsequently I gave an explicit result realizing Teichmüller's principle for the relationship between extremal problems and quadratic differentials. By a process of osmosis I came to call it the General Coefficient Theorem. It is a result which applies to a finite Riemann surface on which is given a positive quadratic differential, a corresponding decomposition into trajectory structure subdomains, and mappings of the latter onto nonoverlapping subdomains of the surface, with these mappings subjected to conditions on preservation of poles, certain coefficient normalizations at the poles (where the same local uniformizing parameter is to be used for variable and image), and certain topological conditions. The result is an inequality for a functional involving coefficients of the quadratic differential at poles of order greater than one and those of the mapping functions at the latter. It went through a number of extensions [**13, 16, 18, 20, 21**] and includes as special cases almost all basic results for univalent functions.

While most methods used in the theory of univalent functions apply only in that context or in slightly modified situations, the method of the extremal metric can readily be used in quite general situations in function theory. There are at least five quite distinct techniques that can be used in this connection.

I. *Direct transfer of a metric induced on the Riemann image.* The Riemann image of a domain by a meromorphic function has a metric induced by its covering of the base surface. Carrying this back to the original domain can give significant results in some cases. See, for example, [**28**].

II. *The method of simple coverings.* Under appropriate geometric or topological conditions it can be advantageous to restrict the metric induced on the Riemann image as above to a subset which covers the base surface simply or perhaps with a fixed finite multiplicity. This method has numerous important applications [**12, 24, 29, 30**].

III. *The method of level sets.* Taking the level sets of a harmonic function as a generalized curve family, its module is readily determined in terms of an integral of the conjugate harmonic differential. Auxiliary techniques then give important results. The proof of the Ahlfors distortion theorem given above is essentially a very simple example of this method. A more sophisticated example is given by [**25**].

IV. *The two constant theorem for modules.* The usual two constant theorem for harmonic measures can be translated into a result for triad modules. By a triad (D, α, P) is meant the configuration consisting of a simply connected domain D of hyperbolic type, an open border arc α of D, and a point P interior to D. By the module $M(P, \alpha, D)$ we mean the module of the family of locally rectifiable open arcs in $D - \{P\}$ running from α back to α and separating P from the closed border arc α^* complementary to α.

For real numbers a, b, $a < b$, we denote by $S(a, b)$ the strip in the (u, v)-plane defined by $a < u < b$, and by $g(a)$ its open border arc determined by $u = a$.

Let (D, α, P) be a triad in the z-sphere with D bounded by a Jordan curve so that α, α^* can be regarded as the corresponding boundary arcs, $U(z)$ a function subharmonic in D, a, b real numbers with $a < b$ such that

$$\varlimsup_{z \to \varsigma} U(z) \leq a, \quad \varsigma \in \alpha, \qquad \varlimsup_{z \to \varsigma} U(z) \leq b, \quad \varsigma \in \alpha^*,$$

and $U(P) > a$. Then

$$M(P, \alpha, D) \leq M(U(P), g(a), S(a, b)).$$

This can be extended to quasiconformal mappings also [**17**]. Applications are found in [**22**]. Also I have recently used this method to give a straightforward proof of Ahlfors' spiral generalization of the Denjoy conjecture.

V. *Symmetrization methods.* The multiple curve family modules $M(a_1, \ldots, a_n)$ discussed above can be utilized also for functions which are not univalent. This is particularly useful for functions which satisfy a valence condition. An important example is given by [**14**].

From its early beginnings as the length-area method, the method of the extremal metric has developed into one of the most powerful and widely used tools in the theory of functions. It would be a major task to list all papers in which it is applied. I will content myself with giving the names of those of my students who have published papers using the method: Sister Barbara Ann Foos [**8**], Arthur Obrock [**32, 33, 34, 35**], Dean Phelps [**37**], Jeffrey Wiener [**42, 43**], Margot Pallmann [**36**], Luby Liao [**31**].

References

1. Lars Ahlfors, *Untersuchungen zur Theorie der konformen Abbildung und der ganzen Funktionen*, Acta Societatis Scientiarum Fennicae, N.S.A. **1** (1930), 1–40.

2. ____, *Conformal invariants*, McGraw-Hill, New York, 1973.

3. Lars Ahlfors and Arne Beurling, *Invariants conformes et problèmes extrémaux*, C.R. Congr. Math. Scand. (Copenhagen, 1946), Jul. Gjellerups Forlag, Copenhagen, pp. 341–351.

4. ____, *Conformal invariants and function-theoretic null sets*, Acta Math. **83** (1950), 100–129.

5. ____, "Conformal invariants," in *Construction and applications of conformal maps: Proceedings of a symposium*, National Bureau of Standards Appl. Math. Series, no. 18, U.S. Government Printing Office, Washington, D.C., 1952, pp. 243–245.

6. H. Bohr, *Über streckentreue und konforme Abbildung*, Math. Z. **1** (1918), 403–420.

7. G. Faber, *Über den Hauptsatz aus der Theorie der konformen Abbildung*, Sitzungsberichte Math.-phys. Klasse der bayerschen Akademie der Wissenschaften München, 1922, pp. 91–100.

8. Sister Barbara Ann Foos, *The values of certain sets of modules*, Duke Math. J. **26** (1959), 467–484.

9. W. Gross, *Zum Verhalten analytischer Funktionen in der Umgebung singulärer Stellen*, Math. Z. **2** (1918), 242–294.

Many of Herbert Grötzsch's important papers are listed in [16]. Here we add only one which was omitted there since explicit reference was not made to it.

10. H. Grötzsch, *Die Werte der Doppelverhältnisses bei schlichter konformer Abbildung*, Sitzungsberichte der preussischen Akademie der Wissenschaften, Phys.-math. Klasse, 1933, pp. 501-515.

11. A. Hurwitz and R. Courant, *Funktionentheorie*, Springer-Verlag, Berlin, 1922.

12. James A. Jenkins, "Some results related to extremal length," in *Contributions to the theory of Riemann surfaces*, Ann. of Math. Stud., no. 30, 1953, pp. 87-94.

13. ____, *A general coefficient theorem*, Trans. Amer. Math. Soc. **77** (1954), 262-280.

14. ____, *On circumferentially mean p-valent functions*, Trans. Amer. Math. Soc. **79** (1955), 423-428.

15. ____, *On the existence of certain general extremal metrics*, Ann. of Math. (2) **66** (1957), pp. 440-453.

16. ____, *Univalent functions and conformal mapping*, Springer-Verlag, Berlin-Göttingen-Heidelberg, 1958.

17. ____, *On the Denjoy conjecture*, Canad. J. Math. **10** (1958), 627-631.

18. ____, *An extension of the General Coefficient Theorem*, Trans. Amer. Math. Soc. **95** (1960), 387-407.

19. ____, *On the global structure of the trajectories of a positive quadratic differential*, Illinois J. Math. **4** (1960), 405-412.

20. ____, *An addendum to the General Coefficient Theorem*, Trans. Amer. Math. Soc. **107** (1963), 125-128.

21. ____, *On normalization in the general coefficient theorem*, Proc. Internat. Congr. Math. (Stockholm, 1962), Inst. Mittag-Leffler, Djursholm, 1963, pp. 347-350.

22. ____, "On the Phragmén-Lindelöf theorem, the Denjoy conjecture and related results," in *Mathematical essays dedicated to A. J. Macintyre*, Ohio University Press, Athens, Ohio, 1970, pp. 183-200.

23. ____, *A topological Three Pole Theorem*, Indiana Univ. Math. J. **21** (1972), 1013-1018.

24. ____, *On results of R. Nevanlinna and Ahlfors*, Bull. London Math. Soc. **7** (1975), 81-83.

25. James A. Jenkins and Kôtaro Oikawa, *On the growth of slowly increasing unbounded harmonic functions*, Acta Math. **124** (1970), 37-61.

26. ____, *On results of Ahlfors and Hayman*, Illinois J. Math. **15** (1971), 664-671.

27. ____, *Conformality and semiconformality at the boundary*, J. Reine Angew. Math. **291** (1977), 92-117.

28. ____, *On the boundary behavior of functions for which the Riemann image has finite spherical area*, Kodai Math. J. (to appear).

29. James A. Jenkins and Nobuyuki Suita, *On regular functions on Riemann surfaces*, Illinois J. Math. **17** (1973), 563-570.

30. ____, *On regular functions on Riemann surfaces*. II, Illinois J. Math. **19** (1973), 122-126.

31. Luby Liao, *Certain extremal problems concerning module and harmonic measure*, J. Analyse Math. **40** (1981), 1-42.

32. Arthur E. Obrock, *Teichmüller inequalities without coefficient normalization*, Trans. Amer. Math. Soc. **159** (1971), 391-416.

33. ____, *Grötzsch domains and Teichmüller inequalities*, Indiana Univ. Math. J. **20** (1970-71), 739-751.

34. ____, *On the use of Teichmüller's principle in conjunction with the continuity method*, J. Analyse Math. **25** (1972), 75-105.

35. ____, *On bounded oscillation and asymptotic expansion of conformal strip mappings*, Trans. Amer. Math. Soc. **173** (1972), 183-201.

36. Margot S. Pallmann, *On level curves of Green's functions*, Kodai Math. Sem. Rep. **29** (1977), 179-185.

37. Dean G. Phelps, *On a coefficient problem in univalent functions*, Trans. Amer. Math. Soc. **143** (1969), 475-485.

38. B. Rodin and S. Warschawski, *Extremal length and boundary behavior of conformal mappings*, Ann. Acad. Sci. Fenn. Ser. A I, Math. **2** (1976), 476–500.

39. O. Teichmüller, *Untersuchungen über konforme und quasikonforme Abbildung*, Deutsche Math. **3** (1938), 621–678.

40. ____, *Ungleichungen zwischen den Koeffizienten schlichter Funktionen*, Sitzungsberichte der preussischen Akademie der Wissenschaften, Phys.-math. Klasse, 1938, pp. 363–375.

41. ____, *Extremale quasikonforme Abbildungen und quadratische Differentiale*, Abhandlungen der preussischen Akademie der Wissenschaften, Math.-naturwiss. Klasse, 1939, no. 22.

42. Jeffrey C. Wiener, *An extremal length problem on a bordered Riemann surface*, Trans. Amer. Math. Soc. **203** (1975), 227–245.

43. ____, *Isolated singularities of a quadratic differential arising from a module problem*, Proc. Amer. Math. Soc. **55** (1976), 47–51.

44. J. Wolff, *Démonstration d'un théorème sur la conservation des angles dans la représentation conforme au voisinage d'un point frontière*, Proc. Nederland. Akad. Wetenschappen, **38** (1935), 46–50.

Some Problems in Complex Analysis

PETER W. JONES

In this note we discuss some open problems in complex analysis. This is a slightly expanded (and in some places contracted) version of a talk given at the Bieberbach Conference. As far as the author knows, only questions 7 and 9 are due to himself.

I. Siegel and Herman phenomena. Let $f(z) = \lambda z + \cdots$ be analytic in a neighborhood of the origin. In the theory of iteration of functions, it is important to understand when f is analytically conjugate (locally) to λz, i.e., when there exists an analytic function $h(z) = z + \cdots$ near zero satisfying $h^{-1} \circ f \circ h = \lambda z$. This is well understood and easy when $|\lambda| \neq 1$. If $\lambda = e^{2\pi i \alpha}$, $\alpha \in \mathbb{Q}$, then conjugation is in general not possible. If α is "very irrational," Siegel proved in the 1940's that h exists. In this last case there is a largest domain $\mathcal{D}$ on which f is conjugate to λz. $\mathcal{D}$ is called a *Siegel disk*.

Another case when one studies conjugation is when f is analytic in a neighborhood of the unit circle, T, and $f: \mathsf{T} \to \mathsf{T}$ is a diffeomorphism. There is then a simple way to associate to f a number $\lambda = e^{2\pi i \alpha}$ (the rotation number) which describes the "average" way in which f rotates T. To do this let $F: \mathsf{R} \to \mathsf{R}$ be a lift of f to R with $\mathsf{T} \leftrightarrow [0, 1]$ and set $\alpha = \lim(1/n) F^n(0)$, where $F^n = F \circ F \circ \cdots \circ F$. Again, if α is "very irrational," there exists an h univalent and analytic near T so that $h^{-1} \circ f \circ h = \lambda z$ (see Herman [7]). There is then a largest topological annulus $\mathcal{R}$ on which f is conjugate to λz. $\mathcal{R}$ is called a *Herman ring*. (Actually, a Herman ring is any largest topological annulus on which an analytic function is topologically conjugate to a rotation.)

Question 1. Let $\{P_n/q_n\}$ be the continued fraction expansion of $\alpha \in (0, 1)$, $P_n/q_n \to \alpha$. Suppose

$$(1) \qquad \sum_{n=1}^{\infty} \frac{\log q_{n+1}}{q_n} < \infty,$$

and suppose $f: \mathsf{T} \to \mathsf{T}$ has rotation number $\lambda = e^{2\pi i \alpha}$. Is f conjugate to λz?

Question 1 has an affirmative answer for Siegel disks [**1**].

Question 2. If (1) fails, is there an $f \leftrightarrow e^{2\pi i \alpha}$ which is not conjugate to λz?

Question 2 is open for both Siegel disks and Herman rings.

Question 3. Suppose f is a rational function with a Siegel disk $\mathcal{D}$ (resp. Herman ring $\mathcal{R}$). Is there a point $z \in \partial \mathcal{D}$ (resp. $\partial \mathcal{R}$) such that $f'(z) = 0$?

Herman [**8**] has shown that, for almost all λ and for $f = \lambda z + z^2$, the answer is affirmative. Carleson and Jones (unpublished) have shown there exist order 3 polynomials for which the answer is affirmative. Nothing is known for Herman rings. It should be remarked that a difficult and beautiful theorem due to Herman [**8**] shows the answer is positive for almost all λ if one knows that f is injective on the boundary of $\mathcal{D}$ (resp. $\mathcal{R}$).

Question 4. Suppose f is a rational function with a Siegel disk $\mathcal{D}$ (resp. Herman ring $\mathcal{R}$). Is $\partial \mathcal{D}$ a quasicircle (resp. $\partial \mathcal{R}$ two quasicircles)?

Nothing is known here. (There are not even any examples!) It is known that if λ is "good" and the boundary is a Jordan curve, there must be a critical point of f on the boundary (Ghys). Thus question 4 is harder than question 3.

II. Corona problems. Let Ω be a planar domain, and let $H^\infty(\Omega)$ be the bounded analytic functions on Ω. The corona theorem is said to be true for Ω if whenever $f_1, \ldots, f_N \in H^\infty(\Omega)$ satisfy $0 < \delta \leq \max_j |f_j(z)| \leq 1$ for all $z \in \Omega$, there are then $g_j \in H^\infty(\Omega)$ such that $\sum_j f_j g_j \equiv 1$.

Carleson proved in the early 60's that the corona theorem is true for unit disk, while Cole proved in the early 70's that the corona theorem fails for general Riemann surfaces.

Question 5. Is the corona theorem true for all planar domains?

Question 5 is "too hard" to answer with our present technology, so we must look at special cases. Garnett and Jones [**6**] have shown the corona theorem holds whenever $\partial \Omega \subset \mathbf{R}$. It is thus natural to ask

Question 6. If $\partial \Omega$ is contained in a Lipschitz curve Γ, is the corona theorem true?

There seems to be some hope of answering question 6 because of Calderón's theorem on Cauchy integrals on Lipschitz curves (see, e.g., [**2, 4**]). One consequence of Calderón's theorem is the solution of the so-called Denjoy conjecture: if $E \subset \Gamma$ has positive length, $l(E) > 0$, then $H^\infty(E^c) \neq$ constants. (This is proved via a duality argument.) It seems that to understand question 6 one must first solve

Question 7. If $E \subset \Gamma$ and $l(E) > 0$, construct *by hand* a nonconstant function in $H^\infty(E^c)$. (Note this is a command and not a question.)

If $\Gamma = \mathbf{R}$ and $E \neq \mathbf{R}$, one simply writes $f(z) = \exp\{\chi_E(z) + i\tilde{\chi}_E(z)\}$ for z in the upper half-plane, and then Schwarz reflects f to all of E^c. This is not available for Lipschitz curves, but it is likely that a similar, though more complicated procedure exists.

Lipschitz curves are clearly very special objects. Another interesting class of domains which have very different function-theoretic properties are complements of homogeneous Cantor sets.

Question 8. Let $E \subset [0,1]$ be the usual Cantor set obtained by taking a fixed ratio of dissection, and let $\Omega^c = E \times E$. Is the corona theorem true for Ω?

Nothing is known here. It should be noted that $H^\infty(\Omega) \neq$ constants if and only if the Hausdorff dimension of $E \times E$ is > 1 (see [**5**]).

III. Projection operators, periods of functions, and harmonic measure.

Let $\mathcal{R}$ be a Riemann surface and let $\pi: \mathcal{U} = \{|z| < 1\} \to \mathcal{R}$ be the universal covering map. Write $\mathcal{R} \cong \mathcal{U}/\Gamma$, where $\Gamma = \{\gamma\}$ is a Fuchsian group, and let Γ^* be the character group of Γ; $\Gamma^* \ni \lambda: \Gamma \to \mathbf{T}$. For each $\lambda \in \Gamma^*$ let $H^\infty_\lambda = \{f \in H^\infty(\mathcal{U}) : f \circ \gamma \equiv \lambda(\gamma) \cdot f, \gamma \in \Gamma\}$ so that H^∞_λ corresponds to some unitary line bundle over $\mathcal{R}$. If $E \subset \partial \mathcal{R}$, let $\omega(z, E, \mathcal{R})$ denote the harmonic measure of E in the surface $\mathcal{R}$, evaluated at the point z.

Unitary line bundles and certain estimates for harmonic measure seem to be closely related to Γ partition functions. A function $P(z) \in H^\infty(\mathcal{U})$ is said to be a Γ *partition function* if $\sum_{\gamma \in \Gamma} P(\gamma z) \equiv 1$ and $\sum_{\gamma \in \Gamma} |P(\gamma(z))| \leq C$ for all z. Γ partition functions are very useful objects because they allow one to construct projection operators P by setting

$$Pf(z) = \sum_{\gamma \in \Gamma} f(\gamma(z)) P(\gamma(z)).$$

Then P maps $H^\infty(\mathcal{U})$ to bounded Γ invariant functions (which are isomorphic to $H^\infty(\mathcal{R})$ under the covering map π). Such projection operators were first studied by Forelli in the 60's. Carleson [**3**] was the first to build Γ partition functions. He showed they exist if $\mathcal{R}$ is a planar domain with boundary $E \subset \mathbf{R}$ satisfying the thickness condition

(2) $\qquad |E \cap (x - t, x + t)| \geq \varepsilon t \quad$ for all $x \in E$ and all $t > 0$.

It turns out that projection operators are closely related to the behavior of the H^∞_λ spaces. Before explaining the connection, we recall an important result due to Widom [**10**]:

$$\inf_{\lambda \in \Gamma^*} \sup\{|f(z_0)| : f \in H^\infty_\lambda, \|f\|_{H^\infty_\lambda} \leq 1, \pi(z_0) = \varsigma_0\} = \exp\left\{-\int_0^\infty \beta_{\varsigma_0}(t)\,dt\right\}.$$

Here $\varsigma_0 \in \mathcal{R}$ and $\beta_{\varsigma_0}(t)$ is the first Betti number of $\{\varsigma \in \mathcal{R} : G(\varsigma, \varsigma_0) > t\}$, where $G(\cdot, \varsigma_0)$ is Green's function on $\mathcal{R}$ with pole at ς_0. We are now ready to state a conjecture relating the above concepts to each other.

Question 9. Are the following conditions equivalent?

(A) There are disjoint crosscuts $\{\mathcal{L}_j\}$ on $\mathcal{R}$ satisfying the following properties. For each j the component of $\mathcal{R} \setminus \bigcup_{k, k \neq j} \mathcal{L}_k$ which contains $\mathcal{L}_j$ (call it $\mathcal{R}_j$) is simply connected, and there is an $\varepsilon > 0$ such that $\omega(z, \partial \mathcal{R}, \mathcal{R}_j) \geq \varepsilon$ for all $z \in \mathcal{L}_j$.

(B) There is a Γ partition function and there is $M < \infty$ such that $\{\varsigma : G(\varsigma, \varsigma_0) > M\}$ is simply connected for all $\varsigma_0 \in \mathcal{R}$.

(C) There is an $\varepsilon > 0$ such that for all $\lambda \in \Gamma^*$ there is an $f \in H_\lambda^\infty$ satisfying $\varepsilon \leq |f(z)| \leq 1$ for all $z \in \mathcal{U}$.

(D) There is a $C < \infty$ such that $\int_0^\infty \beta_\varsigma(t)\, dt \leq C$ for all $\varsigma \in \mathcal{R}$.

(E) There is an $\varepsilon > 0$ such that for all $\varsigma \in \mathcal{R}$ there is a simply connected domain $\Omega = \Omega(\varsigma) \subset \mathcal{R}$ satisfying $\omega(\varsigma, \partial \mathcal{R}, \Omega) \geq \varepsilon$.

The author has partially verified this conjecture in [9]. The result proved is:

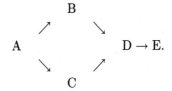

Therefore, all that remains to be proved is that $E \to A$. It is interesting to note that this last implication can be verified if $\mathcal{R}$ is planar and $\partial \mathcal{R} \subset \mathsf{R}$. In that case it turns out that A – E are also equivalent to Carleson's condition (2).

REFERENCES

1. A. D. Brjuno, *Analytical form of differential equations*, Trans. Moscow Math. Soc. **25** (1971), 131–288.

2. A. P. Calderón, *Cauchy integrals on Lipschitz curves and related operators*, Proc. Nat. Acad. Sci. U.S.A. **74** (1977), 1324–1327.

3. L. Carleson, "On H^∞ in multiply connected domains," in *Conference on harmonic analysis in honor of Antoni Zygmund*. Vol. 2, Wadsworth, Belmont, Calif., 1983, pp. 349–372.

4. R. R. Coifman, A. McIntosh, and Y. Meyer, *L'intégrale de Cauchy définit un operateur borné sur L^2 pour les courbes lipschitziennes*, Ann. of Math. (2) **116** (1982), 361–387.

5. J. Garnett, *Positive length but zero analytic capacity*, Proc. Amer. Math. Soc. **24** (1970), 696–699.

6. J. Garnett and P. W. Jones, *The corona theorem for Denjoy domains*, Acta Math. **155** (1985), 27–40.

7. M. Herman, *Sur la conjugaison différentiable des difféomorphismes du cercle à des rotations*, Inst. Hautes Études Sci. Publ. Math. **49** (1979), 5–233.

8. _____, preprint.

9. P. W. Jones, to appear.

10. H. Widom, *H_p sections of vector bundles over Riemann surfaces*, Ann. of Math. (2) **94** (1971), 305–324.

Comments on the Proof of the Conjecture On Logarithmic Coefficients

I. M. MILIN

Editors' Note. This is a modified translation of a letter written in Russian to P. L. Duren in November 1985. It is published here at the editors' suggestion, and with Professor Milin's consent, although time did not permit his approval of the final version. The editors thank Valentin Andreev for his assistance with the translation. By way of background, it should be explained that C. H. FitzGerald raised certain questions in a letter to Professor Milin in the fall of 1984, but received a reply too late for inclusion in his article [1].

Several recently published articles have given accounts of the proof of the Bieberbach conjecture. I cannot refrain from sharing my own view as an eyewitness of these events. By and large I will keep the text of my answers to the questions posed to me by C. FitzGerald (my letter of December 1984). I hope these remarks will help in preparing the Proceedings of the Conference.

1. Let us recall how the conjecture

$$(1) \qquad \sum_{\nu=1}^{n-1}\sum_{k=1}^{\nu} k|\gamma_k|^2 \le \sum_{\nu=1}^{n-1}\sum_{k=1}^{\nu}\frac{1}{k}$$

on logarithmic coefficients of univalent functions arose. In 1966 (but published in 1967) Lebedev and Milin obtained the exponential inequality

$$(2) \qquad \sum_{k=0}^{n-1} |D_k|^2 \le n \exp\left\{\frac{1}{n}\sum_{\nu=1}^{n-1}\sum_{k=1}^{\nu}\left(k|A_k|^2 - \frac{1}{k}\right)\right\},$$

where $\sum_{k=0}^{\infty} D_k z^k = \exp\{\sum_{k=1}^{\infty} A_k z^k\}$. This formally allows the conjecture (1) to be enunciated, because it is easy to prove with the help of (2) that the truth of (1) implies the conjectures of Robertson and Bieberbach.

But for this it was necessary to believe the conjecture (1); to have factual evidence in its favor. At that time I did not have this confidence. Therefore there was a need for a period of accumulation of facts. A concise but correct account of the evolution of the conjecture (1) is given in the papers of A. Z. Grinshpan [7] and P. Duren [12]. I would like to mention some historical details.

(a). *Why did I consider logarithmic coefficients?* First of all because H. Grunsky [8] had studied the expansion

$$\log \frac{z-\varsigma}{F(z)-F(\varsigma)} = \sum_{k,n=1}^{\infty} \alpha_{kn} z^{-k} \varsigma^{-n}$$

for a univalent function $F(z) \in \Sigma$ and had obtained rather easily by contour integration his famous inequality

(3) $$\left| \sum_{k,n=1}^{N} \alpha_{kn} x_k x_n \right| \leq \sum_{k=1}^{N} \frac{1}{k} |x_k|^2$$

as a necessary and sufficient condition for the univalence of $F(z)$. Then in [9] I devoted a lot of time to the expansion

(4) $$\log \frac{z-\varsigma}{F(z)-F(\varsigma)} = \sum_{n=1}^{\infty} A_n(\varsigma) z^{-n}, \qquad F(z) \in \Sigma,$$

and I noticed that the system of coefficients $\{A_n(\varsigma)\}$ has many interesting properties, useful for applications and easy to obtain by means of area theorems.

Finally, I. E. Bazilevich in the paper [10] directly considered the expansion of $\log(f(z)/z)$ and obtained the sharp inequality for $\alpha > 0$

(5) $$\sum_{k=1}^{\infty} k|\gamma_k - \gamma_k^*|^2 \leq \frac{1}{2} \log \frac{1}{\alpha}, \quad \text{where } \alpha = \lim_{r \to 1} (1-r)^2 M_\infty(r, f).$$

This estimates the closeness of the logarithmic coefficients of a given function to those of the Koebe function with the same direction of maximal growth. In this way I developed the conviction that the property of univalence reveals itself rather simply through area theorems or other methods in the form of restrictions on the coefficients of the logarithmic functions (4) and

$$\log \frac{f(z)}{z} = 2 \sum_{n=1}^{\infty} \gamma_n z^n,$$

and that it is necessary to construct an "apparatus of exponentiation" to transfer the restrictions from logarithmic coefficients to coefficients of the functions themselves.

(b). *Growth of logarithmic coefficients and their means.* Although the order of growth of logarithmic coefficients of functions in the class S does not coincide with the order of growth for the Koebe function, the papers of Ch. Pommerenke [11] and I. E. Bazilevich [10] showed that the greatest order of growth of certain means of $k|\gamma_k|^2$ is realized by the Koebe function. Soon afterwards, in the paper [6] I succeeded in proving for each $n = 2, 3, \ldots$ the inequality, sharp in order of magnitude (with respect to n),

$$\sum_{k=1}^{n} k|\gamma_k|^2 \leq \sum_{k=1}^{n} \frac{1}{k} + \delta, \qquad \delta < 0.312,$$

which was inspired by the result of Pommerenke [**11**]. This inequality was the basic fact in favor of the conjecture (1). Later I easily obtained the (unpublished) rough estimate of the means

$$\frac{1}{n}\sum_{\nu=1}^{n-1}\sum_{k=1}^{\nu} k|\gamma_k|^2 \leq \frac{1}{n}\sum_{\nu=1}^{n-1}\sum_{k=1}^{\nu}\frac{1}{k} + \frac{1-C}{2},$$

where $C = 0.577\ldots$ is Euler's constant. I also proved ([**13**], p. 91 (English edition, p. 69), formula (3.49)) the inequality (1) for each function $f(z) \in S$ and for all $n > n_f$ under the condition $\alpha > 0$.

Then in 1969–71, A. Z. Grinshpan [**7**] in his work on a candidate dissertation (Ph.D. thesis) proved among other results the inequality (1) for the initial values $n \leq 3$, and for each function $f(z) \in S$ having $\alpha = 0$ he proved it for $n > n_f$. He also obtained a local theorem analogous to the result of Garabedian-Schiffer [**14**]. This strengthened my belief in the conjecture (1). In 1970, while writing my book [**13**] (p. 72; English edition, p. 55), I did not yet state the conjecture (1) directly, but only noted that the behavior of the sums

$$\sum_{\nu=1}^{n}\sum_{k=1}^{\nu} k|\gamma_k|^2$$

is particularly interesting because the truth of (1) already proves the Bieberbach conjecture. Only in the paper of A. Z. Grinshpan [**7**], written in 1971 and recommended for publication in a joint report by N. A. Lebedev and me to the editorial board of the *Siberian Mathematical Journal*, was the inequality (1) explicitly called Milin's conjecture.

2. In the papers [**1**], [**2**], [**3**], and [**5**] the list of members of the seminar on geometric function theory, who participated in the discussion of L. de Branges's proof, is incomplete. Always present at the seminar, besides those mentioned in these papers, were E. G. Goluzina and A. Z. Grinshpan (together with A. M. Grinshpan, who was of great help to the seminar as interpeter), and some others.

Was it difficult to understand L. de Branges and his proof? How long did it take us to become confident in his proof? Professor L. de Branges arrived in Leningrad at the end of April. By the end of May the participants of the seminar were convinced that L. de Branges had indeed proved the conjecture (1), and we congratulated him on this great achievement.

I will briefly indicate the details of this period. In 1982, L. de Branges sent me (through S. V. Khruschev of LOMI) his paper containing an earlier form of his estimates. Several weeks before the arrival of L. de Branges, E. G. Emel'yanov presented this paper at the LOMI seminar, which made it easier to understand L. de Branges. At the end of May, after discussions by the seminar of the lectures of L. de Branges, E. G. Emel'yanov produced at my request a written translation into Russian of the part of de Branges's book of interest to us, where he replaced the terminology of functional analysis by the usual terminology, which in essence was the initial variant of L. de Branges's proof. On the basis of this I prepared a text of 13 pages for a lecture at the seminar. This text was discussed with A. Z. Grinshpan and E. G. Emel′yanov. Its aim was to help others to recognize the correctness of L. de Branges's proof. In June 1984 I gave two copies of this text to L. de Branges in order to accelerate the recognition of his achievement in his own country. It seems that this was the text which reached C. FitzGerald, Ch. Pommerenke, and others.

References

1. C. H. FitzGerald, *The Bieberbach conjecture: retrospective*, Notices Amer. Math. Soc. **32** (1985), 2–6.

2. Ch. Pommerenke, *The Bieberbach conjecture*, Math. Intelligencer **7** (1985), No. 2, pp. 23–25; 32.

3. J. Oesterlé, *Démonstration de la conjecture de Bieberbach*, Séminaire Bourbaki **37** (1984–85), no. 649; Astérisque No. 133–134 (1986), 319–334.

4. L. de Branges, *A proof of the Bieberbach conjecture*, Acta Math. **154** (1985), 137–152.

5. ____, *A proof of the Bieberbach conjecture*, Steklov Math. Inst. (LOMI) preprint E-5-84 (1984), 1–21.

6. I. M. Milin, *On the coefficients of univalent functions*, Dokl. Akad. Nauk SSSR **176** (1967), 1015–1018; English transl. in Soviet Math. Dokl. **8** (1967), 1255–1258.

7. A. Z. Grinshpan, *Logarithmic coefficients of functions in the class S*, Sibirsk. Mat. Ž. **13** (1972), 1145–1157; English transl. in Siberian Math. J. **13** (1972), 793–801.

8. H. Grunsky, *Koeffizientenbedingungen für schlicht abbildende meromorphe Funktionen*, Math. Z. **45** (1939), 29–61.

9. I. M. Milin, *The area method in the theory of univalent functions*, Dokl. Akad. Nauk SSSR **154** (1964), 264–267; English transl. in Soviet Math. Dokl. **5** (1964), 78–81.

10. I. E. Bazilevich, *On the dispersion of coefficients of univalent functions*, Mat. Sb. **68 (110)** (1965), 549–560; English transl. in Amer. Math. Soc. Transl. (2) **71** (1968), 168–180.

11. Ch. Pommerenke, *Über die Faberschen Polynome schlichter Funktionen*, Math. Z. **85** (1964), 197–208.

12. P. L. Duren, *Coefficients of univalent functions*, Bull. Amer. Math. Soc. **83** (1977), 891–911.

13. I. M. Milin, *Univalent functions and orthonormal systems*, Izdat. "Nauka", Moscow, 1971; English transl., Amer. Math. Soc., Providence, R.I., 1977.

14. P. R. Garabedian and M. Schiffer, *The local maximum theorem for the coefficients of univalent functions*, Arch. Rational Mech. Anal. **26** (1967), 1–32.

Notes on Two Function Models

N. K. NIKOL′SKIĬ AND V. I. VASYUNIN

1. Introduction. These notes reflect some attempts of the authors to understand better the interrelation between two known approaches to the model theory of linear operators. We establish and try to analyze explicit formulae identifying these two function models for Hilbert space contractions, namely, the de Branges-Rovnyak model and the Sz.-Nagy-Foiaş model. The latter, as is well known, represents a contraction as a compression of the multiplication operator $f \to zf$ to its closed coinvariant subspace determined by the characteristic function of the contraction [16, 17]. Originally the de Branges-Rovnyak model was oriented to the investigation of pairs of unitary operators in the spirit of scattering theory [3, 5]. But the most essential point where it differs from all other function models is the use of z-invariant operator ranges (para-closed subspaces, in the terminology of [9]) instead of closed z-invariant subspaces. This approach virtually requires the replacement of the usual L^2-norm by a (nonlocal) differential-integral one which coincides with the L^2-norm in the case of an inner characteristic function only. In principle, the analysis of the compressions of the shift to its coinvariant manifolds associated with such integro-differential metrics looks promising as a source of some operator-theoretic tools more flexible than the classical ones. But it implies a difficulty, namely, the nonlocal character of the above norms. It often restricts effective calculations to the linear hull of the values of the reproducing kernel.

2. Outline. Our general aim mentioned above is to explain a part of the de Branges-Rovnyak model theory, translating it into the usual setting of the Sz.-Nagy-Foiaş function model.

To begin with we briefly describe in §4 the de Branges-Rovnyak model, following [4].

In §6 the basic notion of complementary subspaces is discussed in terms of defect operators.

From the lifting theorem of Sz.-Nagy-Foiaş [18], we easily derive a description of "contractive" z-invariant operator ranges (§7) as subspaces of the form $\theta H^2(E)$ with an analytic operator function θ, $\theta(\varsigma) \colon E \to E_*$. Certain of them are distinguished on which the shift operator S,

$$Sf = zf,$$

is isometric in the range norm. On the complementary space $\mathcal{H}(\theta)$ the conjugate shift S^*,

$$S^*f = \frac{f - f(0)}{z},$$

is considered (§8); its defect operator is unitarily equivalent to the Hermitian square of $f \to f(0)$ iff the analytic polynomials are dense in the L^2-space defined by the operator weight $\Delta = (I - \theta^*\theta)^{1/2}$.

A unitary equivalence of the de Branges-Rovnyak model in the space $\mathcal{D}(\theta)$ (an extension of $\mathcal{H}(\theta)$) and of the Sz.-Nagy-Foiaş model is established in §12 with the help of an explicit formula very similar to that of J. Ball briefly mentioned in [1].

In §10 a criterion of unitary equivalence of (complicated) $\mathcal{D}(\theta)$ with (a little bit simpler) $\mathcal{H}(\theta)$ is proved. In §9 we are concerned with a relation between weighted polynomial density and a description of the extreme points of the unit ball in $H^\infty(E_* \to E)$. In particular, a simple example of an operator function θ is presented which is an extreme point, but the polynomials fail to be dense in L^2_Δ, $\Delta = (I - \theta^*\theta)^{1/2}$.

In concluding §12 we use "a universal form" of the Sz.-Nagy-Foiaş model proposed by the second author in [11]; it is briefly stated in §11.

To conclude we would like to mention once more than many constructions of the paper (e.g., in §§4, 6, 7.4a⇔c⇔e, 8.4) represent a modification of a part of the L. de Branges theory according to another (more traditional) point of view.

3. Acknowledgments. We would like to thankfully mention the lectures of Prof. L. de Branges at the Leningrad Branch of the Steklov Institute of Mathematics during his famous visit to the Institute in the spring 1984. These lectures stimulated us to write this paper. We also are indebted to Prof. L. de Branges for the manuscript of his book [4] kindly placed at our disposal.

We would like to thank the organizers of the Symposium for inviting us to take part in the celebration of the victory over the Bieberbach conjecture.

Lastly, we are grateful to Prof. V. P. Havin and Prof. V. V. Peller for reading the manuscript.

4. The de Branges-Rovnyak model. The model discussed for an arbitrary completely non-unitary contraction, which originated in essence in [3, 5] in connection with problems of the scattering theory for unitary operators, was developed (in particular) in [15, 6] and culminated in [4]. In this section we draft the dBR model following [4]. All properties stated of spaces and operators will be proved in what follows. In [4] only the case $E = E_*$, $\dim E = \infty$, was considered. The general case was considered in [2].

4.1. *Complementation.* Let H be a Hilbert space, and H_1 another Hilbert space contractively embedded into H (in symbols, $H_1 \hookrightarrow H$). Then there exists a unique Hilbert space H_2, $H_2 \hookrightarrow H$, with the properties

(a) $\|x+y\|^2 \leq \|x\|_1^2 + \|y\|_2^2$ for every $x \in H_1$, $y \in H_2$,

(b) each $h \in H$ has a unique decomposition $h = x+y$, $x \in H_1$, $y \in H_2$, with $\|h\|^2 = \|x\|_1^2 + \|y\|_2^2$.

The space H_2 is called the complementary space to H_1.

4.2. *Some notation.* The complementary space to E in H will be denoted by E'.

If E is a Hilbert space, $L^2(E)$ is the usual L^2-space of E-valued functions f on the unit circle $\mathbb{T}$ with respect to the normalized Lebesgue measure m endowed with the norm

$$\|f\|_2^2 = \int_\mathbb{T} \|f(z)\|_E^2 \, dm.$$

The corresponding Hardy space is $H^2(E) = \{f \in L^2(E) : \hat{f}(n) = 0,\ n < 0\}$, where $\hat{f}(n)$ stands for the nth Fourier coefficient.

The symbol $H^2(X)$ has an analogous meaning for any Banach space X. $H^\infty(X) \stackrel{\text{def}}{=} H^2(X) \cap L^\infty(X)$. For the space $(E \to E_*)$ of all (bounded) linear operators from E to E_* we write $H^\infty(E \to E_*)$, $L^\infty(E \to E_*)$.

For $\theta \in L^\infty(E \to E_*)$ let us denote by $\boldsymbol{\theta}$ the operator $f \mapsto \theta f$ (with pointwise application $\theta(\varsigma)f(\varsigma),\ \varsigma \in \mathbb{T}$). It is obvious that the inclusion $\boldsymbol{\theta}H^2(E) \subset H^2(E_*)$ is equivalent to the inclusion $\theta \in H^\infty(E \to E_*)$, and $\|\boldsymbol{\theta}\| \leq 1$ (or $\|\boldsymbol{\theta}|H^2(E)\| \leq 1$) is equivalent to $\|\theta(\varsigma)\| \leq 1$ a.e. on $\mathbb{T}$.

4.3. *Premodel space.* Let $\theta \in H^\infty(E \to E_*)$, $\|\theta\|_\infty \leq 1$, and let $\mathfrak{M}(\theta) = \boldsymbol{\theta} H^2(E)$ be the Hilbert space endowed with the range norm

$$\|f\|_{\mathfrak{M}(\theta)} = \inf\{\|g\|_2 : f = \theta g\}.$$

Then $\mathfrak{M}(\theta) \hookrightarrow H^2(E_*)$. Set $\mathcal{H}(\theta) = \mathfrak{M}(\theta)'$. Then $\mathcal{H}(\theta)$ is S^*-invariant and S^* is a contraction on $\mathcal{H}(\theta)$.

4.4. *Model space.* Let E, E_* be Hilbert spaces, and $\theta \in H^\infty(E \to E_*)$, $\|\theta\|_\infty \leq 1$. Let us consider the following linear set $\mathcal{D}_0(\theta) \subset H^2(E_*) \oplus H^2(E)$, $\mathcal{D}_0(\theta) = \{\{f,g\} : f \in \mathcal{H}(\theta),\ z^{n+1}f - \theta P_+ z^{n+1} Jg \in \mathcal{H}(\theta),\ n \geq 0\}$, where $Jf = \bar{z}f(\bar{z})$ is a unitary operator on any L^2-space over $\mathbb{T}$ which maps H^2 onto $H_-^2 \stackrel{\text{def}}{=} L^2 \ominus H^2$ and vice versa, and $P_+ = P_{H^2}$ is the orthogonal projection on H^2. For every $\{f, g\} \in \mathcal{D}_0(\theta)$ the limit

$$\lim_n (\|z^{n+1}f - \theta P_+ z^{n+1} Jg\|_{\mathcal{H}(\theta)}^2 + \|P_+ z^{n+1} Jg\|_2^2) \stackrel{\text{def}}{=} \|\{f,g\}\|_{\mathcal{D}(\theta)}^2$$

exists. Endowed with the $\|\cdot\|_{\mathcal{D}(\theta)}$-norm $\mathcal{D}(\theta)$,

$$\mathcal{D}(\theta) \stackrel{\text{def}}{=} \{\{f,g\} \in \mathcal{D}_0(\theta) : \|\{f,g\}\|_{\mathcal{D}(\theta)} < \infty\},$$

becomes a Hilbert space, $\mathcal{D}(\theta) \hookrightarrow H^2(E_*) \oplus H^2(E)$. It is called the dBR model space.

4.5. Function model. For every completely nonunitary contraction T acting on a Hilbert space there exists a (essentially unique) function $\theta \in H^\infty(E \to E_*)$ such that T is unitarily equivalent to the model operator $\mathcal{M}_\theta$ acting on $\mathcal{D}(\theta)$ by the formula

$$\mathcal{M}_\theta^*\{f, g\} = \{S^*f, Sg - \theta(\bar{z})^*f(0)\}.$$

$\mathcal{M}_\theta$ is said to be the de Branges-Rovnyak (dBR) model for T.

5. Requisite for contractions.

Let T be a contraction from a Hilbert space H to another one H_*. The positive square root $D_T = (I - T^*T)^{1/2}$ is said to be the defect operator for T (acting on H). It is easy to see that $D_T^2 T^* = T^* D_{T^*}^2$, and hence $D_T^{2n} T^* = T^* D_{T^*}^{2n}$, $n \geq 0$. Approximating $\sqrt{x}$ by polynomials in x we get the well-known intertwining property $D_T T^* = T^* D_{T^*}$. This implies that the following operator A,

$$A = \begin{pmatrix} T & D_{T^*} \\ -D_T & T^* \end{pmatrix},$$

is an isometry on the orthogonal sum $H \oplus H_*$.

5.1. $A^*A = I$. •

5.2. COROLLARY. $\|Tx + D_{T^*}y\|^2 + \|T^*y - D_T x\|^2 = \|x\|^2 + \|y\|^2$ for $x \in H$, $y \in H_*$, and hence $\|Tx + D_{T^*}y\|^2 \leq \|x\|^2 + \|y\|^2$. •

5.3. DOUGLAS'S FACTORIZATION THEOREM [7]. Let A and B be Hilbert space operators defined on the same space. Then (1) $A = CB$ for a contraction C iff $A^*A \leq B^*B$ (equivalently, $\|Ax\| \leq \|Bx\|$ for every x), and (2) $A = UB$ for a partial isometry U with initial space clos Range B iff $A^*A = B^*B$ (equivalently, $\|Ax\| = \|Bx\|$ for every x).

5.4. COROLLARY. A linear operator (α, β) from $H \oplus H$ to H is a contraction iff $\beta = D_{\alpha^*} \gamma$ for a contraction γ.

PROOF.
$$\|\alpha^* h\|^2 + \|\beta^* h\|^2 \leq \|h\|^2, \quad \forall h \in H$$
$$\Leftrightarrow \beta\beta^* \leq D_{\alpha^*}^2 \Leftrightarrow \beta^* = CD_{\alpha^*}, \quad \|C\| \leq 1 \text{ (Theorem 5.3)}. \quad \bullet$$

5.5. COROLLARY. Let A and B be Hilbert space contractions taking values in the same space. Then $A = BC$ for a contraction C iff $D_{B^*} = D_{A^*} K$ for a contraction K.

PROOF. By Theorem 5.3,
$$A = BC, \quad \|C\| \leq 1 \Leftrightarrow AA^* \leq BB^* \Leftrightarrow D_{B^*}^2 \leq D_{A^*}^2$$
$$\Leftrightarrow D_{B^*} = D_{A^*} K, \quad \|K\| \leq 1. \quad \bullet$$

5.6. COROLLARY. Let A, B be contractions on a Hilbert space H. Then $AB = BC$ for a contraction C on H iff there exist contractions X, Y such that $D_{B^*} = X^* D_{B^*} A^* + D_X Y D_{A^*}$.

PROOF. By Theorem 5.3
$$AB = BC, \quad \|C\| \leq 1 \Leftrightarrow ABB^*A^* \leq BB^*$$
$$\Leftrightarrow D_{B^*}^2 \leq D_{B^*A^*}^2 = D_{A^*}^2 + AD_{B^*}^2 A^* = T^*T$$
with $T = \begin{pmatrix} D_{B^*}A^* \\ D_{A^*} \end{pmatrix}$ from H to $H \oplus H$. Again by Theorem 5.3 this is equivalent to $D_{B^*} = CT$ for a contraction C from $H \oplus H$ to H. But $C = (X^*, D_X Y)$ with $\|X\| \leq 1$, $\|Y\| \leq 1$ by Corollary 5.4. •

5.7. LEMMA. *Let $\mathcal{E}, H, H_*$ be Hilbert spaces, and let T be a contraction from H to H_*. The following are equivalent: (1) $\mathcal{E} = TH$ (with the range norm $\|y\| = \inf\{\|x\|: Tx = y\}$); (2) $TT^* = ii^*$ where $i: \mathcal{E} \to H_*$ is the embedding operator.*

PROOF. By the definition of the range norm, T is a partial isometry "onto" (i.e., a coisometry) as an operator from H to $\mathcal{E}$. Denote this coisometry by $T_\mathcal{E}$. Then, $T = iT_\mathcal{E}$ and hence $TT^* = ii^*$.

On the other hand, the equality $TT^* = ii^*$ and Theorem 5.3 imply that $T^* = Ui^*$ for an isometry $U: \mathcal{E} \to H$. Hence, $TH = iU^*H = \mathcal{E}$ and the range norm and the initial norm $\mathcal{E}$ clearly coincide. •

5.8. LEMMA. $T_1 H_1 \subset T_2 H_2$ *iff $T_1 T_1^* \leq T_2 T_2^*$; i.e., iff $T_1 = T_2 C$ for a contraction C.*

PROOF. The inclusion $T_1 H_1 \subset T_2 H_2$ is equivalent to the inequality $i_1 i_1^* \leq i_2 i_2^*$ for the embedding operators. It remains to use 5.7 and 5.3. •

5.9. COROLLARY. *A linear operator A is a contraction from $T_1 H_1$ to $T_2 H_2$ (endowed with the range norms) iff $AT_1 H_1 \subset T_2 H_2$ and iff $AT_1 = T_2 C$ for a contraction C.* •

6. A description of complementary spaces.

6.1. LEMMA. *Let $\mathcal{E}, H, H_*$ be Hilbert spaces. Then $\mathcal{E} \subset H$ iff there exists a contraction $T: H \to H$ such that $\mathcal{E} = TH$ (with the range norm). A contraction T_* from H_* to H generates the same space $\mathcal{E}$ iff $T_* = TU$ with a partial isometry U with final space clos Range T^*.*

PROOF. Note that $\mathcal{E} = i\mathcal{E}$ (with the range norm). Set $T = (ii^*)^{1/2}$ and apply 5.7 and 5.3. •

6.2. LEMMA. *Let $\mathcal{E} \subset H_*$ and $\mathcal{E} = TH$, $\|T\| \leq 1$. Then $\mathcal{E}' = D_{T^*} H_*$ (with the range norm).*

PROOF. Let us define $\mathcal{E}_* = D_{T^*} H_*$ and prove that $\mathcal{E}_*$ is the complementary space to $\mathcal{E}$. Let $x = Ta \in \mathcal{E}$, $y = D_{T^*} b \in \mathcal{E}_*$. Then, by Corollary 5.2, we have $\|x+y\|^2 \leq \|a\|^2 + \|b\|^2$ and hence $\|x+y\|^2 \leq \|x\|_\mathcal{E}^2 + \|y\|_{\mathcal{E}_*}^2$. On the other hand, if

$h \in H_*$ and $a = T^*h$, $b = D_{T^*}h$, we then get $h = x + y$ with $x = Ta$, $y = D_{T^*}b$, and
$$\|h\|^2 = \|T^*h\|^2 + \|D_{T^*}h\|^2 \geq \|x\|_{\mathcal{E}}^2 + \|y\|_{\mathcal{E}_*}^2;$$
i.e., $\|h\|^2 = \|x\|_{\mathcal{E}}^2 + \|y\|_{\mathcal{E}_*}^2$. The uniqueness of the "extremal representation" of such a type also follows from Corollary 5.2. Namely, if $h = Ta + D_{T^*}b$ and $\|h\|^2 = \|a\|^2 + \|b\|^2$, then one gets $P_2 A\binom{a}{b} = 0$ where P_2 is the orthogonal projection of $H \oplus H_*$ onto its second coordinate. Hence, by 5.1,
$$\binom{a}{b} = A^*A\binom{a}{b} = A^*(P_1 + P_2)A\binom{a}{b} = A^*P_1A\binom{a}{b} = A^*\binom{h}{0};$$
i.e., $a = T^*h$, $b = D_{T^*}h$. ●

6.3. COROLLARY. $\mathcal{E}'' = \mathcal{E}$ for $\mathcal{E} \subsetneq H_*$.

In fact, $D_{D_{T^*}} = (TT^*)^{1/2}$ and $TH = (TT^*)^{1/2}H_*$. ●

6.4. COROLLARY. $\mathcal{E} \subsetneq \mathcal{E}_1 \subsetneq H_*$ iff $\mathcal{E}'_1 \subsetneq \mathcal{E}' \subsetneq H_*$.

For the proof apply 6.1, 5.8, 6.2, 5.5, and 6.3 ●

6.5. LEMMA. *Let* $\mathcal{E} = TH \subsetneq H_*$ *and let* $\mathcal{E}'$ *be the complementary space. Then for* $x \in \mathcal{E}$ *and* $y \in \mathcal{E}'$
$$\|x\|_{\mathcal{E}}^2 = \sup\{\|x + D_{T^*}b\|^2 - \|b\|^2 : b \in H_*\},$$
$$\|y\|_{\mathcal{E}'}^2 = \sup\{\|y + Ta\|^2 - \|a\|^2 : a \in H\}.$$

PROOF. In view of 6.2 and 6.3 it is enough to prove one of these formulae. By the definition of $\mathcal{E}'$ and the range norm, $\|y + Ta\|^2 \leq \|y\|_{\mathcal{E}'}^2 + \|Ta\|_{\mathcal{E}}^2 \leq \|y\|_{\mathcal{E}'}^2 + \|a\|^2$ for every $a \in H$. Then, $\|y\|_{\mathcal{E}'}^2 \geq \sup\{\cdots\}$. For $y = D_{T^*}^2 c$, $c \in H_*$, the converse is also true, because putting $x = T^*c$ we get
$$\|y + Tx\|^2 = \|D_{T^*}^2 c + TT^*c\|^2 = \|D_{T^*}c\|^2 + \|x\|^2 = \|y\|_{\mathcal{E}'}^2 + \|x\|^2.$$
Here we use the fact that D_{T^*} is an isometry from $(\operatorname{Ker} D_{T^*})^\perp \supset D_{T^*}H_*$ to $\mathcal{E}'$. On the other hand, $\operatorname{clos} D_{T^*}H_* = (\operatorname{Ker} D_{T^*})^\perp$ and hence $\operatorname{clos}_{\mathcal{E}'} D_{T^*}^2 H_* = \mathcal{E}'(= D_{T^*}H_*)$. So, the desired equality being valid on a dense set (namely, on $D_{T^*}^2 H_*$), it is valid everywhere. ●

6.6. LEMMA. *Let* T *be a contraction* $H \to H_*$ *and let* $x, y \in H_*$. *Then* $x \in \mathcal{E}$, $y \in \mathcal{E}'$, *provided the right-hand side suprema which appear in Lemma 6.5 are finite*.

PROOF. Taking 6.2 and 6.3 into account, we can restrict ourselves to one of the assertion stated. Let us deal with y. Then,
$$\sup\{\|y + Ta\|^2 - \|a\|^2 : a \in H\}$$
$$= \sup\{\|y\|^2 + 2\lambda|(y, Ta)| - \lambda^2\|D_Ta\|^2 : a \in H, \lambda \in \mathbb{R}\} < \infty.$$

From the last inequality one can easily derive that $|(y, Ta)| \leq \text{const} \, \|D_T a\|$, $a \in H$. Hence, there exists a vector $u \in H$ (and even $u \in \text{clos}\, D_T H$) such that $(y, Ta) = (u, D_T a)$ for every $a \in H$. This yields $T^* y = D_T u$, and consequently

$$y = D_{T^*}^2 y + TT^* y = D_{T^*}^2 y + TD_T u = D_{T^*}(D_{T^*} y + Tu) \in D_{T^*} H_* = \mathcal{E}'. \quad \bullet$$

7. Shift invariant operator ranges and complementary spaces. Just as the base of the SzNF model is formed by the description of all closed shift invariant subspaces, a similar role is played for the dBR model by the computation of all shift invariant operator ranges or, equivalently (§6.2), of all shift invariant Hilbert spaces continuously embedded into a Hardy space $H^2(E)$. We keep our mind on contractively embedded spaces reserving a remark below (§7.10) for the general case.

Let us start with the following elementary lemma.

7.1. LEMMA. *Let E and E_* be Hilbert spaces, and let $\theta \in H^\infty(E \to E_*)$, $\|\theta\|_\infty \leq 1$. Set*

$$\mathfrak{M}(\theta) = \theta H^2(E).$$

Then $\mathfrak{M}(\theta) \subsetneq H^2(E_*)$ and $S\mathfrak{M}(\theta) \subsetneq \mathfrak{M}(\theta)$.

PROOF. The last embedding means that the restriction of the shift operator S to $\mathfrak{M}(\theta)$ is a contraction. Indeed,

$$\|f\|_{\mathfrak{M}(\theta)} = \inf\{\|g\|_2 : \theta g = f\}$$

for $f \in \mathfrak{M}(\theta)$, and

$$\|Sf\|_{\mathfrak{M}(\theta)} = \inf\{\|h\|_2 : \theta h = zf\} \leq \inf\{\|g\|_2 : \theta z g = zf\} = \|f\|_{\mathfrak{M}(\theta)}. \quad \bullet$$

7.2. LEMMA. *S is an isometry on $\mathfrak{M}(\theta)$ iff $\text{Ker}\,\theta$ is a reducing subspace for S; i.e., $\text{Ker}\,\theta = H^2(E_{**})$, $E_{**} \subset E$.*

PROOF. If $f = \theta g$ with $g \in H^2(E)$, $\|f\|_{\mathfrak{M}(\theta)} = \|Pg\|_2$, where P is the orthogonal projection onto $(\text{Ker}\,\theta)^\perp$, the orthogonal complement of the kernel of θ as an operator from $H^2(E)$ to $H^2(E_*)$. Hence

$$\|Sf\|_{\mathfrak{M}(\theta)} = \|z\theta g\|_{\mathfrak{M}(\theta)} = \|z\theta Pg\|_{\mathfrak{M}(\theta)} = \|\theta z Pg\|_{\mathfrak{M}(\theta)} = \|PzPg\|_2.$$

The last norm is equal to $\|Pg\|_2$ for every g iff $S(\text{Ker}\,\theta)^\perp \subset (\text{Ker}\,\theta)^\perp$. But, in any case, $S(\text{Ker}\,\theta) \subset \text{Ker}\,\theta$. The description of reducing subspaces for S as $H^2(E_{**})$, $E_{**} \subset E$, is well known; cf., e.g., [17, 12]. $\bullet$

7.3. REMARK. It is possible that $\text{Ker}\,\theta = \{0\}$ although $\text{Ker}\,\theta(\varsigma) \neq \{0\}$ for every $\varsigma \in \mathbb{D}$ and even if there exist analytic functions f taking values in E such that $\theta(\varsigma)f(\varsigma) \equiv 0$, $\varsigma \in \mathbb{D}$. An example: $E = H^2$, $\theta = S^* - zI$, $S^*f = \bar{z}(f - f(0))$.

On the other hand, it is easy to see that without loss of generality we can always assume that $\theta(\varsigma)e \neq 0$ for every $e \neq 0$, $e \in E$.

7.4. THEOREM. *Let E_* be a Hilbert space and let $\mathcal{E} \subset H^2(E_*)$. The following are equivalent.*

(a) $S\mathcal{E} \subset \mathcal{E}$ (*i.e., $\mathcal{E}$ is S-invariant and $S|\mathcal{E}$ is a contraction*).

(b) $\mathcal{E} = TH$ *and* $ST = TC$ *for a contraction C and a Hilbert space H.*

(c) $\mathcal{E} = \mathfrak{M}(\theta)$ *for a contractive function $\theta \in H^\infty(E \to E_*)$.*

(d) $\mathcal{E} = TH$ *and there exists a contraction X such that $S^*D_{T^*} = D_{T^*}X$ and* $\|(D_{T^*}f)(0)\| \leq \|D_X f\|_2$ *for every $f \in H^2(E_*)$.*

(e) $S^*\mathcal{E}' \subset \mathcal{E}'$ *and*

$$(7.4.1) \qquad \|f(0)\|^2 \leq \|f\|^2_{\mathcal{E}'} - \|S^*f\|^2_{\mathcal{E}'}, \qquad f \in \mathcal{E}'.$$

7.5. COMMENT. The equivalence (a)⇔(e) shows that from $S\mathcal{E} \subset \mathcal{E}$ it follows that $S^*\mathcal{E}' \subset \mathcal{E}'$, but (at least formally) not vice-versa. The following simple construction supplies us with an example of $\mathcal{E}', \mathcal{E}' \subset H^2$, for which $S^*\mathcal{E}' \subset \mathcal{E}'$ but (7.4.1) fails to be valid. For some more comments see §7.9 below.

7.6. EXAMPLE. Set $\mathcal{E}' = \text{span}(\mathbb{1}, z)$ endowed with a new scalar product such that $\|\mathbb{1}\|_{\mathcal{E}'} = \sqrt{2}$, $\|z\|_{\mathcal{E}'} = \sqrt{3}$, and $1 < (\mathbb{1}, z)_{\mathcal{E}'} < \sqrt{2}$. Then it is easy to see that $S^*\mathcal{E}' \subset \mathcal{E}' \subset H^2$, but $1 > \|\mathbb{1} - z\|^2_{\mathcal{E}'} - \|\mathbb{1}\|^2_{\mathcal{E}'} = 3 - 2(\mathbb{1}, z)_{\mathcal{E}'}$. •

7.7. PROOF OF THEOREM 7.4. (a)⇔(b) by 6.1 and 5.9.

(c)⇒(a) is Lemma 7.1.

(b)⇒(c): We find a function θ with $TH = \mathfrak{M}(\theta)$ in a lifting theorem [18] for interwining contractions. For the reader's convenience a more elementary and independent proof will be given in §7.8.

Let $\mathcal{U}$ be a unitary operator which transforms C into its canonical model; i.e., $\mathcal{U}: H \to \mathcal{K}_1$, $T_1 = \mathcal{U}C\mathcal{U}^{-1}$, and

$$\mathcal{K}_1 = \mathcal{H}_1 \ominus \{\theta_1 f \oplus \Delta_1 f : f \in H^2(E_1)\}, \qquad \mathcal{H}_1 = H^2(E_{1*}) \oplus \text{clos}\,\Delta_1 L^2(E_1),$$

$\Delta_1 = (I - \theta_1^*\theta_1)^{1/2}$, $T_1 F = P_1 SF$, $F \in \mathcal{K}_1$, where $P_1 = P_{\mathcal{K}_1}$ is the orthogonal projection and θ_1 is a contractive function $\theta_1 \in H^\infty(E_1 \to E_{1*})$. The model T_2 for S on $H^2(E_*)$ is trivial: $E_2 = \{\mathbb{0}\}$, $E_{2*} = E_*$, $\mathcal{K}_2 = \mathcal{H}_2 = H^2(E_*)$, $\theta_2 = \mathbb{0}: \{\mathbb{0}\} \to E_*$, $T_2 = S$. Now the equation $ST = TC$ from (b) can be rewritten as $T_2 X = X T_1$, where $X = T\mathcal{U}^{-1}$. Then, by Lemma 2.1 of [18], p. 235, there exists a function $\theta \in H^\infty(E_{1*} \to E_*)$ such that

$$(7.7.1) \qquad \theta\theta_1 = \mathbb{0},$$

and $X = Y|\mathcal{K}_1$, where Y is an operator from $\mathcal{H}_1$ to $\mathcal{H}_2$ with $\|Y\| = \|X\|$ of the form

$$Y = (\boldsymbol{\theta}, \mathbb{0}).$$

Condition (7.7.1) implies $Y|\mathcal{K}_1^\perp = 0$; hence $YY^* = XX^*$ and

$$\boldsymbol{\theta\theta}^* = YY^* = XX^* = TT^*.$$

Therefore $TH = \mathfrak{M}(\theta)$.

(b)⇔(d): By Corollary 5.6 with $A = S$, $B = T$, we see that (b) is equivalent to

$$(7.7.2) \qquad D_{T^*} = X^* D_{T^*} S^* + D_X Y P_0$$

for some contractions X, Y, where $P_0 = D_{S^*}$ is the projection of $H^2(E_*)$ onto the constants. Multiplying (7.7.2) from the right by S, one can easily see that (7.7.2) is equivalent to the pair of equations

$$D_{T^*}S = X^*D_{T^*}, \qquad D_{T^*}P_0 = D_X Y P_0,$$

or

$$S^* D_{T^*} = D_{T^*} X, \qquad P_0 D_{T^*} = P_0 Y^* D_X$$

(with a contraction Y^*), or by Theorem 5.3 to an equation and an inequality

$$S^* D_{T^*} = D_{T^*} X; \qquad \|P_0 D_{T^*} f\| \le \|D_X f\| \quad \text{for } f \in H^2(E_*).$$

(d)$\Leftrightarrow$(e): This equivalence can be reached by simple and natural computations too. Namely, if (d) holds and if $f = D_{T^*} \cdot g$ with $\|f\|_{\mathcal{E}'} = \|g\|_2$, then

$$\|f(0)\|^2 + \|S^* f\|_{\mathcal{E}'}^2 = \|(D_{T^*} \cdot g)(0)\|^2 + \|D_{T^*} \cdot Xg\|_{\mathcal{E}'}^2$$
$$\le \|D_X g\|_2^2 + \|Xg\|_2^2 = \|g\|_2^2 = \|f\|_{\mathcal{E}'}^2.$$

Conversely, if (e) holds, we can represent $\mathcal{E}'$ as $\mathcal{E}' = D_{T^*} \cdot H^2(E_*)$ with a contraction $T \colon H \to H^2(E_*)$. Since $S^* \mathcal{E}' \subset \mathcal{E}'$, we get $S^* D_{T^*} = D_{T^*} \cdot X$ with $\|X\| \le 1$ (see Corollary 5.9). Moreover, without loss of generality one can suppose that

(7.7.3) $$\text{Range } X \subset (\text{Ker } D_{T^*})^\perp.$$

Then, for $g \in H^2(E_*)$ and $f = D_{T^*} \cdot g$, one gets

$$\|(D_{T^*} \cdot g)(0)\|^2 = \|f(0)\|^2 \le \|f\|_{\mathcal{E}'}^2 - \|S^* f\|_{\mathcal{E}'}^2 \quad \text{(by (7.4.1))}$$
$$= \|D_{T^*} \cdot g\|_{\mathcal{E}'}^2 - \|D_{T^*} \cdot Xg\|_{\mathcal{E}'}^2 = \|D_{T^*} \cdot g\|_{\mathcal{E}'}^2 - \|Xg\|_2^2 \quad \text{(by (7.7.3))}$$
$$\le \|g\|_2^2 - \|Xg\|_2^2 = \|D_X g\|_2^2. \quad \bullet$$

7.8. An elementary proof of (b)$\Rightarrow$(c) of Theorem 7.4. Without loss of generality one can assume that T is an injection (if not replace it by the restriction $T|(\text{Ker } T)^\perp$). Let us consider the minimal isometric dilation V of C; $V \colon \mathcal{H} \to \mathcal{H}$, $H \subset \mathcal{H}$, and $C^n = PV^n|H$, $n \ge 0$, where P is the orthogonal projection onto H. It is very easy to see (and is well known) that $V^* H \subset H$, and hence

(7.8.1) $$PVP = PV.$$

From (b) it follows that

$$S^n T = TPV^n|H; \quad \text{i.e. } T^* S^{*n} = V^{*n} = T^*, \qquad n \ge 0.$$

This implies

$$\lim_n \|V^{*n} V^m T^* f\| = \lim_n \|V^{*(n-m)} T^* f\| = \lim_n \|T^* S^{*(n-m)} f\| = 0$$

for every $f \in H^2(E_*)$ and every $m \ge 0$. But span $(V^m T^* f \colon m \ge 0, f \in H^2(E_*)) = \mathcal{H}$ in view of the minimality of V and the equality clos Range $T^* = (\text{Ker } T)^\perp = H$. So V is a completely non-unitary operator, i.e., a shift. Let $\mathcal{U}$ be a unitary operator, $\mathcal{U}\colon H^2(E) \to \mathcal{H}$, such that $\mathcal{U}^{-1}V\mathcal{U} = S$. From (7.8.1) we get $STP = TPVP = TPV$ on $\mathcal{H}$, and hence $STPU = TPUS$ on $H^2(E)$. But an operator intertwining the shifts coincides with $\boldsymbol{\theta}$ for a function $\theta \in H^\infty(E \to E_*)$. Thus, $TPU = \boldsymbol{\theta}$ and $\boldsymbol{\theta}H^2(E) = TP\mathcal{H} = TH$. The range norms also obviously coincide. •

7.9. *Some more comments on Theorem 7.4.* Let us start with the following amusing question. What operator ranges are both S- and S^*-invariant? Our answer is the following.

7.9.1. THEOREM. *Let $\mathcal{E} \hookrightarrow H^2(E_*)$ and $\mathcal{E} = \mathfrak{M}(\theta) = \boldsymbol{\theta}H^2(E)$. If there exists an operator function F, $F(\varsigma)\colon E_* \to E$ such that $F\phi e \in H^2(E)$ for $e \in E$ and $\phi \in H^\infty(E \to E_*)$ and $\boldsymbol{\theta}F = P_R$ (the orthogonal projection of E_* onto $R = \operatorname{span}\{\theta(\varsigma)E\colon \varsigma \in \mathbb{D}\}$, then $S\mathcal{E} \hookrightarrow \mathcal{E}$, $S^*\mathcal{E} \subset \mathcal{E}$. If $\dim E_* < \infty$, the converse is also true.*

PROOF. $S^*\boldsymbol{\theta}SH^2(E) = \boldsymbol{\theta}H^2(E)$, and hence to prove that $S^*\mathcal{E} \subset \mathcal{E}$ it is necessary and sufficient that $S^*\boldsymbol{\theta}e \in \boldsymbol{\theta}H^2(E)$ for every $e \in E$.

If such a function F exists,

$$S^*\boldsymbol{\theta}e = \boldsymbol{\theta}F\frac{\theta - \theta(0)}{z}e \in \boldsymbol{\theta}H^2(E).$$

Conversely, if $S^*\mathcal{E} \subset \mathcal{E}$, then for every $e \in E$ and $n \geq 0$ there exists $e' \in E$ with $(S^{*n}\theta e)(0) = \theta(0)e'$. So, for the Taylor coefficients of θ, $\theta = \sum_{n\geq 0}\theta_n z^n$, we get $\theta_n E \subset \theta(0)E$, $n \geq 0$. Therefore, $\theta(0)E$, being a closed subspace of E_* ($\dim E_* < \infty$!), contains $\theta(\varsigma)E$ for every $\varsigma \in \mathbb{D}$. Set $F \equiv \mathbb{0}$ on the orthogonal complement of $R = \theta(0)E$. Let A be a right inverse of $\theta(0)\colon E \to R$, and let f_e be the unique function for which $S^*\boldsymbol{\theta}Ae = \boldsymbol{\theta}f_e$, $f_e \in (\operatorname{Ker}\boldsymbol{\theta})^\perp$, $e \in R$. Define $F(\varsigma)$ on R by

$$F(\varsigma)e = Ae - \varsigma f_e(\varsigma), \qquad \varsigma \in \mathbb{D}.$$

Since the operator $e \mapsto f_e$ is linear (and bounded) on R, $F(\varsigma)$ is a bounded $E_* \to E$ operator for $\varsigma \in \mathbb{D}$, and $Fe \in H^2(E)$ for every $e \in E_*$. If $\{e_i\}_{1 \leq i \leq n}$ is an orthonormal basis in E_*, then $\|F(\varsigma)\| \leq (\sum_{i=1}^n \|F(\varsigma)e_i\|_E^2)^{1/2}$, $\varsigma \in \mathbb{D}$, and hence $F\phi e \in H^2(E)$ for every $\phi \in H^\infty(E \to E_*)$ and $e \in E$. The identity

$$\theta(\varsigma)F(\varsigma)e \equiv P_R e, \qquad e \in E_*$$

holds by the definition of F. •

7.9.2. *A partial case.* If $\mathcal{E} = \mathfrak{M}(\theta) = \boldsymbol{\theta}H^2 \neq \{\mathbb{0}\}$ (the scalar case, $E = E_* = \mathbb{C}$), then $\mathcal{E}$ is S^*-invariant iff $1/\theta \in H^2$. •

Theorem 7.4 easily yields a new proof for the following results of C. Foiaş [9] on the so-called "strange S-invariant operator ranges."

7.9.3. COROLLARY. *There exists an operator range $\mathcal{E} \hookrightarrow H^2$ which is invariant under any operator commuting with S but which is not the range of any such operator.*

In fact, set $\mathcal{E} = \boldsymbol{\theta}H^2(\mathbb{C}^2)$, $\theta = (1/\sqrt{2})(\vartheta_1, \vartheta_2)$ where ϑ_1 is a Blaschke product with real zeros λ_k, $\lambda_k \to 1$, and $\vartheta_2 = \exp(z+1)(z-1)^{-1}$. It is easy to see that

$\mathcal{E}$ is invariant under any multiplication operator. If $\mathcal{E} = \vartheta H^2$, $\vartheta \in H^\infty$, we get $\vartheta_1, \vartheta_2 \in \vartheta H^2$. Hence ϑ is an outer function, and therefore
$$\lim_{r \to 1}(1-r)\log|\vartheta(r)| = 0.$$
On the other hand, $\vartheta \in \mathcal{E}$, i.e. $\vartheta = \vartheta_1 f_1 + \vartheta_2 f_2$ with $f_i \in H^2$, $i = 1, 2$. Thus,
$$\varliminf_k (1-\lambda_k)\log|\vartheta(\lambda_k)| = \varliminf_k (1-\lambda_k)\log|\vartheta_2(\lambda_k)f_2(\lambda_k)|$$
$$\leq \varliminf_k (1-\lambda_k)\log|\vartheta_2(\lambda_k)| = -2.$$
Contradiction. •

Now, let us discuss when $S^*\mathcal{E} \subsetneq \mathcal{E}$ implies $S\mathcal{E}' \subset \mathcal{E}'$. From §§7.5–7.6 we know that there exist $\mathcal{E}$'s for which $S^*\mathcal{E} \subsetneq \mathcal{E}$ but $S\mathcal{E}' \not\subset \mathcal{E}'$ (cf. §§7.5–7.6). The following supplies us with information on the question we are now interested in; in particular, §§7.9.10–7.9.13 contain some examples of $\mathcal{E}$'s with $S^*\mathcal{E} \subsetneq \mathcal{E}$, $S\mathcal{E}' \not\subset \mathcal{E}'$.

7.9.4. LEMMA. *Let $\mathcal{E} = TH \subsetneq H^2(E_*)$. Then $S^*\mathcal{E} \subsetneq \mathcal{E}$ iff*

(7.9.5) $\qquad \|D_{T^*}S^*g\|_2 \leq \|D_{T^*}g\|_2, \qquad g \in H_0^2(E_*),$

or equivalently, $\|D_{T^*}f\|_2 \leq \|D_{T^*}Sf\|_2$, $f \in H^2(E_*)$.

PROOF. Put $T_1 = T_2 = T$ and $A = S^*$ in Corollary 5.9, apply Corollary 5.6 with $A = S^*$, $B = T$, and use Theorem 5.3. •

7.9.6. LEMMA. *In the notation of Lemma 7.9.4, $S\mathcal{E}' \subset \mathcal{E}'$ iff*

(7.9.7) $\qquad \|D_{T^*}S^*g\|_2 \leq \lambda\|D_{T^*}g\|_2$ *for some $\lambda > 0$ and every* $g \in H^2(E_*)$.

PROOF. Similar reasoning, using that $S\mathcal{E}' \subset \mathcal{E}' \Leftrightarrow (\frac{1}{\lambda}S)\mathcal{E}' \subsetneq \mathcal{E}'$ for some $\lambda > 0$. •

Convention. From now on (7.9.5) is supposed to be valid.

7.9.8. SUFFICIENT CONDITION. $S\mathcal{E}' \subset \mathcal{E}'$ *if (in the notation of Lemma 7.9.4)*

(7.9.9) $\qquad \|D_{T^*}g(0)\|_2 \leq \mu\|D_{T^*}g\|_2$ *for some $\mu > 0$ and every* $g \in H^2(E_*)$.

PROOF.
$$\|D_{T^*}S^*g\|_2 = \|D_{T^*}S^*(g - g(0))\|_2 \leq \|D_{T^*}(g - g(0))\|_2$$
$$\leq (1+\mu)\|D_{T^*}g\|_2. \quad \bullet$$

7.9.10. CRITERION. *Suppose that (in addition to (7.9.5))*

(7.9.11) $\qquad \|D_{T^*}Sg\|_2 \leq C\|D_{T^*}g\|_2$ *for some $C > 0$ and every* $g \in H^2(E_*)$.

Then (7.9.7) $\Leftrightarrow$ (7.9.9).

PROOF.
$$\|D_{T^*}g(0)\|_2 \leq \|D_{T^*}g\|_2 + \|D_{T^*}SS^*g\|_2 \leq \|D_{T^*}g\|_2 + C\|D_{T^*}S^*g\|_2$$
$$\leq (1+C\lambda)\|D_{T^*}g\|_2. \quad \bullet$$

7.9.12. EXAMPLE. Let $H = H^2(E)$, and let $T: H \to H^2(E_*)$ be an operator such that $\|T^*f\|_2 = \|\Delta f\|_2$ for every f, where $\Delta \in L^\infty(E_* \to E)$, $\|\Delta\|_\infty \leq 1$. Then $D_{T^*}^2 = P_+\delta^2$, $\delta^2 = I - \Delta^*\Delta$, and

$$\|D_{T^*}\cdot g\|_2^2 = (\delta^2 g, g) = (\delta^2 Sg, Sg) = \|D_{T^*}\cdot Sg\|_2^2 \quad \text{for } g \in H^2(E_*).$$

Hence both (7.9.5) and (7.9.11) are satisfied, and so (7.9.7)$\Leftrightarrow$ (7.9.9). But in the case under consideration, (7.9.9) means that $\|\delta g(0)\|_2 \leq \mu\|\delta g\|_2$, $g \in H^2(E_*)$, or equivalently

$$\mu \inf\{\|\delta(e + zf)\|_2 : f \in H^2(E_*)\} \geq \|\delta e\|_2, \quad e \in E_*.$$

In particular, putting $E = E_* = \mathbb{C}$, one can obtain the following criterion: the S^*-invariant operator range TH^2 has S-invariant complement iff either $\delta = 0$ or $\int_\mathbb{T} \log \delta\, dm > -\infty$. •

7.9.13. ANOTHER SUFFICIENT CONDITION. *(7.9.7) is satisfied if*

(7.9.14) $\qquad D_{T^*} = VA, \qquad S^*A = AS^*, \qquad$ *and V is left invertible.*

PROOF. Let $\mathcal{U}V = I$, $g \in H^2(E)$. Then

$$\|D_{T^*}\cdot S^*g\|_2 \leq \|V\| \cdot \|AS^*g\|_2 = \|V\| \cdot \|S^*Ag\|_2 \leq \|V\| \cdot \|Ag\|_2$$
$$\leq \|V\| \cdot \|\mathcal{U}\| \cdot \|VAg\|_2 = \|V\| \cdot \|\mathcal{U}\| \cdot \|D_{T^*}\cdot g\|_2. \quad \bullet$$

7.9.15. FINAL REMARKS. There exist operators T with property (7.9.5), which satisfy (7.9.14) but not (7.9.9), and on the other hand which satisfy (7.9.9) but not (7.9.14). Of course, (7.9.14) is satisfied by the orthogonal projection T onto a (closed) S^*-invariant subspace of $H^2(E_*)$.

Every $D_{T^*}\cdot H^2(E_*)$ is contractively embedded into some $D_*H^2(E_*)$ with D_* satisfying (7.9.14) (For one of them, put $\|D_*f\|_2 = \lim_n \|D_{T^*}\cdot S^n f\|_2$, $f \in H^2(E_*)$.)

7.10. *S-invariant operator ranges.* Suppose $\mathcal{E}$ is an arbitrary Hilbert space continuously (and wlog contractively) embedded into $H^2(E_*)$ and such that $S\mathcal{E} \subset \mathcal{E}$ (not necessarily contractively). Then $(rS)\mathcal{E} \hookrightarrow \mathcal{E}$ for a suitable $0 < r < 1$, and if $\mathcal{E} = TH$ we get $rST = TC$ with a contraction C. Now, a description of S-invariant operator ranges follows.

7.10.1. LEMMA. *Let $\mathcal{E} \hookrightarrow H^2(E_*)$. Then $\mathcal{E}$ is S-invariant iff there exist r, $0 < r < 1$, a Hilbert space E, and a strong H^2 operator function F from E to E^* (i.e., $F(\cdot)e \in H^2(E_*)$ for every $e \in E$) such that $\mathcal{E} = \mathbf{F}(C_r H^2(E))$, where $(C_r f)(\varsigma) = f(r\varsigma)$, $\varsigma \in \mathbb{D}$.*

PROOF. Choosing a number r, $0 < r < 1$, such that $rST = TC$ with a contraction C, one can proceed just as in §7.8. Then, keeping the notation of §7.8 we arrive at the equation $rST P\mathcal{U} = TP\mathcal{U}S$ on $H^2(E)$. Set $F(\varsigma)e = (TP\mathcal{U}e)(\varsigma)$ for $z \in \mathbb{D}$, $e \in E$. The function F is a strong $E \to E_*$ valued H^2

operator function. From the equation we obtain for any E-valued polynomial $f = \sum_{n \geq 0} z^n e_n$:
$$(TP\mathcal{U}f)(\varsigma) = \sum_{n \geq 0} r^n \varsigma^n (TP\mathcal{U}e_n)(\varsigma) = \sum_{n \geq 0} r^n \varsigma^n F(\varsigma) e_n = F(\varsigma)f(r\varsigma).$$

The operator FC_r, being continuous,
$$\left(\|FC_r f\|_2 = \left\| \sum_{n \geq 0} r^n z^n F e_n \right\|_2 \leq \sum_{n \geq 0} r^n \|F e_n\|_2 \right.$$
$$\left. \leq \text{const} \sum_{n \geq 0} r^n \|e_n\| \leq \frac{\text{const}}{(1-r^2)^{1/2}} \|f\|_2 \right)$$

has to coincide with $TP\mathcal{U}$ on $H^2(E)$. Thus,
$$\mathcal{E} = TH = TP\mathcal{U}H^2(E) = FC_r H^2(E). \quad \bullet$$

8. Defect operator for S^* on $\mathcal{H}(\theta)$. As before, $\mathcal{H}(\theta)$ stands for the complementary space to $\mathfrak{M}(\theta)$: $\mathcal{H}(\theta) = \mathfrak{M}(\theta)' = D_{\theta^*} \cdot H^2(E_*)$. Then $S^*\mathcal{H}(\theta) \subset \mathcal{H}(\theta)$, and one can consider S^* as an operator on $\mathcal{H}(\theta)$. To distinguish between the latter and S^* on $H^2(E_*)$, let us use the notation $S_\theta^* = S^*|\mathcal{H}(\theta)$. Inequality (7.4.1) means that $\|f(0)\| \leq \|D_{S_\theta^*} f\|_{\mathcal{H}(\theta)}$ for $f \in \mathcal{H}(\theta)$. Now, we compute both S_θ^{**} and $\|D_{S_\theta^*} f\|_{\mathcal{H}(\theta)}$ explicitly.

8.1. LEMMA. *Let $\theta \in H^\infty(E \to E_*)$, $\|\theta\|_\infty \leq 1$, and $P_0 f = f(0)$ for $f \in H^2(E_*)$. Then*
$$D_{\theta^*}^2 + \theta P_0 \theta^* = P_0 + S D_{\theta^*}^2 S^*, \qquad S^* D_{\theta^*}^2 + S^* \theta P_0 \theta^* = D_{\theta^*}^2 \cdot S^*.$$

PROOF.
$$D_{\theta^*}^2 + \theta P_0 \theta^* = D_{\theta^*}^2 + \theta D_S^2 \cdot \theta^* = D_{S \cdot \theta^*}^2 = D_{\theta^* S^*}^2$$
$$= D_{S^*}^2 + S D_{\theta^*}^2 S^* = P_0 + S D_{\theta^*}^2 S^*. \quad \bullet$$

8.2. LEMMA. *For every $e \in E$, $S^*\theta e \in \mathcal{H}(\theta)$ and*
$$\|S^*\theta e\|_{\mathcal{H}(\theta)}^2 = \|D_{\theta(0)} e\|^2 - \inf\{\|D_\theta(e + zx)\|_2^2 : x \in H^2(E_*)\}.$$

PROOF. By Lemmas 6.6 and 6.5 it is enough to compute the following sup:
$$\sup\{\|S^*\theta e + \theta x\|_2^2 - \|x\|_2^2 : x \in H^2(E_*)\}$$
$$= \sup\{\|\theta e - \theta(0) e + z \theta x\|_2^2 - \|x\|_2^2 : x \in H^2(E_*)\}$$
$$= \sup\{\|\theta(e + zx)\|_2^2 - \|\theta(0) e\|^2 - \|zx\|_2^2 : x \in H^2(E_*)\}$$
$$= \sup\{\|\theta(e + zx)\|_2^2 + \|e\|^2 - \|\theta(0) e\|^2 - \|e\|^2 - \|zx\|_2^2 : x \in H^2(E_*)\}$$
$$= \|D_{\theta(0)} e\|^2 - \inf\{\|D_\theta(e + zx)\|_2^2 : x \in H^2(E_*)\}. \quad \bullet$$

8.3. COROLLARY. $\|S^*\theta e\|_{\mathcal{H}(\theta)} \leq \|D_{\theta(0)} e\| \leq \|e\|$ *for every $e \in E$, and hence the mapping $j: e \to S^*\theta e$ is a contraction from E to $\mathcal{H}(\theta)$.* $\bullet$

8.4. THEOREM. $S_\theta^{**} f = Sf - \theta j^* f$ *for every $f \in \mathcal{H}(\theta)$.*

PROOF. Let $h \in H^2(E_*)$, $g = P_+\theta^*h$, $\varphi = D^2_{\theta^*}h$, and $f \in \mathcal{H}(\theta)$. Then

(8.5) $\quad (S^*_\theta \varphi, f)_{\mathcal{H}(\theta)} = (D^2_{\theta^*}h, S^{**}_\theta f)_{\mathcal{H}(\theta)} = (h, S^{**}_\theta f)_2$

because

(8.6) $\quad (y_1, y_2)_{\mathcal{H}(\theta)} = (x_1, x_2)_2$

provided $y_i \in \mathcal{H}(\theta)$ and one of the x_i's belongs to $D_{\theta^*} H^2(E_*)$. On the other hand, by Lemma 8.1,

$$(S^*_\theta \varphi, f)_{\mathcal{H}(\theta)} = (S^*_\theta D^2_{\theta^*}h, f)_{\mathcal{H}(\theta)} = (D^2_{\theta^*} S^* h, f)_{\mathcal{H}(\theta)} - (jg(0), f)_{\mathcal{H}(\theta)}$$
$$= (S^* h, f)_2 - (g(0), j^* f) \quad \text{(by (8.6))}$$
$$= (h, Sf)_2 - (P_+\theta^* h, j^* f)_2 = (h, Sf - \theta j^* f)_2.$$

Together with (8.5) this yields that $S^{**}_\theta f = Sf - \theta j^* f$. •

8.7. THEOREM. *For functions $f = D^2_{\theta^*} h$ from the dense subset $D^2_{\theta^*} H^2(E_*)$ of $\mathcal{H}(\theta)$ the following equality holds:*

$$\|D_{S^*_\theta} f\|^2_{\mathcal{H}(\theta)} = \|f(0)\|^2 + \inf\{\|D_\theta(g(0) + zx)\|^2_2 : x \in H^2(E)\},$$

where $g = P_+\theta^* h$.

PROOF.

$$\|D_{S^*_\theta} f\|^2_{\mathcal{H}(\theta)} = ((I - S^{**}_\theta S^*_\theta)f, f)_{\mathcal{H}(\theta)} = (f(0) + \theta j^* S^* f, D^2_{\theta^*} h)_{\mathcal{H}(\theta)}$$
(by Theorem 8.4)
$$= (f(0) + \theta^*_j S^* f, h)_2 \quad \text{(by (8.6))}$$
$$= (f(0), h(0)) + (j^* S^* f, P_0\theta^* h)$$
$$= (f(0), f(0) + \theta(0)g(0)) + (S^* D^2_{\theta^*}h, jg(0))_{\mathcal{H}(\theta)}$$
$$= (f(0), f(0) + \theta(0)g(0)) + (S^*_\theta f, jg(0))_{\mathcal{H}(\theta)}$$
$$= \|f(0)\|^2 + (f(0), \theta(0)g(0)) + (f, S^{**}_\theta S^* \theta g(0))_{\mathcal{H}(\theta)}$$
$$= \|f(0)\|^2 + (f(0), \theta(0)g(0)) + (h, SS^*\theta g(0) - \theta j^* S^* \theta g(0))_2$$
(by Theorem 8.4. and (8.6))
$$= \|f(0)\|^2 + (f(0), \theta(0)g(0)) + (S^* h, S^* \theta g(0))$$
$$\quad - (P_0\theta^* h, j^* S^* \theta g(0))_2$$
$$= \|f(0)\|^2 + (f(0), \theta(0)g(0)) + (h, \theta g(0))_2 - (h(0), \theta(0)g(0))$$
$$\quad - \|S^*\theta g(0)\|^2_{\mathcal{H}(\theta)}$$
$$= \|f(0)\|^2 + (f(0), \theta(0)g(0)) + \|g(0)\|^2$$
$$\quad - (f(0) + \theta(0)g(0), \theta(0)g(0)) - \|S^*\theta g(0)\|^2_{\mathcal{H}(\theta)}$$
$$= \|f(0)\|^2 + \|D_{\theta(0)}g(0)\|^2 - \|S^*\theta g(0)\|^2_{\mathcal{H}(\theta)}$$
$$= \|f(0)\|^2 + \inf\{\|D_\theta(g(0) + zx)\|^2_2 : x \in H^2(E)\}$$
(by Lemma 8.2). •

8.8. COROLLARY. *Given $\aleph(\theta)$, suppose (without loss of generality) that $\{e \in E : \theta(\varsigma)e \equiv 0\} = \{0\}$. (This does not mean that $\mathrm{Ker}\,\theta = \{0\}$!) Then $\|D_{S_\theta^*}f\|^2_{\aleph(\theta)} = \|f(0)\|^2$ for every $f \in \aleph(\theta)$ iff*

(8.9) $\qquad \inf\{\|D_\theta(e+zx)\|_2 : x \in H^2(E)\} = 0 \quad \text{for every } e \in E;$

i.e., the polynomials in z are dense in the weighted space $L^2(E, \Delta)$ endowed with the norm

$$\|g\|^2_\Delta = \int_\mathbb{T} \|\Delta(\varsigma)g(\varsigma)\|^2\, dm(\varsigma), \qquad \Delta(\varsigma) = (1 - \theta(\varsigma)^*\theta(\varsigma))^{1/2}.$$

In fact,

$$E \ominus \mathrm{clos}\{g(0) : g = P_+\theta^*h,\ h \in H^2(E_*)\}$$
$$= \{e \in E : 0 = (g(0), e) = (P_+\theta^*h, e) = (h, \theta e),\ \forall h \in H^2(E_*)\}$$
$$= \{e \in E : \theta e \equiv 0\} = \{0\},$$

and hence (8.9) holds. Moreover, is easy to see that $\|D_\theta g\|_2 = \|\Delta g\|_2 = \|g\|_\Delta$ for any $g \in H^2(E)$. •

9. Remarks on polynomial density and extreme points.

The knowledge of the defect operator, for an operator which is intended to be the model, is crucial for constructing the function model. In particular, this is the case for S_θ^* in the framework of the dBR model. The simplest case is when $D_{S_\theta^*}$ is metrically equivalent to the evaluation operator $f \to f(0)$ on $H^2(E_*)$ (compare Theorem 8.7 with §10, especially with §§10.7, 10.9). By Corollary 8.8 this simplest case occurs if and only if condition (8.9) is fullfilled. This section contains some comments on the meaning of this condition.

9.1. LEMMA. *Let $\Delta \in L^\infty(E \to E)$, $\Delta(\varsigma) \geq 0$ a.e. The following are equivalent.*
1. $\inf\{\|\Delta(e+zx)\|_2 : x \in H^2(E)\} = 0$ *for every $e \in E$.*
2. $\mathrm{clos}\,\Delta H^2(E)$ *is a reducing subspace of the shift operator on $L^2(E)$.*
3. $\mathrm{clos}\,\Delta H^2(E) = \mathrm{clos}\,\Delta L^2(E).$
4. *The polynomials are dense in $L^2(E, \Delta)$.*
5. $\Delta L^2(E) \cap H^2_-(E) = \{0\}.$

PROOF. The equivalence of 1–4 can be ascertained in the usual way. E.g., $1 \Rightarrow 2$ since for $f \in H^2(E)$

$$\inf\{\|\Delta(\bar{z}f + g)\|_2 : g \in H^2(E)\} = \inf\{\|\Delta(f(0) + zx)\|_2 : x \in H^2(E)\} = 0,$$

if 1 holds.

On the other hand, 3 is equivalent to the coincidence of the orthogonal complements $\mathrm{Ker}(\Delta|H^2(E))^*$ and $\mathrm{Ker}\,\Delta^*$ ($= \mathrm{Ker}\,\Delta = (\mathrm{clos}\,\Delta L^2)^\perp$). But the first of them is equal to $\{f \in L^2(E) : P_+\Delta f = 0\}$. Thus, 3 is reduced to the following implication: $f \in L^2(E),\ P_+\Delta f = 0 \Rightarrow \Delta f = 0$. This is just property 5. •

Note that property 5 is satisfied provided the range function $\varsigma \mapsto \Delta(\varsigma)E$ contains no coanalytic direction; i.e., from $f \in H^2_-(E)$, $f(\varsigma) \in \Delta(\varsigma)E$ a.e. it follows that $f \equiv \mathbb{0}$.

Now, let us use the well-known description of shift invariant subspaces of $L^2(E)$ (Beurling, Helson, Lax, Halmos; see, e.g., [12]). Then, there exist a projection-valued function P ($P(\varsigma): E \to E$), a Hilbert space E_{**}, and an isometry-valued function ϑ ($\vartheta(\varsigma): E_{**} \to E$) such that

(9.2) $$\mathcal{E} \stackrel{\text{def}}{=} \text{clos}\,\mathbf{\Delta} H^2(E) = \boldsymbol{P}L^2(E) \oplus \boldsymbol{\vartheta}H^2(E_{**}).$$

Statements 1–5 of Lemma 9.1 are equivalent to $E_{**} = \{\mathbb{0}\}$. The subspace $\mathcal{E}$ is completely nonreducing iff $P \equiv \mathbb{0}$. By the famous Wold-Kolmogorov theorem this is the case iff $\bigcap_{n \geq 0} z^n \mathcal{E} = \{\mathbb{0}\}$. The following includes a result by Sz.-Nagy–Foiaş, which shows that $1 \Leftrightarrow 2$ (see [17, Proposition V.4.2 and a remark about it in Comments for Chapter V]). Our proof is easier.

9.3. THEOREM. *The following are equivalent.*

1. *Any statement of Lemma 9.1 holds.*
2. *If $F \in H^\infty(E \to E_{**})$, E_{**} is a Hilbert space, and if $F^*F \leq \Delta^2 = I - \theta^*\theta$ a.e. on* T, *then $F \equiv \mathbb{0}$.*
3. *If Q is a measurable projection-valued function on* T *and if $\Delta Q \Delta = F^*F$ with $F \in H^\infty(E \to E_{**})$, then $\Delta Q \equiv \mathbb{0}$.*

PROOF. It is clear that $2 \Rightarrow 3$.

$1 \Rightarrow 2$: Let $F^*F \leq \Delta^2$, $F = \sum_{n \geq 0} \hat{F}(n) z^n$. Then for every $x \in H^2(E)$ we have $\|Fx\|_2 \leq \|\Delta x\|_2$, and from 1, $\inf\{\|F(e + zx)\|_2 : x \in H^2(E)\} = 0$ for every $e \in E$. Hence, $F(0)e = \mathbb{0}$, $e \in E$, and $F(0) = \hat{F}(0) = \mathbb{0}$. By induction, $\hat{F}(n) = \mathbb{0}$ for $n \geq 0$; i.e., $F \equiv \mathbb{0}$.

$3 \Rightarrow 1$: Let P be a function from the representation (9.2). Set $Q = I - P$, and let us prove that $\Delta Q \Delta$ is factorable as F^*F. (This yields $Q\Delta \equiv \mathbb{0}$ and $E_{**} = \mathbb{0}$ since $\vartheta(\varsigma)E_{**} \subset Q(\varsigma)\Delta(\varsigma)E$ a.e.).

For every $e \in E$ there exists a unique $f = f_e \in H^2(E_{**})$ such that $Q\Delta e = \vartheta f$. Define a function F by the equality $F(\varsigma)e = f_e(\varsigma)$ (for $\varsigma \in \mathbb{D}$ or for a.e. $\varsigma \in \mathsf{T}$). Then, $F(\cdot)e \in H^2(E_{**})$ and

$$\|F(\varsigma)e\|_{E_{**}} = \|f_e(\varsigma)\|_{E_{**}} = \|\vartheta(\varsigma)f_e(\varsigma)\|_E = \|Q(\varsigma)\Delta(\varsigma)e\|_E \leq \|\Delta(\varsigma)e\|_E.$$

Hence, $F \in H^\infty(E \to E_{**})$ and $Q\Delta = \vartheta F$, which implies $\Delta Q \Delta = F^*F$. ∎

9.4. REMARKS. In fact, we also proved that if $\mathcal{E} = \text{clos}\,\mathbf{\Delta} H^2(E)$ is completely nonreducing (i.e., $P \equiv \mathbb{0}$) then $\Delta^2 = F^*F$, $F \in H^\infty(E \to E_{**})$.

The converse is also true (and known). In fact, $\Delta^2 = F^*F$ implies $\Delta(\varsigma) = c(\varsigma)F(\varsigma)$ a.e. with operators $c(\varsigma)$ which are isometries on $\text{clos}\,F(\varsigma)E$. Hence,

$$\bigcap_{n \geq 0} z^n \mathcal{E} = \bigcap_{n \geq 0} z^n \,\text{clos}\,\boldsymbol{c}\boldsymbol{F}H^2(E) = \boldsymbol{c} \bigcap_{n \geq 0} z^n \,\text{clos}\,\boldsymbol{F}H^2(E)$$

$$\subset \boldsymbol{c} \bigcap_{n \geq 0} z^n H^2(E_{**}) = \{\mathbb{0}\}$$

and the Wold-Kolmogorov theorem yields the result. •

It is worth mentioning a corollary of the last remark.

9.5. COROLLARY. *If* rank $\Delta = 1$ *a.e. on* T, *then the statements of Theorem 9.3 are equivalent to a new one:*

$$\Delta^2 \neq F^*F \quad \text{for any } F \in H^\infty(E \to E_{**}).$$

In fact, (9.2) implies that $\operatorname{clos}\Delta(\varsigma)E = P(\varsigma)E \oplus \vartheta(\varsigma)E_{**}$ a.e. on T. So, if $E_{**} \neq \{0\}$, then $P \equiv 0$ and by the previous remark $\Delta^2 = F^*F$. •

One more comment to Theorem 9.3 consists of the following link between our problem and that of the extreme points in $H^\infty(E \to E_*)$.

9.6. COROLLARY. *If* $\Delta = (I - \theta^*\theta)^{1/2}$ *and if all (one) of the statements of 9.1 hold(s), then* θ *is an extreme point of the unit ball of* $H^\infty(E \to E_*)$.

Indeed, if $F \in H^\infty(E \to E_*)$ and $\|\theta \pm F\|_\infty \leq 1$, then

$$\|F(\varsigma)e\|^2 + \|\theta(\varsigma)e\|^2 = \tfrac{1}{2}(\|F(\varsigma)e + \theta(\varsigma)e\|^2 + \|F(\varsigma)e - \theta(\varsigma)e\|^2) \leq \|e\|^2, \qquad e \in E;$$

i.e., $F^*F + \theta^*\theta \leq I$ and hence (by Theorem 9.3) $F \equiv 0$. •

Example 9.9 (below) shows that the converse is not true in general. Of course, if $E = E_* = \mathbb{C}$ it is well known (and results from what follows) that the polynomials in z are dense in $L^2(\Delta dm)$ iff θ is an extreme point, and iff $\int_T \log(1 - |\theta|^2)\,dm = -\infty$.

It should be noted that the conditions under which a positive operator function (like $\Delta Q\Delta$ from Statement 3 of Theorem 9.3) admits a factorization of the form F^*F are known, at all events for the matrix case (dim $E < \infty$). For the reader's convenience we recall these conditions both in the Helson-Lowdenslager (see [**10**]) and in the Matveev-Rozanov form (see [**14**]).

9.7. THEOREM. *Let* $W \in L^\infty(E \to E)$, $W(\varsigma) \geq 0$ *a.e. on* T, $W \neq 0$, *and* dim $E < \infty$. *The following are equivalent.*

1. $W = F^*F$ for some $F \in H^\infty(E \to E^*)$, E^* being a Hilbert space.

2. *The range function* $E(\varsigma) \overset{\text{def}}{=} W(\varsigma)E$ *is coanalytic; i.e.,* $E(\varsigma) =$ *linear hull of* $\{f(\varsigma): f \in H^2_-(E),\ f(\varsigma) \in W(\varsigma)E \text{ a.e.}\}$ *for almost all* $\varsigma, \varsigma \in$ T, *and*

$$\int_T \log \det(W(\varsigma)|E(\varsigma))\,dm(\varsigma) > -\infty.$$

3. rank $W(\varsigma)$ *is constant a.e. (say, r), and if* $e_1, \ldots, e_n$ *is an orthonormal basis in* E, *then there exists a principal minor* $\delta = \{(W(\cdot)e_{i_p}, e_{i_q})\}_{1 \leq p, q \leq r}$ *such that*

$$\int_T \log \det \delta\,dm > -\infty,$$

and the quotients

$$\det \delta_{kp}/\det \delta$$

are boundary values of Nevanlinna meromorphic functions (Nev_) in the outside of the disc $\{\varsigma : |\varsigma| > 1\}$; here $1 \leq p \leq r$, $1 \leq k \leq n$, and δ_{kp} stands for the minor of W which results by replacing the pth line by the kth one in δ.

PARTIAL CASES. A. $\operatorname{rank} W = 1$ a.e. Then $W = (\cdot, g)f$ with bounded $f = \sum_{i=1}^{n} f_i e_i$, $g = \sum_{i=1}^{n} g_i e_i$. Now W is factorable if and only if

$$\int_{\mathsf{T}} \log |f_p \bar{g}_p| dm > -\infty \quad \text{for some } p,\ 1 \leq p \leq n$$

and

$$f_k/f_p \in \operatorname{Nev}_- \quad \text{for any } k,\ 1 \leq k \leq n.$$

B. $\dim E = 1$ or 2. For scalar functions we have the well-known criterion $\int_{\mathsf{T}} \log W > -\infty$. For $\dim E = 2$, W is factorable iff $\operatorname{rank} W = \operatorname{const}$ a.e. and $\int_{\mathsf{T}} \log \det W > -\infty$ for $\operatorname{rank} W = 2$ or the conditions of the previous case are fullfilled for $\operatorname{rank} W = 1$.

9.8. EXAMPLE. Let $\dim E = n < \infty$ and $\theta \in H^\infty(E \to E_*)$, $\|\theta\|_\infty \leq 1$. Suppose that θ is an "almost inner" function; i.e., $\theta(\varsigma)$ is an isometry onto a subspace $E(\varsigma)$ of codimension 1 but not on all of E. Then $\operatorname{rank}(I - \theta^*\theta) = 1$ a.e. and $\operatorname{Trace}(I - \theta^*\theta) > 0$ a.e. on T, and

$$\Delta = (I - \theta^*\theta)^{1/2} = (I - \theta^*\theta)/(\operatorname{Trace}(I - \theta^*\theta))^{1/2} \quad \text{a.e.}$$

Thus, the criterion in Partial Case A for $W = \Delta$ can be stated as follows: there exists p, $1 \leq p \leq n$, such that

$$\int_{\mathsf{T}} \log \frac{1 - \|\theta(\varsigma)e_p\|^2}{\left(\sum_{i=1}^{n}(1 - \|\theta(\varsigma)e_i\|^2)\right)^{1/2}} dm(\varsigma) > -\infty,$$

and for every $k \neq p$

$$\frac{(\theta(\cdot)e_p, \theta(\cdot)e_k)}{1 - \|\theta(\cdot)e_p\|^2} \in \operatorname{Nev}_-.$$

9.9. EXAMPLE. Let $E_* = E = \mathbb{C}^2$ and $\theta = (\cdot, f)h$ with $h = h_1 e_1 + h_2 e_2$, $f = f_1 e_1 + f_2 e_2$. Suppose $h_i, \bar{f}_i \in H^\infty$ $(i = 1, 2)$ and $|h_1|^2 + |h_2|^2 = 1$ a.e., $|f_1|^2 + |f_2|^2 = 1$ a.e. Then $\theta^*\theta = (\cdot, f)f$ are orthogonal projections a.e., and so are $\Delta = (I - \theta^*\theta)^{1/2} = I - \theta^*\theta = I - (\cdot, f)f = (\cdot, f_*)f_*$, where $f_* = \bar{f}_2 e_1 - \bar{f}_1 e_2$. One of the f_i (say f_1) is nonzero a.e. Then, from Partial Case A it is clear that Δ is factorable if and only if $\bar{f}_2/\bar{f}_1 \in \operatorname{Nev}_-$. The two conceivable cases are attainable. If Δ is factorable, the space $\operatorname{clos} \Delta H^2(\mathbb{C}^2)$ is completely nonreducing (and all the more $\operatorname{clos} \Delta H^2(\mathbb{C}^2) \neq \operatorname{clos} \Delta L^2(\mathbb{C}^2)$). It is easy to see that in the opposite case (i.e., in the case $\bar{f}_2/\bar{f}_1 \notin \operatorname{Nev}_-$), $\operatorname{clos} \Delta H^2(\mathbb{C}^2)$ is a reducing subspace of $L^2(\mathbb{C}^2)$.

On the other hand, whether or not θ is an extreme point of the unit ball of H^∞ depends on the function h only. Indeed, for every function F from $H^\infty(E \to E_*)$, the inequalities $\|\theta \pm F\|_\infty \leq 1$ imply both $\theta^*\theta + F^*F \leq I$ and $\theta\theta^* + FF^* \leq I$. Hence, $F^*F \leq \Delta^2$ and $FF^* \leq \Delta_*^2$ a.e., where $\Delta = (I - \theta^*\theta)^{1/2}$,

$\Delta_* = (I - \theta\theta^*)^{1/2}$. The factorization Theorem 5.3 gives $F(\varsigma) = C(\varsigma)\Delta(\varsigma)$, $F(\varsigma)^* = C_*(\varsigma)\Delta_*(\varsigma)$ almost everywhere for some contractions $C(\varsigma), C_*(\varsigma)$. Hence

$$F(\varsigma)^* = \Delta(\varsigma)C(\varsigma)^*, \qquad F(\varsigma) = \Delta_*(\varsigma)C_*(\varsigma)^*.$$

9.10. COROLLARY. *If $\|\theta \pm F\|_\infty \leq 1$, then $F(\varsigma)E \subset \Delta_*(\varsigma)E_*$, $F(\varsigma)^*E_* \subset \Delta(\varsigma)E$ a.e.* •

In the example under consideration, $\Delta_* = \Delta_*^2 = I - (\cdot, h)h = (\cdot, h_*)h_*$, where $h_* = \overline{h}_2 e_1 - \overline{h}_1 e_2$. This yields $\Delta_*(\varsigma) = (\cdot, h_*(\varsigma))h_*(\varsigma)$ and $\Delta_*(\varsigma)\mathbb{C}^2 = h_*(\varsigma)\mathbb{C}$ a.e. on $\mathbb{T}$. Since one of the h_i is nonzero, let us suppose $h_1 \not\equiv 0$ (in fact, $h_1 \neq 0$ a.e.). Now, if $\varphi \in H^2(\mathbb{C}^2)$, $\varphi = \varphi_1 e_1 + \varphi_2 e_2$, and $\varphi(\varsigma) \in \Delta_*(\varsigma)\mathbb{C}^2$ a.e., then either $\varphi = \mathbb{0}$ a.e. or $\varphi_2 \neq 0$ a.e. and $\varphi_1/\varphi_2 = -\overline{h}_2/\overline{h}_1$ a.e. Corollary 9.10 implies that θ is an extreme point provided $h_2/h_1 \notin \text{Nev}_-$.

CONCLUSION: If $\overline{f}_2/\overline{f}_1 \in \text{Nev}_-$ and $h_2/h_1 \notin \text{Nev}_-$, then the function θ is an extreme point of the unit ball of $H^\infty(\mathbb{C}^2 \to \mathbb{C}^2)$, but $\text{clos}\,\Delta H^2(\mathbb{C}^2) \neq \text{clos}\,\Delta L^2(\mathbb{C}^2)$. •

10. Function models: non-coisometric case.

Let us turn to the function models for completely nonunitary contractions. Trying to understand an ability of S_θ^{**} to serve as a model operator, we start with the following observation.

10.1. LEMMA. *Any operator S_θ^* has no isometric parts; i.e., if $\mathcal{E}$ is an S_θ^*-invariant subspace of $\mathcal{H}(\theta)$ and if the restriction $S_\theta^*|\mathcal{E}$ is an isometry, then $\mathcal{E} = \{\mathbb{0}\}$.*

PROOF. Let $f \in \mathcal{E}$. Then (7.4.1) implies that for every $n \geq 0$

$$\|(S^{*n}f)(0)\|^2 \leq \|D_{S_\theta^*}S^{*n}f\|_{\mathcal{H}(\theta)}^2 = \|S^{*n}f\|_{\mathcal{H}(\theta)}^2 - \|S^{*(n+1)}f\|_{\mathcal{H}(\theta)}^2 = 0,$$

and hence $f = \mathbb{0}$. •

10.2. DEFINITION. An operator A is said to be completely nonisometric if there is no nonzero A-invariant subspace on which A is an isometry.

Below (in §10.5), we give a (known) formula for the largest invariant subspace on which A acts as an isometry, but let us mention first an obstacle for S_θ^{**} to serve as a universal operator.

10.3. COROLLARY. *If an operator T is unitarily equivalent to an operator of the form S_θ^{**}, then T^* is completely nonisometric.*

10.4. LEMMA. *Let A be a Hilbert space contraction (acting on H). Then the set*

$$\{x : \|x\| = \|A^n x\|,\ n \geq 0\} = \bigcap_{n \geq 0} \operatorname{Ker} D_A A^n$$

is the largest subpsace on which A acts isometrically. A is completely nonisometric if and only if $H = \text{span}(A^{*n}D_A H : n \geq 0)$.

PROOF.
$$\|x\| = \|A^n x\|, \quad n \geq 0 \Leftrightarrow \|A^n x\| = \|A^{n+1} x\|, \quad n \geq 0$$
$$\Leftrightarrow x \in \bigcap_{n \geq 0} \operatorname{Ker} D_A A^n.$$

Moreover,
$$\left(\bigcap_{n \geq 0} \operatorname{Ker} D_A A^n \right)^\perp = \operatorname{span}(A^{*n} D_A H : n \geq 0),$$

and the result follows. ●

Now, we turn to the Sz.-Nagy–Foiaş model.

10.5. *The Sz.-Nagy–Foiaş model* is constructed by starting from a completely non-unitary operator T and its characteristic function $\theta = \theta_T$, see [17]. One sets

$$\theta_T = -T + z D_{T^*}(I - zT^*)^{-1} D_T$$

and considers θ_T as an operator function from $E = \operatorname{clos} D_T H$ to $E_* = \operatorname{clos} D_{T^*} H$. ($H$ stands for the Hilbert space on which T is defined.) θ, being a contractive analytic function, produces the model space $\mathcal{K}_\theta$ by the formula

$$\mathcal{K}_\theta = (H^2(E_*) \oplus \operatorname{clos} \Delta L^2(E)) \ominus \{\theta u \oplus \Delta u : u \in H^2(E)\},$$

where $\Delta(\varsigma) = (I - \theta(\varsigma)^* \theta(\varsigma))^{1/2}$ a.e. on T. The initial operator T is unitarily equivalent to the model operator M_θ acting on $\mathcal{K}_\theta$ as the compression of the shift

$$M_\theta \{f, g\} = P_{\mathcal{K}_\theta} \{zf, zg\}.$$

Moreover,
$$M_\theta^* \{f, g\} = \{P_+ \bar{z} f, \bar{z} g\} = \{S^* f, \bar{z} g\} \quad \text{for } \{f, g\} \in \mathcal{K}_\theta.$$

Conversely, every contractive function θ, $\theta \in H^\infty(E \to E_*)$, which is not an isometric constant on any nonzero subspace of E (a so-called pure contractive function) is the characteristic function of a (essentially unique) completely nonunitary contraction. Lemma 10.4 yields immediately the following.

10.6. LEMMA. *The largest M_θ^*-invariant subspace of $\mathcal{K}_\theta$ on which M_θ^* is an isometry is $\mathbb{0} \oplus (\operatorname{clos} \Delta L^2(E) \ominus \operatorname{clos} \Delta H^2(E))$. Hence, M_θ^* is completely nonisometric iff $\operatorname{clos} \Delta L^2(E) = \operatorname{clos} \Delta H^2(E)$.*

PROOF. Let $\{f, g\} \in \mathcal{K}_\theta$. Then
$$\|M_\theta^{*n} \{f, g\}\|_2^2 = \|S^{*n} f\|_2^2 + \|\bar{z}^n g\|_2^2 = \|S^{*n} f\|_2^2 + \|g\|_2^2,$$

and hence $\|\{f, g\}\|_2 = \|M_\theta^{*n} \{f, g\}\|_2$ for every $n \geq 0$ if and only if $f = \mathbb{0}$. But,

$$\{\mathbb{0}, g\} \in \mathcal{K}_\theta \Leftrightarrow g \in \operatorname{clos} \Delta L^2(E) \ominus \operatorname{clos} \Delta H^2(E). \quad ●$$

Now, from Lemmas 10.1 and 10.6 and the Sz.-Nagy–Foiaş theory drafted in §10.5 one can derive the following.

10.7. COROLLARY 1. *A contraction T^* is completely nonisometric iff* clos $\Delta L^2(E) =$ clos $\Delta H^2(E)$ *where* $\Delta = (I - \theta_T^*\theta_T)^{1/2}$.

2. *If a SzNF model operator M_θ is unitarily equivalent to an operator of the form S_ϑ^{**} (ϑ is a contractive H^∞-function), then* clos $\Delta L^2(E) =$ clos $\Delta H^2(E)$. •

Now we are in a position to identify the operator S_θ^{**} with a maximal part (restriction) of M_θ having completely nonisometric adjoint.

10.8. THEOREM. *Let $\theta \in H^\infty(E \to E_*)$ be a contractive operator function, let $\Delta = (I - \theta^*\theta)^{1/2}$ and*

$$\mathcal{K}_\theta^* = (H^2(E_*) \oplus \operatorname{clos} \Delta H^2(E)) \ominus \{\theta u \oplus \Delta u : u \in H^2(E)\}.$$

Then $M_\theta \mathcal{K}_\theta^* \subset \mathcal{K}_\theta^*$ and $M_\theta | \mathcal{K}_\theta^*$ is unitarily equivalent to S_θ^{**}.

PROOF. Set

$$VD_{\theta^*}^2 h = \{D_{\theta^*}^2 h, -\Delta P_+ \theta^* h\}, \qquad h \in H^2(E_*).$$

Then, it is obvious that $VD_{\theta^*}^2 H \in H^2(E_*) \oplus \operatorname{clos}\Delta H^2(E)$. On the other hand, for $u \in H^2(E)$ we have

$$(D_{\theta^*}^2 h, \theta u)_2 - (\Delta P_+ \theta^* h, \Delta u)_2 = (\theta^* D_{\theta^*}^2 h - \Delta^2 P_+ \theta^* h, u)_2$$
$$= (D_\theta^2 P_+ \theta^* h - P_+ \Delta^2 P_+ \theta^* h, u)_2 = (\mathbb{0}, u)_2 = 0.$$

This means that $VD_{\theta^*}^2 h \in \mathcal{K}_\theta^*$ for any $h \in H^2(E_*)$. Moreover

$$\|VD_{\theta^*}^2 h\|_2^2 = \|D_{\theta^*}^2 h\|_2^2 + \|\Delta P_+ \theta^* h\|_2^2$$
$$= (\|D_{\theta^*} h\|_2^2 - \|\theta^* D_{\theta^*} h\|_2^2) + (\|P_+ \theta^* h\|_2^2 - \|\theta P_+ \theta^* h\|_2^2)$$
$$= (\|D_{\theta^*} h\|_2^2 - \|D_\theta P_+ \theta^* h\|_2^2) + (\cdots)$$
$$= (\|D_{\theta^*} h\|_2^2 - \|P_+ \theta^* h\|_2^2 + \|\theta P_+ \theta^* h\|_2^2) + (\|P_+ \theta^* h\|_2^2 - \|\theta P_+ \theta^* h\|_2^2)$$
$$= \|D_{\theta^*} h\|_2^2 = \|D_{\theta^*}^2 h\|_{\mathcal{H}(\theta)}^2.$$

Hence, V extends from $D_{\theta^*}^2 H^2(E_*)$ to an operator (we keep the same notation) isometrically mapping $\mathcal{H}(\theta)$ into $\mathcal{K}_\theta^*$.

In fact, V is unitary because if $\{\varphi, \psi\} \in \mathcal{K}_\theta^* \ominus V\mathcal{H}(\theta)$, then $(\theta u, \varphi)_2 + (\Delta u, \psi)_2 = 0$ for every $u \in H^2(E)$ and $(D_{\theta^*}^2 h, \varphi)_2 - (\Delta P_+ \theta^* h, \psi)_2 = 0$ for every $h \in H^2(E)$, which implies (putting $u = P_+ \theta^* h$) that $(h, \varphi)_2 = 0$ for every $h \in H^2(E_*)$. Thus, $\varphi = \mathbb{0}$ and therefore $\psi \perp \Delta H^2(E)$; i.e., $\psi = \mathbb{0}$.

It is clear that $\mathcal{K}_\theta \ominus \mathcal{K}_\theta^* = \{\mathbb{0}\} \oplus (\operatorname{clos}\Delta L^2(E) \ominus \operatorname{clos}\Delta H^2(E))$, and hence $M_\theta^*(\mathcal{K}_\theta \ominus \mathcal{K}_\theta^*) \subset \mathcal{K}_\theta \ominus \mathcal{K}_\theta^*$; i.e., $M_\theta \mathcal{K}_\theta^* \subset \mathcal{K}_\theta^*$.

It remains to check the identity $P_* M_\theta^* V = V S_\theta^*$, where P_* stands for the orthogonal projection onto $\mathcal{K}_\theta^*$. To this end it is enough to prove that

(10.8.1) $$(VS_\theta^* f, Vg) = (M_\theta^* Vf, Vg)$$

for all f, g from $D_{\theta^*}^2 H^2(E_*)$. Let $f = D_{\theta^*}^2 h$, $g = D_{\theta^*}^2 k$. Then

$$(VS_\theta^* f, Vg) = (S_\theta^* f, g)_{\mathcal{H}(\theta)} = (S^* f, k)_2.$$

On the other hand,

$$\begin{aligned}(M_\theta^* Vf, g) &= (S^*f, D_{\theta^*}^2 k)_2 + (\bar{z}\Delta P_+\theta^* h, \Delta P_+\theta^* k)_2 \\ &= (S^*f, k)_2 - (S^*f, \theta P_+\theta^* k)_2 + (\Delta^2 \bar{z} P_+\theta^* h, P_+\theta^* k)_2 \\ &= (S^*f, k)_2 - (D_{\theta^*}^2 S^* h, \theta P_+\theta^* k)_2 + (S^*\theta P_0\theta^* h, \theta P_+\theta^* k)_2 \\ &\quad + (\Delta^2 P_+\bar{z} P_+\theta^* h, P_+\theta^* k)_2 + (\Delta^2 P_-\bar{z} P_+\theta^* h, P_+\theta^* k)_2\end{aligned}$$

(using Lemma 8.1)

$$\begin{aligned}&= (S^*f, k)_2 - (D_\theta^2 \theta^* S^* h, P_+\theta^* k)_2 + (\bar{z}\theta P_0\theta^* h, \theta P_+\theta^* k)_2 \\ &\quad + (\Delta^2 \theta^* S^* h, P_+\theta^* k)_2 - (\theta^*\theta P_-\bar{z}P_+\theta^* h, P_+\theta^* k)_2 \\ &= (S^*f, k)_2.\end{aligned}$$

This completes the proof of (10.8.1) and the theorem itself. •

The converse to Corollary 10.3 follows.

10.9. COROLLARY. *Let T be a contraction with completely nonisometric adjoint T^*, and let $\theta = \theta_T$ be its characteristic function. Then T is unitarily equivalent to S_θ^{**} on $\mathcal{H}(\theta)$.* •

11. Digression. Now we can see that the space $\mathcal{H}(\theta)$ has to be enlarged to serve as a model space for general operators (with adjoints having isometric parts). This aim can be reached by using the de Branges-Rovnyak space $\mathcal{D}(\theta)$ defined in §4.4.

At first sight the space $\mathcal{D}(\theta)$ looks a little bit complicated because like the space $\mathcal{K}_\theta$ it is "two-storied" and its nature as a complement of a graph-space with respect to another space is more or less veiled. The use of the range inner product from $\mathcal{H}(\theta)$ does not simplify matters. But, following [4], for the sake of brevity, we introduce the space $\mathcal{D}(\theta)$ in what may be too straightforward a way, although a simpler but somewhat formal way to describe $\mathcal{D}(\theta)$ consists in the explicit writing down of the linear hull of the reproducing kernels of $\mathcal{D}(\theta)$, followed by the completion; cf. [4, 1]. We think it is an advantage of $\mathcal{D}(\theta)$ that it is more symmetric with respect to coordinate permutation than $\mathcal{K}_\theta$. One more advantage of $\mathcal{D}(\theta)$ is that it falls into a "one-storied" space more often (under the condition clos $\Delta H^2(E) = $ clos $\Delta L^2(E)$) than $\mathcal{K}_\theta$ (under the condition $\Delta \equiv \mathbb{O}$).

There exists a symmetric transcription for the SzNF model space too. It was introduced by B. S. Pavlov [13] and is very useful for non-selfadjoint scattering problems. This model space $\mathcal{K}$ can be cursorily described as the space of pairs $f \oplus f_*$ from $L^2(E \oplus E_*, \overline{\Delta})$ as follows:

$$\mathcal{K} = \{f \oplus f_* : f + \theta^* f_* \in H_-^2(E_*),\ \theta f + f_* \in H^2(E_*)\},$$

where $L^2(E \oplus E_*, \overline{\Delta})$ stands for the completion of the space $L^2(E \oplus E_*)$ with respect to the seminorm $\|f \oplus f_*\|_{\overline{\Delta}} = \|\overline{\Delta}(f \oplus f_*)\|_2$,

$$\overline{\Delta}^2 \stackrel{\text{def}}{=} \begin{pmatrix} I & \theta^* \\ \theta & I \end{pmatrix}.$$

Moreover, there exists "the most symmetric" coordinate-free transcription of the SzNF model space introduced by the second author (see [**11**]) as a development of Pavlov's idea mentioned above. We use this "universal" form of the SzNF model in what follows, and that is why we describe it briefly, referring for all details to [**11**].

In fact, any model depends only on two isometric embeddings of some L^2-spaces into the space of minimal unitary dilation, say $\mathcal{H}$. Symmetry and universality mentioned above consist not in fixing a special choice of these embeddings but in using common properties of such embeddings. To be more specific, let us consider Hilbert spaces $E, E_*, \mathcal{H}$ and linear maps

$$\pi: L^2(E) \to \mathcal{H}, \qquad \pi_*: L^2(E_*) \to \mathcal{H}$$

such that
  (i) π and π_* are isometries; i.e., $\pi^*\pi = I$, $\pi_*^*\pi_* = I$.
  (ii) $\pi_*^*\pi$ commutes with the shift operator and maps $H^2(E)$ into $H^2(E_*)$; hence $\pi_*^*\pi = \boldsymbol{\theta}$, θ being a contractive operator function, $\theta \in H^\infty(E \to E_*)$.
  (iii) $\mathcal{H} = \mathrm{clos}(\pi L^2(E) + \pi_* L^2(E_*))$.
Set $\Delta = (I - \theta^*\theta)^{1/2}$, $\Delta_* = (I - \theta\theta^*)^{1/2}$. Since

$$\|(\pi - \pi_*\theta)f\|_\mathcal{H} = \|\Delta f\|_{L^2(E_*)}, \qquad \|(\pi_* - \pi\theta^*)g\|_\mathcal{H} = \|\Delta_* g\|_{L^2(E)}$$

for every f and g, the equalities

$$\tau\Delta = \pi - \pi_*\theta, \qquad \tau_*\Delta_* = \pi_* - \pi\theta^*$$

determine the partial isometries

$$\tau: L^2(E) \to \mathcal{H}, \qquad \tau_*: L^2(E_*) \to \mathcal{H}$$

with initial spaces clos $\boldsymbol{\Delta} L^2(E)$ and clos $\boldsymbol{\Delta}_* L^2(E_*)$ respectively. It is easy to see that

(11.1) $$\begin{cases} I = \pi\pi^* + \tau_*\tau_*^* = \pi_*\pi_*^* + \tau\tau^*, \\ \tau^*\pi = \boldsymbol{\Delta}, \qquad \tau^*\pi_* = \mathbb{0}; \qquad \tau_*^*\pi = \mathbb{0}, \qquad \tau_*^*\pi_* = \boldsymbol{\Delta}_*. \end{cases}$$

Now set
  (iv) $K = \{x \in \mathcal{H} : \pi^*x \in H^2_-(E),\ \pi_*^*x \in H^2(E_*)\}$, and
  (v) $Ux = \pi z\pi^*x + \tau_* z\tau_*^*x$, $x \in \mathcal{H}$,
  (vi) $T_K = P_K U|K$.

From definitions (i)–(vi) and formula (11.1) it is not hard to derive successively that

$$P_K = I - \pi P_+ \pi^* - \pi_* P_- \pi_*^*,$$

that U is a unitary operator on $\mathcal{H}$, that T_K is a completely nonunitary contraction (for this it is very useful to prove first the dual representation $U = \pi_* z\pi_*^* = \tau z\tau^*$), and, finally, that U is a minimal unitary dilation of T_K.

Conversely, every completely nonunitary contraction is unitarily equivalent to an operator of the form (vi). An operator T_K from (vi) is called the model operator, and K is called the model space. For explicit formulae of the model space we refer to [17, 8, 13], and for a quick self-contained construction to [19].

We conclude by indicating formulae connecting the abstract form of the model described above with Sz.-Nagy–Foiaş's and Pavlov's forms:

(SzNF): $\quad \mathcal{H} = L^2(E_*) \oplus \text{clos}\, \Delta L^2(E),$

$$\pi f = \theta f \oplus \Delta f, \qquad \pi^*(h \oplus k) = \theta^* h + \Delta k,$$

$$\pi_* g = g \oplus \mathbb{0}, \qquad \pi_*^*(h \oplus k) = h,$$

$$\tau f = \mathbb{0} \oplus f, \qquad \tau^*(h \oplus k) = k,$$

$$\tau_* g = \Delta_* g - \theta^* g, \qquad \tau_*^*(h \oplus k) = \Delta_* h - \theta k.$$

(P): $\quad \mathcal{H} = L^2(E \oplus E_*, \overline{\Delta}), \quad \overline{\Delta}^2 = \begin{pmatrix} I & \theta^* \\ \theta & I \end{pmatrix},$

$$\pi f = f \oplus \mathbb{0}, \qquad \pi^*(h \oplus k) = h + \theta^* k,$$

$$\pi_* g = \mathbb{0} \oplus g, \qquad \pi_*^*(h \oplus k) = \theta h + k,$$

$$\tau f = \Delta^{-1} f \oplus (-\theta \Delta^{-1} f), \qquad \tau^*(h \oplus k) = \Delta h,$$

$$\tau_* g = (-\theta^* \Delta_*^{-1} g) \oplus \Delta_*^{-1} g, \qquad \tau_*^*(h \oplus k) = \Delta_* k.$$

12. Function models. To begin with, we give a straightforward proof of the fact that $\mathcal{D}(\theta)$ is an extension of $\mathcal{H}(\theta)$. The definition of $\mathcal{D}(\theta)$ is in §4.4. Recall that $Jf = \bar{z} f(\bar{z})$.

12.1. LEMMA. *The operator i,*

$$i(D_{\theta^*}^2 h) = \{D_{\theta^*}^2 h, JP_-\theta^* h\}, \qquad h \in H^2(E_*),$$

isometrically embeds the space $\mathcal{H}(\theta)$ into $\mathcal{D}(\theta)$. The adjoint is given by $i^*\{f, g\} = f$.

PROOF. For $n \geq 0$ we have

$$z^{n+1} D_{\theta^*} h - \theta P_+ z^{n+1} P_- \theta^* h$$
$$= z^{n+1} h - z^{n+1} \theta P_+ \theta^* h - \theta P_+ z^{n+1} \theta^* h + \theta z^{n+1} P_+ \theta^* h = D_{\theta^*}^2 z^{n+1} h.$$

Moreover,

$$\|D_{\theta^*}^2 z^{n+1} h\|_{\mathcal{H}(\theta)}^2 = \|D_{\theta^*} z^{n+1} h\|_2^2 = \|h\|_2^2 - \|P_+ \theta^* z^{n+1} h\|_2^2,$$

and the last difference has a finite limit as $n \to \infty$. Hence, $i(D_{\theta^*}^2 h) \in \mathcal{D}(\theta)$. If $\{f, g\} \in \mathcal{D}(\theta)$, we get

$$(\{f, g\}, i D_{\theta^*}^2 h)_{\mathcal{D}(\theta)} = \lim_n \{(z^{n+1} f - \theta P_+ z^{n+1} Jg, D_{\theta^*}^2 z^{n+1} h)_{\mathcal{H}(\theta)}$$
$$+ (P_+ z^{n+1} Jg, P_+ z^{n+1} P_- \theta^* h)_2\}$$
$$= \lim_n (f - \theta \bar{z}^{n+1} P_+ z^{n+1} Jg + \theta P_- \bar{z}^{n+1} P_+ z^{n+1} Jg, h)_2$$
$$= \lim_n (f - \theta P_+ \bar{z}^{n+1} P_+ z^{n+1} Jg, h)_2$$
$$= (f - \theta P_+ Jg, h)_2 = (f, h)_2 = (f, D_{\theta^*}^2 h)_{\mathcal{H}(\theta)}.$$

Thus $i^*\{f,g\} = f$, and from $i^*iD_{\theta*}^2.h = D_{\theta*}^2.h$ it follows that $i^*i = I$; i.e., i is an isometry. •

12.2. THEOREM. *Let $\theta \in H^\infty(E \to E_*)$ be a contractive operator function. Then the operator $\mathcal{U}$ defined by*

$$\mathcal{U}\{f,g\} = \{f, J(\theta^*f + \Delta g)\}$$

is a unitary mapping from K_θ to $\mathcal{D}(\theta)$; $\Delta = (I - \theta^*\theta)^{1/2}$. The operator $\mathcal{U}$ transforms the model operator M_θ into the operator $\mathcal{M}_\theta = \mathcal{U}M_\theta\mathcal{U}^{-1}$ on $\mathcal{D}(\theta)$ such that

(12.3) $$\mathcal{M}_\theta^*\{f,g\} = \{S_\theta^*f,\ Sg - \theta(\bar{z})^*g(0)\}$$

for $\{f,g\} \in \mathcal{D}(\theta)$. On a coordinate-free form of the model space, say $\mathcal{K}$ (see [11]), the operator $\mathcal{U}$ acts by the formula

(12.4) $$\mathcal{U}x = \{\pi_*^*x,\ J\pi^*x\}, \qquad x \in \mathcal{K}.$$

The proof of the theorem breaks into lemmas. We use a coordinate-free form of the model; see §11 for the notation. The dissatisfied reader can transform our proofs into Sz.-Nagy–Foiaş's or Pavlov's form using formulae (SzNF) and (P) from §11.

12.5. LEMMA. $\pi_*^*\mathcal{K} = \mathcal{H}(\theta)$.

PROOF.

$$\pi_*^*P_\mathcal{K}\pi_* = \pi_*^*(I - \pi P_+\pi^* - \pi_*P_-\pi_*^*)\pi_* = P_+ - \theta P_+\theta^*$$

on $L^2(E_*)$. Hence $\pi_*^*P_\mathcal{K}\pi_*|H^2(E_*) = D_{\theta*}^2$ and in view of Theorem 5.3 there exists a partial isometry U from $H^2(E_*)$ to $\mathcal{K}$ with initial space clos Range $D_{\theta*}$ and such that $P_\mathcal{K}\pi_*|H^2(E_*) = UD_{\theta*}$. Now, we have

$$\pi_*^*\mathcal{K} = P_+\pi_*^*\mathcal{K} = D_{\theta*}U^*\mathcal{K} = D_{\theta*}(\text{clos } D_{\theta*}H^2(E_*)) = \mathcal{H}(\theta). \quad \bullet$$

12.6. LEMMA. *Let $n \geq 0$ and*

$$A_n = z^{n+1}\pi_*^* - \theta P_+ z^{n+1}\pi^*$$

be an operator from $\mathcal{K}$ to $L^2(E_*)$. Then $A_k\mathcal{K} = \mathcal{H}(\theta)$.

PROOF.

$$A_nA_n^* = (z^{n+1}\pi_*^* - \theta P_+ z^{n+1}\pi^*)(\pi_*\bar{z}^{n+1} - \pi\bar{z}^{n+1}P_+\theta^*)$$
$$= I - \theta P_+ z^{n+1}\theta^*\bar{z}^{n+1} - z^{n+1}\theta\bar{z}^{n+1}P_+\theta^* + \theta P_+\theta^* = I - \theta P_+\theta^*.$$

Hence, $A_nA_n^*|H^2(E_*) = D_{\theta*}^2$, and in view of Theorem 5.3 there exists a partial isometry U_n from $H^2(E_*)$ to $\mathcal{K}$ with initial space clos Range $D_{\theta*}$ and such that $A_n^*|H^2(E_*) = U_nD_{\theta*}$. This yields

$$A_n\mathcal{K} = P_+A_n\mathcal{K} = D_{\theta*}U_n^*\mathcal{K} = D_{\theta*}(\text{clos } D_{\theta*}H^2(E_*)) = \mathcal{H}(\theta). \quad \bullet$$

12.7. LEMMA. $\mathcal{U}$ *is an isometry.*

PROOF. Lemmas 12.5 and 12.6 yield $\pi_*^* x \in \mathcal{H}(\theta)$ and $z^{n+1}\pi_*^* x - \theta P_+ z^{n+1}\pi^* x = A_n x \in \mathcal{H}(\theta)$ for every $x \in K$. Hence, a necessary condition in order that $\mathcal{U}x$ belong to $\mathcal{D}(\theta)$ is satisfied. Let us compute the limit involved in the definition of the norm $\|\mathcal{U}x\|_{\mathcal{D}(\theta)}$, namely, $\lim_n(a_n + b_n)$, where

$$a_n = \|A_n x\|_{\mathcal{H}(\theta)}^2, \qquad b_n = \|P_+ z^{n+1}\pi^* x\|_2^2.$$

By Lemmas 6.5–6.6 one obtains

$$a_n = \sup\{\|A_n x + \theta f\|_2^2 - \|f\|_2^2 : f \in H^2(E)\}$$
$$= \|A_n x\|^2 + \sup\{2\operatorname{Re}(\theta f, A_n x)_2 - \|\Delta f\|_2^2 : f \in H^2(E)\},$$

where (recall) $\Delta = (I - \theta^* \theta)^{1/2}$. Next, let us remark that

$$P_+ \theta^* A_n = P_+ z^{n+1} \theta^* \pi_*^* - P_+ \theta^* \theta P_+ z^{n+1} \pi^*$$
$$= P_+ \Delta^2 P_+ z^{n+1}\pi^* - P_+ z^{n+1}(\pi^* - \theta^* \pi_*^*)$$
$$= P_+ \Delta^2 P_+ z^{n+1}\pi^* - P_+ z^{n+1} \Delta \tau^*,$$

and hence, by the definition of A_n,

$$a_n = \|z^{n+1}\pi_*^* x - \theta P_+ z^{n+1}\pi^* x\|_2^2$$
$$+ \sup\{2\operatorname{Re}(f, \Delta^2 P_+ z^{n+1}\pi^* x - \Delta z^{n+1}\tau^* x)_2 - \|\Delta f\|_2^2 : f \in H^2(E)\}$$
$$= \|\pi_*^* x\|_2^2 - 2\operatorname{Re}(z^{n+1}\pi_*^* x, \theta P_+ z^{n+1}\pi^* x)_2 + \|\theta P_+ z^{n+1}\pi^* x\|_2^2$$
$$+ \sup\{-\|\Delta f - (\Delta P_+ z^{n+1}\pi^* x - z^{n+1}\tau^* x)\|_2^2$$
$$+ \|\Delta P_+ z^{n+1}\pi^* x - z^{n+1}\tau^* x\|_2^2 : f \in H^2(E)\}$$
$$= (\|\pi_*^* x\|_2^2 + \|\tau^* x\|_2^2) + (\|\theta P_+ z^{n+1}\pi^* x\|_2^2 + \|\Delta P_+ z^{n+1}\pi^* x\|_2^2)$$
$$- 2\operatorname{Re}(z^{n+1}\pi^* x, P_+ z^{n+1}(\theta^* \pi_*^* + \Delta \tau^*)x)$$
$$- \inf\{\|\Delta f - \Delta P_+ z^{n+1}\pi^* x + z^{n+1}\tau^* x\|_2^2 : f \in H^2(E)\}$$
$$= \|x\|^2 + \|P_+ z^{n+1}\pi^* x\|_2^2 - 2\|P_+ z^{n+1}\pi^* x\|_2^2$$
$$- \inf\{\|\Delta f + z^{n+1}\tau^* x\|_2^2 : f \in H^2(E)\} \quad \text{(using the formulae of §11).}$$

Since the last inf is just equal to $\operatorname{dist}(\tau^* x, \Delta \bar{z}^{n+1} H^2(E))$ and $\tau^* x \in \operatorname{clos} \Delta L^2(E_*)$, we get

$$\lim_n(a_n + b_n) = \|x\|^2.$$

Thus, $\mathcal{U}x \in \mathcal{D}(\theta)$ and $\|\mathcal{U}x\|_{\mathcal{D}(\theta)} = \|x\|$ for every $x \in K$. ●

12.8. LEMMA. $\mathcal{U}K = \mathcal{D}(\theta)$.

PROOF. Let $\{f, g\} \in \mathcal{D}(\theta) \ominus \mathcal{U}K$. Then $\{f, g\}$ is orthogonal to $\mathcal{U}x$, $x = P_K \pi_* f$. By the definitions of $\mathcal{U}$ and A_n and of the norm in $\mathcal{D}(\theta)$, this yields

$$0 = \lim_n\{(A_n x, z^{n+1} f - \theta P_+ z^{n+1} Jg)_{\mathcal{H}(\theta)} + (P_+ z^{n+1}\pi^* P_K \pi_* f, P_+ z^{n+1} Jg)\}.$$

Now, using the formula for P_K from §11, we can see that

$$P_K \pi_* = \pi_* P_+ - \pi P_+ \theta^*,$$

and hence
$$A_n x = A_n P_K \pi_* f = (z^{n+1}\pi_*^* - \theta P_+ z^{n+1}\pi^*)(\pi_* P_+ - \pi P_+ \theta^*)f$$
$$= z^{n+1}f - z^{n+1}\theta P_+ \theta^* f - \theta P_+ z^{n+1}\theta^* f + \theta P_+ z^{n+1} P_+ \theta^* f$$
$$= z^{n+1}f - \theta P_+ z^{n+1}\theta^* f = D_{\theta^*}^2 z^{n+1}f.$$

Substituting in the previous scalar product and taking into account that
$$\pi^* P_K \pi_* f = \theta^* f - P_+ \theta^* f = P_- \theta^* f,$$
we obtain
$$0 = \lim_n \{(z^{n+1}f, z^{n+1}f - \theta P_+ z^{n+1}Jg)_2 + (z^{n+1}P_-\theta^* f, P_+ z^{n+1}Jg)_2$$
$$= \|f\|_2^2 - \lim_n (z^{n+1}P_+\theta^* f, z^{n+1}Jg)_2 = \|f\|^2.$$

Thus, $f = \mathbb{0}$.

In particular, this means that $\{\mathbb{0}, g\}$ is orthogonal to $\mathcal{U}P_K \pi Jg$. A reasoning similar to the previous one shows that

(12.9)
$$0 = \lim_n \{(A_n P_K \pi Jg, -\theta P_+ z^{n+1} Jg)_{\mathcal{H}(\theta)}$$
$$+ (P_+ z^{n+1}\pi^* P_K \pi Jg, P_+ z^{n+1} Jg)_2\},$$
$$P_K \pi = \pi P_- - \pi_* P_- \theta, \quad \pi^* P_K \pi = P_- - \theta^* P_- \theta,$$
$$A_n P_K \pi Jg = z^{n+1} P_+ \theta Jg - \theta P_+ z^{n+1} Jg + \theta P_+ z^{n+1} \theta^* P_- \theta Jg$$
$$= (z^{n+1}\theta Jg - z^{n+1}P_-\theta Jg) - \theta P_+ z^{n+1}Jg + \theta P_+ \theta^* z^{n+1} P_- \theta Jg$$
$$= -D_{\theta^*}^2 P_+ z^{n+1} P_- \theta Jg + \theta P_- z^{n+1} Jg - P_- \theta z^{n+1} Jg.$$

The last two members of this expression can easily be estimated because $\theta P_- - P_- \theta = \theta P_- - P_- \theta P_- = P_+ \theta P_-$, and taking the Hermitian square we obtain
$$P_+ \theta P_- \theta^* P_+ = P_+ \theta\theta^* P_+ - P_+\theta P_+ \theta^* P_+ = D_{\theta^*}^2 P_+ - P_+(I - \theta\theta^*)P_+ \le D_{\theta^*}^2 P_+.$$
In view of Theorem 5.3 this means that $P_+\theta P_- = D_{\theta^*} C$ for a contraction C, and hence
$$\lim_n \|\theta P_- z^{n+1} Jg - P_- \theta z^{n+1} Jg\|_{\mathcal{H}(\theta)} = \lim_n \|D_{\theta^*} C P_- z^{n+1} Jg\|_{\mathcal{H}(\theta)}$$
$$\le \lim_n \|C P_- z^{n+1} Jg\|_2 = 0.$$

Now, one can finish the calculation of the right-hand side of (12.9):
$$0 = \lim_n \{(D_{\theta^*}^2 P_+ z^{n+1} P_- \theta Jg, \theta P_+ z^{n+1} Jg)_{\mathcal{H}(\theta)}$$
$$+ (P_+ z^{n+1} Jg - P_+ z^{n+1}\theta^* P_- \theta Jg, P_+ z^{n+1} Jg)_2\}$$
$$= \lim_n \{(P_+ z^{n+1} P_- \theta Jg, \theta P_+ z^{n+1} Jg)_2 + \|P_+ z^{n+1} Jg\|_2^2$$
$$- (P_+ z^{n+1}\theta^* P_- \theta Jg, P_+ z^{n+1} Jg)_2\}$$
$$= \lim_n \|P_+ z^{n+1} Jg\|_2^2 = \|g\|^2.$$

Thus, $g = \mathbb{0}$ and the lemma follows. ∎

12.10. LEMMA. $\mathcal{M}_\theta^*\{f,g\} = \{S_\theta^* f, S_\theta g - \theta(\bar{z})^* g(0)\}$.

PROOF. Following the general transcription of the model (see §11, we get $\mathcal{M}_\theta = \mathcal{U} T_K \mathcal{U}^{-1}$, where $\mathcal{U}$ is from (12.4), and hence

$$T_K^* x = -U^* x - \pi_* \bar{z}(\pi_*^* x)(0), \quad x \in K,$$

because $T_K^* = P_K U^* | K$ and $U^* = \pi \bar{z} \pi^* + \tau_* \bar{z} \tau_*^*$ (and, in the dual form, $U^* = \pi_* \bar{z} \pi_*^* + \tau \bar{z} \tau^*$). Let

$$\mathcal{U} x \stackrel{\text{def}}{=} \{\pi_*^* x, J\pi^* x\} = \{f, g\}, \quad \{f, g\} \in \mathcal{D}(\theta).$$

Then $T_K^* x = U^* x - \pi_* \bar{z} f(0)$,

$$\mathcal{U} T_K^* x = \{\pi_*^* U^* x - \bar{z} f(0), J(\pi^* U^* x - \theta^* \bar{z} f(0))\}$$

$$= \{\bar{z} f - \bar{z} f(0), z J\pi^* x - \theta(\bar{z})^* f(0)\} \quad \text{(using both forms of } U^* \text{ to check}$$

$$\text{that } \pi_*^* U^* = \bar{z} \pi_*^*, \ \pi^* U^* = \bar{z} \pi^*)$$

$$= \{S_\theta^* f, Sg - \theta(\bar{z})^* f(0)\}. \quad \bullet$$

REFERENCES

1. J. A. Ball, *Models for noncontractions*, J. Math. Anal. Appl. **52** (1975), 235–254.
2. ___, *Unitary perturbations of contractions*, Dissertation, Univ. of Virginia, 1973.
3. L. de Branges, *Perturbations of selfadjoint transformations*, Amer. J. Math. **84** (1962), 543–560.
4. ___, *Square summable power series*, Springer-Verlag (to appear).
5. L. de Branges and J. Rovnyak, "Canonical models in quantum scattering theory," in *Perturbation theory and its applications in quantum mechanics*, edited by C. H. Wilcox, Wiley, New York, 1966, pp. 295–392.
6. L. de Branges and L. Schulman, *Perturbations of unitary transformations*, J. Math. Anal. Appl. **23** (1968), 294–326.
7. R. G. Douglas, *On majorization, factorization and range inclusion of operators on Hilbert spaces*, Proc. Amer. Math. Soc. 17 (1966), 413–415.
8. ___, "Canonical models," in *Topics in operator theory*, edited by C. Pearcy, Math. Surveys, no. 13, Amer. Math. Soc., Providence, R. I., 1974, pp. 161–218.
9. C. Foiaş, *Invariant para-closed subspaces*, Indiana Univ. Math. J. **21** (1972), 887–906.
10. H. Helson, *Lectures on invariant subspaces*, Academic Press, New York, 1964.
11. N. G. Makarov and V. I. Vasyunin, "A model for noncontractions and stability of the continuous spectrum," in *Complex analysis and spectral theory*, Lecture Notes in Math., vol. 864, Springer-Verlag, 1981, pp. 365–412.
12. N. K. Nikol'skiĭ, *Treatise on the shift operator*, Springer-Verlag, 1985.
13. B. S. Pavlov, *On separation conditions for spectral components of a dissipative operator*, Izv. Akad. Nauk SSSR Ser. Mat. **39** (1975), 123–148. (Russian)
14. Yu. A. Rozanov, *Stationary stochastic processes*, Gusudarstv. Izdat. Fiz.-Mat. Lit., Moscow, 1963. (Russian)
15. L. A. Shulman, *Perturbations of unitary transformations*, J. Math. Anal. Appl. **28** (1969), 231–254.
16. B. Sz.-Nagy and C. Foiaş, *Sur les contractions de l'espace de Hilbert. VIII: Fonctions charactéristiques: Modèles fonctionnels*, Acta Sci. Math. (Szeged) **25** (1964), 38–71.
17. ___, *Analyse harmonique des opèrateurs de l'espace de Hilbert*, Masson, Paris, Akad. Kiadó, Budapest, 1967.
18. ___, *On the structure of intertwining operators*, Acta Sci. Math. (Szeged) **35**, (1973), 225–254.

19. V. I. Vasyunin, "Construction of the function model of B. Sz.-Nagy and C. Foiaş," in *Investigations on linear operators and the theory of functions.* VIII, Zap. Naučn. Sem. Leningrad, Otdel. Mat. Inst. Steklov (LOMI) **73** (1977), 16–23.

Added in proof. In September 1985 Donald Sarason kindly supplied us with two interesting manuscripts ("Shift-Invariant Spaces from the Brangesian Point of View" and "Doubly Shift-Invariant Spaces in H^2").[1] They are devoted to calculation of the characteristic function of S_θ^{**}, to the question of the S-invariance of the space $\mathcal{H}(\theta)$, and to some other questions (all for the scalar case $\dim E = \dim E_* = 1$).

These papers suggested the following complements to our constructions.

1. *The characteristic function of S_θ^{**}*. Put $\Phi = \vartheta^* \Delta$. In view of (9.2) this is a contractive outer H^∞-operator function (in fact, the maximal analytic function such that $\Phi^*\Phi \leq \Delta^2$ a.e. on T). The factorization $\theta = (I, \mathbb{O})\binom{\theta}{\Phi}$ is regular (in the sense of Sz.-Nagy–Foiaş), and its second factor $\Psi \stackrel{\text{def}}{=} \binom{\theta}{\Phi}$ corresponds to the M_θ-invariant subspace K_θ^*. Therefore (cf. Theorem 10.8) the pure part of Ψ is the characteristic function of S_θ^{**}.

2. By the way, one can easily rewrite some identities of the paper in terms of Φ and Ψ. For instance, the infimum from Lemma 8.2 is equal to $\|\Phi(0)e\|^2$, and so the lemma itself and Theorem 8.7 can be rewritten as follows: $\|je\|^2 = \|e\|^2 - \|\Psi(0)e\|^2$ (or $j^*j = I - \Psi^*\Psi$) and $\|D_{S_\theta^*} f\|^2_{\mathcal{H}(\theta)} = \|f(0)\|^2 + \|\Phi(0)g(0)\|^2$ (in the notation of Theorem 8.7).

3. *S-invariance of $\mathcal{H}(\theta)$*. By Theorem 8.4, $S\mathcal{H}(\theta) \subset \mathcal{H}(\theta)$ iff $\theta j^* \mathcal{H}(\theta) \subset \mathcal{H}(\theta)$. Let E' be the subspace of E such that the restriction $\Psi|E'$ is the pure part of Ψ. Then in the case $\dim E < \infty$, $j^*\mathcal{H}(\theta) = j^*jE = E'$, and $\mathcal{H}(\theta)$ is S-invariant iff $\theta E' \subset \mathcal{H}(\theta)$, or—in terms of Φ—iff $E' \subset \Phi(0)^* E_{**}$. This criterion can be stated as follows:
$$\int_T \log \det \Delta \, dm > -\infty.$$

4. *Similarity of S_θ^{**} and S*. A criterion of similarity to an isometry is given in Theorem 2.4 [**18**]: a contraction is similar to an isometry if and only if its characteristic function has a bounded analytic left-inverse. It is clear that this isometry is a unilateral shift iff the given operator is of the class C_{10}. So if $\dim E = \dim E_* = 1$, S_θ^{**} is similar to S iff $\inf_\mathbb{D}\{|\theta| + |\Phi|\} > 0$.

---

[1] The first manuscript is included in these Proceedings.

The Growth of the Derivative of a Univalent Function

CH. POMMERENKE

1. Introduction. Let the function f be analytic and univalent in the unit disk D. It is not an essential restriction to assume that $f \in S$, i.e., that $f(z) = z + a_2 z^2 + z_3 z^3 + \cdots$ for $z \in \mathrm{D}$.

de Branges [9] proved in 1984 the famous Bieberbach conjecture that $|a_n| \leq n$ for $n = 2, 3, \ldots$ with equality if and only if f is a rotation of the Koebe function. One classical way of estimating coefficients is through the integral mean

$$(1.1) \quad n|a_n| \leq \frac{1}{2\pi r^{n-1}} \int_0^{2\pi} |f'(re^{it})|\, dt, \qquad r = \frac{n}{n+1}.$$

This method gives the correct order $a_n = O(n)$ but is too crude to prove $|a_n| \leq n$.

Baernstein [1] has recently used (1.1) to extend the relation

$$(1.2) \quad f(z) = O\left(\frac{1}{(1-z)^\alpha}\right) \quad \Leftrightarrow \quad a_n = O(n^{\alpha-1})$$

from the range $1/2 < \alpha \leq 2$ to $1/2 - 1/320 < \alpha \leq 2$. This is surprising because $1/2$ had been suspected to be a natural limit for (1.2) to be true.

We shall report on some recent progress on two problems, namely,

(A) the growth of f' on almost all radii, and

(B) the growth of the integral means of f'.

Makarov [19, 20, 22] has obtained remarkable results on problem (A) and has applied these results to give almost complete answers to the interesting and difficult question of how harmonic measure is connected with Hausdorff measure. There are only partial answers to (B); in particular there is an interesting conjecture of Brennan [4].

2. Bloch functions. The function g is called a Bloch function if it is analytic in D and if

$$(2.1) \quad \|g\|_\mathcal{B} \equiv |g(0)| + \sup_{|z|<1} (1 - |z|^2)|g'(z)| < \infty.$$

If $f \in S$, then

$$(2.2) \quad g(z) = \log f'(z)$$

is a Bloch function with $g(0) = 0$ and

$$\|g\|_\mathcal{B} = \sup_{|z|<1} (1-|z|^2) \left|\frac{f''(z)}{f'(z)}\right| \leq 6 \tag{2.3}$$

by a classical result of Bieberbach [3]. In the converse direction, the Becker univalence criterion [12, 2] shows that, with $g = \log f'$,

$$\|g\|_\mathcal{B} \leq 1, \quad g(0) = 0 \quad \Rightarrow \quad f \in S. \tag{2.4}$$

The following result is essentially due to Makarov [19].

THEOREM 1. *If f is a Bloch function with $g(0) = 0$, then*

$$\frac{1}{2\pi} \int_0^{2\pi} |g(re^{it})|^{2n}\, dt \leq n! \|g\|_\mathcal{B}^{2n} \left(\log \frac{1}{1-r^2}\right)^n \tag{2.5}$$

for $0 \leq r < 1$ and $n = 1, 2, \ldots$.

Clunie and MacGregor [7] independently proved a weaker version of (2.5), and Makarov's result has a factor A^n on the right-hand side. The present form of (2.5) was proved by Korenblum (unpublished) by a method related to that of [7].

PROOF. We write

$$\lambda(r) = \log \frac{1}{1-r^2} \tag{2.6}$$

and prove by induction that

$$J_n(r) \equiv \frac{1}{2\pi} \int_0^{2\pi} |g(re^{it})|^{2n}\, dt \leq n! \lambda(r)^n \|g\|^{2n}. \tag{2.7}$$

The case $n = 0$ is trivial. Let $n \geq 1$.

Assume that (2.6) holds with n replaced by $n-1$. The Hardy identity shows that

$$\frac{d}{dr}[rJ_n'(r)] = \frac{4n^2 r}{2\pi} \int_0^{2\pi} |g|^{2n-2} |g'|^2\, dt.$$

Since $|g'(re^{it})| \leq \|g\|/(1-r^2)$, we obtain from the induction hypothesis that

$$\frac{d}{dr}[rJ_n'(r)] \leq \frac{4n^2 r}{(1-r^2)^2} J_{n-1}(r) \|g\|^2 \tag{2.8}$$

$$\leq \frac{4n \cdot n! r}{(1-r^2)^2} \lambda(r)^{n-1} \|g\|^{2n}$$

$$\leq n! \frac{d}{dr}\left[r \frac{d}{dr} \lambda(r)^n\right] \|g\|^{2n},$$

and (2.7) follows by integration because $J_n(0) = \lambda(0) = 0$.

THEOREM 2. *If g is a Bloch function, then, for almost all t,*

$$\limsup_{r \to 1} \frac{|g(re^{it})|}{\sqrt{\log \frac{1}{1-r} \log\log\log \frac{1}{1-r}}} \leq \|g\|_\mathcal{B}. \tag{2.9}$$

This theorem is due to Makarov [19] except that the estimate of Makarov has an absolute constant on the right-hand side. Korenblum [18] independently proved the same estimate with log log log replaced by log log.

PROOF. Let $\|g\|_B = 1$, and let

(2.10) $\qquad g^*(r,t) = \max_{0 \le \rho \le r} |g(\rho e^{it})| \qquad (0 \le t \le 2\pi, \ 0 \le r < 1).$

Let $\lambda(r)$ be defined by (2.6), and let $K_1, K_2, \ldots$ denote constants. It follows from the Hardy-Littlewood maximal theorem [10, p. 12–13] for analytic functions and from Theorem 1 that

(2.11) $\qquad \int_0^{2\pi} g^*(re^{it})^{2n} \, dt \le K_1 \int_0^{2\pi} |g(re^{it})|^{2n} \, dt \le K_2 n! \lambda(r)^n.$

We write

(2.12) $\qquad \psi_n(r) = n\dfrac{d}{dr}(\log \lambda(r))^{-1/n} = -\dfrac{2r}{1-r^2}\lambda(r)^{-1}(\log \lambda(r))^{-1-1/n}.$

It follows from (2.11) by Fubini's theorem that

$$\int_0^{2\pi}\int_{1/2}^1 g^*(r,t)^{2n} \frac{\psi_n(r)}{\lambda(r)^n}\, dr\, dt \le K_3 n! n.$$

Hence there exists a set $A_n \subset \partial D$ with

(2.13) $\qquad \operatorname{meas} A_n > 2\pi - 1/n^2$

such that, for $e^{it} \in A_n$,

$$\int_{1/2}^1 g^*(r,t)^{2n} \frac{\psi_n(r)}{\lambda(r)^n}\, dr \le K_3 n! n^3.$$

It follows that, for $e^{it} \in A_n$ and $1/2 < r < 1$,

$$\frac{g^*(r,t)^{2n}}{\lambda(r)^n(\log \lambda(r))^{1+1/n}} \le K_4 g^*(r,t)^{2n} n \int_r^1 \frac{\psi_n(\rho)}{\lambda(\rho)^n}\, d\rho \le K_5 n! n^4$$

because $g^*(r,t)$ increases with r, by (2.10). Hence we obtain from Stirling's formula that

(2.14) $\qquad |g(re^{it})|^{2n} \le g^*(r,t)^{2n} \le K_6 e^{-n} n^{n+5} \lambda(r)^n (\log \lambda(r))^{1+1/n}.$

Now let $A = \bigcup_{k=1}^\infty \bigcap_{n=k}^\infty A_n$. We see from (2.13) that $\operatorname{meas} A = 2\pi$. Let $e^{it} \in A$. Then $e^{it} \in A_n$ for $n \ge k$ for suitable k. Let $r < 1$ be so large that $r > r_0$ and $\log \log \lambda(r) > k + 1$. We consider

(2.15) $\qquad n \equiv [\log \log \lambda(r)] \ge k$

and obtain from (2.14) that

$$|g(re^{it})|^{2n} \le K_6 e^{-n} n^{n+5} \lambda(r)^n \exp[(1+1/n)(n+1)].$$

It follows that

$$\limsup_{r \to 1} \frac{|g(re^{it})|^2}{\lambda(r)n} \le e^{-1} \cdot e = 1,$$

and this is equivalent to (2.9) in view of (2.15) and (2.6).

Bloch functions of extreme behavior are given by functions with Hadamard gaps, for instance by

$$(2.16) \qquad g_q(z) = \frac{1}{B_q} \sum_{k=1}^{\infty} z^{q^k}, \qquad q = 2, 3, \ldots,$$

where B_q is chosen such that $\|g_q\|_\mathcal{B} = 1$. It follows from the law of the iterated logarithm for gap series of M. Weiss [30] that

$$(2.17) \qquad \limsup_{r \to 1} \frac{|g_q(re^{it})|}{\sqrt{\log \frac{1}{1-r} \log \log \log \frac{1}{1-r}}} = \frac{1}{B_q \sqrt{\log q}} \quad \text{for almost all } t;$$

(Makarov [19], except for the constant). The optimal choice is $q = 15$. Since $B_q < 0.886$ [14], we obtain that the right-hand side of (2.17) is > 0.685. Thus (2.9) is off by this factor at most.

3. Hausdorff measure: compression. Let $h(t)$ ($t \geq 0$) be a continuous strictly increasing function with $h(0) = 0$. For $E \subset \mathbb{C}$, the outer Hausdorff measure with respect to h is defined by

$$(3.1) \qquad \Lambda_h(E) = \lim_{\delta \to 0} \inf_{(D_j)} \sum_j h(\operatorname{diam} D_j),$$

where the infimum is taken over all countable collections of disks D_j such that

$$(3.2) \qquad E \subset \bigcup_j D_j, \qquad \operatorname{diam} D_j < \delta \qquad (j = 1, 2, \ldots).$$

In particular, $\Lambda_h(E) = 0$ if and only if, for every $\varepsilon > 0$, there exist disks D_j of radii r_j such that

$$(3.3) \qquad E \subset \bigcup_j D_j, \qquad \sum_j \operatorname{diam} D_j < \varepsilon.$$

Perhaps the most interesting case is $h(t) = t$, linear Hausdorff measure. A curve has finite linear measure if and only if it is rectifiable. We write Λ_α instead of Λ_h if $h(t) = t^\alpha$.

The following remarkable result is due to Makarov [19].

THEOREM 3. *Let f map $\mathbb{D}$ conformally onto a Jordan domain, and let*

$$(3.4) \qquad h(t) = t \exp\left(K \sqrt{\log \frac{1}{t} \log \log \log \frac{1}{t}}\right),$$

where K is a certain absolute constant. Then, for $A \subset \partial \mathbb{D}$,

$$(3.5) \qquad \Lambda_h(f(A)) = 0 \quad \Rightarrow \quad \operatorname{meas} A = 0.$$

This theorem is amazingly precise. In a paper of Carleson [5], a breakthrough in 1973, it was shown that (3.5) holds with $h(t) = t^\beta$ for some $\beta > \frac{1}{2}$, whereas Makarov's theorem shows that (3.5) holds for $h(t) = t^\beta$ and *all* $\beta < 1$.

Makarov [19] (see also Kaufman and Wu [17]) has used the gap series (2.16) to show that there exist f and $A \subset \partial D$ such that

$$\Lambda_h(f(A)) = 0, \quad \text{meas } A = 2\pi,$$

where h is given by (3.4) with a different constant K. Makarov has therefore solved the problem (3.5) completely except for the determination of the best possible value of K.

The proof of Theorem 3 uses Theorem 2 as one basic tool. It follows from (2.9) and (2.3) applied to $g = \log f'$ that

$$\liminf_{r \to 1} \frac{\log |f'(re^{it})|}{\sqrt{\log \frac{1}{1-r} \log \log \log \frac{1}{1-r}}} \geq -6 \quad \text{for almost all } t.$$

This is a lower estimate for $|f'|$ on almost all radii. We refer to Makarov's paper [19] for the proof of Theorem 3.

4. Hausdorff measure: expansion. Let f be analytic and univalent in D. Then the angular limit $f(\varsigma)$ exists at almost all points $\varsigma \in \partial D$. Øxendal [25] posed the question whether, for all $\alpha > 1$, there exists a measurable set $A \subset \partial D$ such that

(4.1) $\quad\quad\quad\quad \text{meas } A = 2\pi, \quad \Lambda_\alpha(f(A)) = 0.$

This was proved by Carleson [6] for snowflake curves.

Makarov [19] has now completely settled the Øxendal problem and has even proved much more. The only assumption he makes is that $h(t) = o(t)$ at $t \to 0$. This includes the case $h(t) = t^\alpha$, $\alpha > 1$, but excludes the case $h(t) = t$ of linear measure where (4.1) is in general false.

Makarov [20] has also stated a result (see Theorem 5 below) that deals with the case of linear measure. This result can be extended [29] to the general case.

THEOREM 4. *There exists a measurable set $A \subset \partial D$ with meas $A = 2\pi$ such that $f(A)$ has σ-finite linear measure, i.e., the set $f(A)$ is the union of countably many sets of finite linear measure.*

This implies the Makarov result mentioned above because if E is of σ-finite linear measure, then $\Lambda_h(E) = 0$ if $h(t) = o(t)$ $(t \to 0)$.

The proof of Theorem 4 is based on two lemmas which will not be proved here. The first lemma is essentially due to Makarov [20].

LEMMA 1. *Let E be the set of $\varsigma \in \partial D$ such that*

(4.2) $\quad\quad\quad\quad \liminf_{r \to 1} |f'(r\varsigma)| = 0.$

Then there is a set $A_0 \subset E$ such that

(4.3) $\quad\quad\quad\quad \text{meas } A_0 = \text{meas } E, \quad \Lambda_1(f(A_0)) = 0.$

It was Makarov [19, 20] who realized that a limit inferior condition suffices to construct sets with not-too-large image. Plessner's theorem and the Koebe distortion theorem show that, for almost all $\varsigma \in \partial D$, one of the following two alternatives holds:

(i) $f'(\varsigma) = \lim_{r \to 1} f'(r\varsigma) \neq 0, \infty$ exists, or
(ii) $\liminf_{r \to 1} |f'(r\varsigma)| = 0$.

Hence Lemma 1 implies the following result of Makarov [20].

THEOREM 5. *If $f'(\varsigma)$ exists almost nowhere, then there is a set $A_0 \subset \partial D$ such that*

(4.4) $$\operatorname{meas} A_0 = 2\pi, \quad \Lambda_1(f(A_0)) = 0.$$

Thus harmonic measure is concentrated on a set of linear measure zero if the boundary is sufficiently bad. The first example of a function f with a set E_0 satisfying (4.4) was constructed by McMillan and Piranian [24].

Theorem 5 explains why many computer-plotted curves of bad conformal mappings (e.g. in iteration theory) are so incomplete: almost all points plotted lie in a set of zero linear measure, and only the limited resolution of the output improves the picture somewhat.

LEMMA 2. *If*

(4.5) $$A_1 = \{\varsigma \in D : f'(\varsigma) \text{ exists } \neq 0, \infty\},$$

then $f(A_1)$ has σ-finite linear measure.

PROOF OF THEOREM 4. Let E and A_1 be defined by (4.2) and (4.5). We construct $A_0 \subset E$ as in Lemma 1. If $A = A_0 \cup A_1$, then $f(A) = f(A_0) \cup f(A_1)$ has σ-finite linear measure by Lemma 1 and Lemma 2. Furthermore $\operatorname{meas} A = \operatorname{meas} A_0 + \operatorname{meas} A_1 = 2\pi$ by Plessner's theorem as we remarked above.

The function $f(\varsigma)$ is absolutely continuous on A in the sense that

$$B \subset A, \quad \operatorname{meas} B = 0 \quad \Rightarrow \quad \Lambda_1(f(B)) = 0.$$

Indeed, if $B \subset A$ and $\operatorname{meas} B = 0$, then

$$\Lambda_1(f(B)) \leq \Lambda_1(f(A_0)) + \Lambda_1(B \cap A_1) = 0$$

by Lemma 1 and by a result of McMillan [23] about A_1.

5. Integral means: estimates. We come now to Problem (B). For $f \in S$ and $\lambda \in \mathbb{R}$, let

(5.1) $$I_\lambda(r, f') = \frac{1}{2\pi} \int_0^{2\pi} |f'(re^{it})|^\lambda \, dt \quad (0 \leq r < 1).$$

Since $(1-r)/8 < |f'(re^{it})| < 2/(1-r)^3$, we see that

$$I_\lambda(r, f') \leq \begin{cases} 2^\lambda (1-r)^{-3\lambda} & \text{if } \lambda > 0, \\ 8^{|\lambda|}(1-r)^{-|\lambda|} & \text{if } \lambda < 0. \end{cases}$$

We define

(5.2) $$\beta(\lambda) = \limsup_{r \to 1-0} \max_{f \in S} \frac{\log I_\lambda(r, f')}{\log \frac{1}{1-r}},$$

so that trivially $\beta(\lambda) \leq 3\lambda$ for $\lambda > 0$ and $\beta(\lambda) \leq |\lambda|$ for $\lambda < 0$. It follows from Hölder's inequality that $\beta(\lambda)$ is convex.

Feng and MacGregor [13] have proved that

(5.3) $$\beta(\lambda) = 3\lambda - 1 \quad \text{for } 2/5 < \lambda < +\infty.$$

Brennan [4] has made the very interesting conjecture that

$$\iint_D |f'(z)|^{-\alpha}\, dx\, dy < \infty \quad \text{for } \alpha < 2;$$

this is (except for the uniformity involved in (5.2)) equivalent to

(5.4) $$\beta(-2) = 1 \quad \text{(Brennan conjecture)}.$$

He proved that $\beta(-2) < 2$. The following result is proved in [27].

THEOREM 6. *For all λ,*

(5.5) $$\beta(\lambda) \leq -1/2 + \lambda + \sqrt{1/4 - \lambda + 4\lambda^2};$$

furthermore,

(5.6) $$\beta(\lambda) < |\lambda| - 0.399 \quad \text{for } -\infty < \lambda \leq -1.$$

The right-hand side of (5.5) is $\sim 3\lambda^2$ as $\lambda \to 0$; this is a slight improvement of the estimate $\beta(\lambda) \leq 9\lambda^2$ proved in [8]. It follows from (5.6) that

$$1 \leq \beta(-2) < 1.601.$$

Proof of (5.6). Since $|f'(z)| \geq (1 - |z|)/8$, it is sufficient to prove (5.6) for $\lambda = -1$, that is, to show that $\beta(-1) < 0.601$.

We consider the analytic function

(5.7) $$w(z) = \frac{1}{\sqrt{f'(z)}} = \sum_{n=0}^{\infty} c_n z^n \quad (z \in D).$$

It follows from Parseval's formula that

(5.8) $$I(r) \equiv I_{-1}(r, f') = \sum_{n=0}^{\infty} |c_n|^2 r^{2n}.$$

Hence

(5.9) $$I''''(r) \leq 16 \sum_{n=2}^{\infty} n^2(n-1)^2 |c_n|^2 r^{2n-4} = \frac{16}{2\pi} \int_0^{2\pi} |w''(re^{it})|^2\, dt.$$

Computation shows that

(5.10) $$w'' = -w S_f/2, \quad \text{where } S_f \text{ is the Schwarzian derivative};$$

this simple relation is the reason why we choose $\lambda = -1$. Since

$$|S_f(z)| \leq \frac{6}{(1-|z|^2)^2} \quad \text{for } z \in D,$$

we obtain from (5.9) and (5.10) that

$$\text{(5.11)} \qquad I''''(r) \leq \frac{144}{(1-r^2)^4} I(r) < \frac{9.01}{(1-r)^4} I(r)$$

if $r_0 < r < 1$, where r_0 is an absolute constant < 1. The differential equation

$$v''''(r) = \frac{9.01}{(1-r)^4} v(r) \qquad (r_0 \leq r < 1)$$

has the positive solution

$$v(r) = c(1-r)^{-\alpha}, \qquad \alpha(\alpha+1)(\alpha+2)(\alpha+3) = 9.01,$$

and calculation shows that $\alpha < 0.601$. Now a comparison theorem (related to one of Herold [16]) shows that

$$I(r) < c(1-r)^{-\alpha} \quad \text{for } r_0 \leq r < 1$$

if c is chosen large enough. Hence $\beta(-1) \leq \alpha < 0.601$.

In a new paper, Makarov [22] has proved that, for $0 < p < 1$, there exists $q > p/2$ such that

$$A \subset \partial D, \quad \Lambda_p(A) > 0 \quad \Rightarrow \quad \Lambda_q(f(A)) > 0.$$

For $0.601 < p < 1$ this follows from (5.6); the general case $0 < p < 1$ is proved by an argument of Carleson [5]. Makarov also remarks that the result would more easily follow if (in our notation) $\beta(-\gamma) = \gamma - 1$ holds for some $\gamma > 0$, in particular if the Brennan conjecture $\beta(-2) = 1$ is true. This increases the likelihood that the Brennan conjecture is true.

6. Integral means: examples.

At the moment, there are two types of examples with large $I_\lambda(r, f')$. One is the Koebe function $z(1-z)^{-2}$. It shows that

$$\beta(\lambda) \geq \begin{cases} 3\lambda - 1 & \text{for } \lambda > 1/3, \\ |\lambda| - 1 & \text{for } \lambda < -1, \end{cases}$$

as we see from (5.2); compare (5.3). The Brennan conjecture $\beta(-2) = 1$ is that, roughly speaking, the Koebe function is not too far from being extremal in the case $\lambda = -2$.

The other type of example comes from Hadamard gap series. Let

$$f'(z) = \exp\left(-\frac{i}{5} \sum_{k=0}^{\infty} z^{2^k}\right), \qquad f(0) = 0.$$

Then $f \in S$ by the Becker univalence criterion (2.4).

Makarov [21] has shown that, if $|\lambda| \leq 1$,

$$\text{(6.1)} \qquad I_\lambda(r, f') \geq (1-r)^{-c\lambda^2} \qquad (0 \leq r < 1),$$

where $c > 0.0003$. His ingenious proof uses ergodic theory and an extension of a lemma of Hawkes [15].

It follows from (6.1) that (5.3) cannot be extended to the full range $1/3 < \lambda < +\infty$. This gives a negative answer to questions by Gaier and Hayman [26, Problem 4.3] and Duren [11, p. 229].

A less sophisticated method of proof can be used with random gap series [28]. Let $q = 2, 3 \ldots$ and let

(6.2) $$f'(z) = \exp\left[\frac{1}{B_q} \sum_{k=0}^{\infty} \pm z^{q^k}\right], \qquad f(0) = 0,$$

where the signs are independent random variables with equal probability for each sign and where B_q is chosen as in (2.16). Thus $\|[\cdots]\| \leq 1$ for each choice of signs, and therefore $f \in S$ by (2.4).

THEOREM 7. *For all $\lambda \in \mathbb{R}$ and $q = 2, 3, \ldots,$*

(6.3) $$\beta(\lambda) \geq \frac{\log \cosh(\lambda/B_q)}{2 \log q}.$$

This estimate is obtained using the functions (6.2) with suitable choices of signs. Unfortunately the choice depends on r, and the method does not prove Makarov's result (6.1) which gives a function f independent of r and λ.

It follows from (6.3) together with (5.5) that

$$0.117 < \liminf_{\lambda \to 0} \frac{\beta(\lambda)}{\lambda^2} \leq \limsup_{\lambda \to 0} \frac{\beta(\lambda)}{\lambda^2} \leq 3;$$

here we choose $q = 15$ and use $B_{15} < 0.886$ [14]. We also obtain (with $q = 11$, $B_{11} < 0.946$) that $0.099 < \beta(-1) < 0.601$ where the upper estimate comes from (5.6).

References

1. A. Baernstein II, *Coefficients of univalent functions with restricted maximum modulus*, Complex Variables Theory Appl. **5** (1986), 225–236.
2. J. Becker, *Löwnersche Differentialgleichung und quasikonform fortsetzbare schlichte Funktionen*, J. Reine Angew. Math. **225** (1972), 23–43.
3. L. Bieberbach, *Über die Koeffizienten derjenigen Potenzreihen, welche eine schlichte Abbildung des Einheitskreises vermitteln*, S.-B. Preuss. Akad. Wiss., 1916, 940–955.
4. J. E. Brennan, *The integrability of the derivative in conformal mapping*, J. London Math. Soc. (2) **18** (1978), 261–272.
5. L. Carleson, *On the distortion of sets on a Jordan curve under conformal mapping*, Duke Math. J. **40** (1973), 547–559.
6. ____, *On the support of harmonic measure for sets of Cantor type*, Ann. Acad. Sci. Fenn. Ser. A I. Math. **10** (1985), 113–123.
7. J. G. Clunie and T. H. MacGregor, *Radial growth of the derivative of univalent functions*, Comment. Math. Helv. **59** (1984), 362–375.
8. J. Clunie and Ch. Pommerenke, *On the coefficients of univalent functions*, Michigan Math. J. **14** (1967), 71–78.
9. L. de Branges, *A proof of the Bieberbach conjecture*, Acta Math. **154** (1985), 137–152.
10. P. L. Duren, *Theory of H^p spaces*, Academic Press, 1970.
11. ____, *Univalent functions*, Springer-Verlag, 1983.
12. P. L. Duren, H. S. Shapiro, and A. L. Shields, *Singular measures and domains not of Smirnov type*, Duke Math. J. **33** (1966), 247–254.

13. J. Feng and T. H. MacGregor, *Estimates on integral means of the derivatives of univalent functions*, J. Analyse Math. **29** (1976), 203–231.

14. D. Gnuschke-Hauschild and Ch. Pommerenke, *On Bloch functions and gap series*, Preprint TU, Berlin, 1985.

15. J. Hawkes, *Probabilistic behaviour of some lacunary series*, Z. Wahrsch. Verw. Gebiete **53** (1980), 21–33.

16. H. Herold, *Ein Vergleichssatz für komplexe lineare Differentialgleichungen*, Math. Z. **126** (1972), 91–94.

17. R. Kaufman and J.-M. Wu, *Distortion of the boundary under conformal mapping*, Michigan Math. J. **29** (1982), 267–280.

18. B. Korenblum, *BMO estimates and radial growth of Bloch functions*, Bull. Amer. Math. Soc. **12** (1985), 99–102.

19. N. G. Makarov, *On the distortion of boundary sets under conformal mappings*, Proc. London Math. Soc. **51** (1985), 369–384.

20. ___, *Defining subsets, the support of harmonic measure, and perturbations of the spectra of operators in Hilbert space*, Soviet Math. Dokl. **29** (1984), 103–106.

21. ___, *A note on integral means of the derivative in conformal mapping*, Proc. Amer. Math. Soc. **96** (1986), 233–236.

22. ___, *Two remarks on integral means of the derivative in conformal mapping*, Preprint, Leningrad, 1985.

23. J. E. McMillan, *Boundary behavior of a conformal mapping*, Acta Math. **123** (1969), 43–67.

24. J. E. McMillan and G. Piranian, *Compression and expansion of boundary sets*, Duke Math. J. **40** (1973), 599–605.

25. B. Øxendal, *Brownian motion and sets of harmonic measure zero*, Pacific J. Math. **95** (1981), 179–192.

26. Ch. Pommerenke (ed.), *Problems in complex function theory*, Bull. London Math. Soc. **4** (1972), 354–366.

27. ___, *On the integral means of the derivative of a univalent function*, J. London Math. Soc. **32** (1985), 254–258.

28. ___, *On the integral means of the derivative of a univalent function. II*, Bull. London Math. Soc. **17** (1985), 565–570.

29. ___, *On conformal mapping and linear measure*, J. Analyse Math. (to appear).

30. M. Weiss, *The law of the iterated logarithm for lacunary trigonometric series*, Trans. Amer. Math. Soc. **91** (1959), 444–469.

Shift-Invariant Spaces from the Brangesian Point of View

DONALD SARASON

1. Introduction. In the manuscript [2], L. de Branges establishes in Theorem 15 an extension of A. Beurling's theorem on shift-invariant subspaces of the Hardy space H^2 of the unit disk [1]. Beurling's theorem states that the nontrivial shift-invariant subspaces of H^2 are the subspaces bH^2 with b an inner function. In de Branges's theorem, the shift-invariant subspace of Beurling's theorem is replaced by a shift-invariant Hilbert space contained contractively in H^2; that change entails the replacement of the inner function of Beurling's theorem by a general function in the unit ball of H^∞. de Branges's Theorem 15 in fact is stated and proved for vector-valued H^2 spaces and so extends not only Beurling's original theorem but also its vector-valued generalization due to P. Lax [8] and P. R. Halmos [5].

According to the theory developed in [2], a Hilbert space contained contractively in another has associated with it a complementary space, which is just the ordinary orthogonal complement in the special case where the first space is an ordinary subspace of the second one. If the first space is one of the de Branges shift-invariant spaces, its complementary space is invariant under the backward shift, and the backward shift acts as a contraction in it. The restrictions of the backward shift to such subspaces thus comprise a natural class of contractions that could possibly be useful as model operators.

The main purpose of this article is to analyze, for scalar-valued H^2, the structure of the complementary spaces of de Branges shift-invariant spaces and the structure of the restrictions of the backward shift to those complementary spaces. An examination of the same spaces and operators is carried out in the book [3] of de Branges and J. Rovnyak, but the approach here, emphasizing the role of Toeplitz operators, seems to lead a bit farther in some directions.

The paper is reasonably self-contained, at the expense of some duplication (from an altered viewpoint) of preliminary material from [2] and [3]. §§2 and 4 review the basic facts about contractively contained Hilbert spaces and their complementary spaces, respectively. §3 concerns de Branges's extension of Beurling's theorem. As already mentioned, this result is established in [2] in a vector-valued context. A simple proof of the scalar-valued version will be sketched here. The remainder of the paper, §§5-10, is devoted to the analysis mentioned in the last paragraph. The end result is an explication of how the model operators in de Branges's theory, for the scalar-valued situation, relate to the model operators of B. Sz.-Nagy and C. Foiaş [9].

The author is grateful to Professor de Branges for an advance copy of [2].

NOTATIONS AND CONVENTIONS. The open unit disk will be denoted by D. The space H^2 will be thought of either as the space of holomorphic functions in D with square-summable Maclaurin coefficients or as the corresponding space of boundary functions, according to convenience. The space H^∞ of bounded holomorphic functions in D will also be identified with its space of boundary functions. The norm and inner product in H^2 (or in L^2 of ∂D) will be denoted by $\|\ \|_2$ and $\langle\ ,\ \rangle$.

The unilateral shift operator in H^2 will be denoted by S $((Sf)(z) = zf(z))$. For u in L^∞ of ∂D, the Toeplitz operator with symbol u will be denoted by T_u; this is the operator on H^2 that sends f to $P(uf)$, where P is the orthogonal projection of L^2 onto H^2. The adjoint of T_u is $T_{\bar u}$.

If x and y are vectors in a Hilbert space, then $x \otimes y$ will denote the rank-one operator on that space that sends the vector w to the vector $\langle w, y\rangle x$.

2. Contractively contained Hilbert spaces.

The Hilbert space $\mathcal{M}$ is said to be contained contractively in the Hilbert space $\mathcal{K}$ if $\mathcal{M}$ is a vector subspace of $\mathcal{K}$ and if the inclusion map of $\mathcal{M}$ into $\mathcal{K}$ is a contraction (i.e., $\|x\|_\mathcal{K} \leq \|x\|_\mathcal{M}$ for all x in $\mathcal{M}$). Given the Hilbert space $\mathcal{K}$, there is a Hilbert space contained contractively in it associated with each contraction operator B from some Hilbert space $\mathcal{J}$ into $\mathcal{K}$. The space associated with B is denoted by $\mathcal{M}(B)$; as a vector space it equals the range of B, and one defines the norm in it by setting $\|Bx\|_{\mathcal{M}(B)} = \|x\|_\mathcal{J}$ whenever x belongs to the orthogonal complement of the kernel of B. If $\mathcal{M}$ is any Hilbert space contained contractively in $\mathcal{K}$, then clearly $\mathcal{M} = \mathcal{M}(B)$ with B equal to the inclusion map of $\mathcal{M}$ into $\mathcal{K}$.

If A and B are two contraction operators from Hilbert spaces into $\mathcal{K}$, one easily checks that $\mathcal{M}(A)$ and $\mathcal{M}(B)$ are equal (as Hilbert spaces) if and only if $AA^* = BB^*$. In particular, $\mathcal{M}(B) = \mathcal{M}((BB^*)^{1/2})$. The space $\mathcal{M}(B)$ is an ordinary subspace of $\mathcal{K}$ if and only if B is a partial isometry.

3. Brangesian shift-invariant spaces.

If b is a function in the unit ball of H^∞, the Hilbert space $\mathcal{M}(T_b)$ will be denoted by $\mathcal{M}(b)$. It is evident that $\mathcal{M}(b)$ is S-invariant and that S acts as an isometry in $\mathcal{M}(b)$. The converse is de Branges's extension of Beurling's theorem (the scalar-valued version): *Let $\mathcal{M}$ be a Hilbert space contained contractively in H^2 such that $S\mathcal{M} \subset \mathcal{M}$ and such that S acts as an isometry in $\mathcal{M}$. Then there is a function b in the unit ball of H^∞ such that $\mathcal{M} = \mathcal{M}(b)$.*

One can obtain a simple proof of this result by using Halmos's method of analyzing isometries [5]. Here is a sketch. Let $\mathcal{M}$ be as in the statement above and let U denote the restriction of S to $\mathcal{M}$. Since obviously $\bigcap_1^\infty U^n \mathcal{M} = \{0\}$, the isometry U is pure, in other words, $\mathcal{M}$ is spanned by the orthogonal complement of $U\mathcal{M}$ in $\mathcal{M}$ and its images under the positive powers of U. Let b be any unit vector in $\mathcal{M}$ that is orthogonal to $U\mathcal{M}$. The sequence $b, Ub, U^2b, \ldots$ is then an orthonormal sequence in $\mathcal{M}$, from which one easily deduces that the map $f \to bf$ is an isometry of H^2 into $\mathcal{M}$ and hence a contraction of H^2 into itself. It follows that b is in the unit ball of H^∞ and that $\mathcal{M}(b)$ is contained isometrically in $\mathcal{M}$. If $\mathcal{M}(b)$ were not all of $\mathcal{M}$, then $\mathcal{M}$ would contain a unit vector b_1 that is orthogonal both to $\mathcal{M}(b)$ and to $U\mathcal{M}$. In that case the map $f \oplus f_1 \to bf + b_1 f_1$ would be an isometry of $H^2 \oplus H^2$ into $\mathcal{M}$ and so an injective contraction of $H^2 \oplus H^2$ into H^2. (Injectivity would hold because the inclusion map of $\mathcal{M}$ into H^2 is injective.) This is a contradiction, as the map in question sends $(-b_1) \oplus b$ to 0. Hence $\mathcal{M} = \mathcal{M}(b)$, and the proof is complete.

From the preceding proof it is clear that b is uniquely determined by $\mathcal{M}(b)$ to within a multiplicative constant of unit modulus.

Incidentally, it is simple to construct an S-invariant Hilbert space $\mathcal{M}$ contained contractively in H^2 such that S acts as a contraction but not as an isometry in $\mathcal{M}$. For example, one can let the powers of z be an orthogonal basis for $\mathcal{M}$, taking $\|z^n\|_\mathcal{M}$ to be greater than or equal to 1, nonincreasing with respect to n, and not constant.

4. Complementary spaces. If $\mathcal{K}$ is a Hilbert space and B is a contraction operator from a Hilbert space into $\mathcal{K}$, then the complementary space of $\mathcal{M}(B)$, denoted $\mathcal{H}(B)$, is defined to be $\mathcal{M}((1 - BB^*)^{1/2})$. It is a simple exercise in operator theory to prove that, if x is in $\mathcal{M}(B)$, y is in $\mathcal{H}(B)$, and $w = x + y$, then $\|w\|_\mathcal{K}^2 \leq \|x\|_{\mathcal{M}(B)}^2 + \|y\|_{\mathcal{H}(B)}^2$. Moreover, given w in $\mathcal{K}$, the preceding inequality becomes an equality when $x = BB^*w$ and $y = w - BB^*w$, and only in that case. These properties of $\mathcal{H}(B)$ will play no role here but are mentioned because they form the basis for the definition of complementary space given in [2].

If $\mathcal{M}(B)$ is an ordinary subspace of $\mathcal{K}$, then clearly $\mathcal{H}(B)$ is its orthogonal complement.

It seems appropriate to mention here, in passing, that de Branges's notion of complementary space played a basic role in the discovery of his proof of the Bieberbach conjecture.

5. Complementary spaces of Brangesian shift-invariant spaces. For the remainder of this paper, b will denote a nonconstant function in the unit ball of H^∞. The space $\mathcal{H}(T_b)$ will be denoted by $\mathcal{H}(b)$, and the norm and inner product in $\mathcal{H}(b)$ will be denoted by $\|\ \|_b$ and $\langle\ ,\ \rangle_b$. Thus, if f and g are in H^2, then
$$\langle (1 - T_b T_{\bar{b}})^{1/2} f, (1 - T_b T_{\bar{b}})^{1/2} g \rangle_b = \langle f, g \rangle.$$
From this one easily deduces that, if f is in H^2 and g is in $\mathcal{H}(b)$, then
$$\langle (1 - T_b T_{\bar{b}}) f, g \rangle_b = \langle f, g \rangle,$$
an equality that will be used repeatedly below.

Our first task is to identify a few functions that belong to $\mathcal{H}(b)$. The space $\mathcal{H}(b)$ is a Hilbert space of holomorphic functions in D, and the evaluation functionals at the points of D are clearly continuous in $\mathcal{H}(b)$ (since they are continuous in H^2). The kernel function k_w^b for the functional in $\mathcal{H}(b)$ of evaluation at the point w of D (defined by the relation $\langle f, k_w^b \rangle_b = f(w)$), is easily seen to be given by $k_w^b = (1 - T_b T_{\bar{b}}) k_w$, where k_w is the kernel function for evaluation at w in H^2 ($k_w(z) = (1 - \bar{w}z)^{-1}$). As k_w is an eigenvector of $T_{\bar{b}}$ with eigenvalue $\overline{b(w)}$, we find that $k_w^b = (1 - \overline{b(w)}b) k_w$.

The function S^*b belongs to $\mathcal{H}(b)$. In fact, S^*b spans the range of the rank-one operator $S^*T_b - T_b S^*$, so to prove S^*b lies in $\mathcal{H}(b)$ (that is, in the range of $(1 - T_b T_{\bar{b}})^{1/2}$) it will suffice, by a known criterion [4], to establish the inequality

$$(1) \qquad (S^*T_b - T_b S^*)(S^*T_b - T_b S^*)^* \leq 1 - T_b T_{\bar{b}}.$$

A short calculation reveals that the left side of (1) equals $S^*T_b T_{\bar{b}} S - T_b T_{\bar{b}}$, so (1) does in fact hold.

The space $\mathcal{H}(b)$ is S^*-invariant and S^* acts as a contraction in it. This assertion is equivalent to the assertion that there is a factorization $S^*(1 - T_b T_{\bar{b}})^{1/2} = (1 - T_b T_{\bar{b}})^{1/2} R$ with $\|R\| \leq 1$, which, by the criterion of [4], is true if and only if the inequality

$$(2) \qquad S^*(1 - T_b T_{\bar{b}}) S \leq 1 - T_b T_{\bar{b}}$$

holds. A calculation shows that the difference between the right and left sides of (2) can be written as $S^*T_b(1 - SS^*)T_{\bar{b}} S$, so (2) is in fact true.

The restriction of S^* to $\mathcal{H}(b)$ will be denoted by X. We shall need a formula from [3] for X^*. It can be derived as follows. As the functions of the form $(1 - T_b T_{\bar{b}}) f$ are dense in $\mathcal{H}(b)$, we need only to determine how X^* acts on such functions. Let g be in $\mathcal{H}(b)$, let f be in H^2, and let $h = (1 - T_b T_{\bar{b}}) f$. We have

$$\langle Xg, h \rangle_b = \langle S^*g, h \rangle_b = \langle S^*g, f \rangle = \langle g, Sf \rangle = \langle g, (1 - T_b T_{\bar{b}}) Sf \rangle_b$$

so that $X^*h = (1 - T_b T_{\bar{b}}) Sf$. Moreover,

$$(1 - T_b T_{\bar{b}}) Sf = Sh - T_b(T_{\bar{b}} S - S T_{\bar{b}}) f.$$

Since $T_{\bar{b}} S - S T_{\bar{b}} = 1 \otimes S^*b$ (an equality one can most easily derive by first showing $S^*T_b - T_b S^* = S^*b \otimes 1$), we have $(T_{\bar{b}} S - S T_{\bar{b}}) f = \langle f, S^*b \rangle 1 = \langle h, S^*b \rangle_b 1$. Consequently

$$X^*h = Sh - \langle h, S^*b \rangle_b b,$$

a formula valid for all h in $\mathcal{H}(b)$ since the reasoning above establishes it for a dense subset of $\mathcal{H}(b)$.

Does the constant function 1 belong to $\mathcal{H}(b)$? That key question is answered by the first proposition.

PROPOSITION 1. *For $|w| < 1$, the kernel function k_w belongs to $\mathcal{H}(b)$ if and only if either $b(w) = 0$ or b is not an extreme point of the unit ball of H^∞.*

It is known [7, p. 138] that b is an extreme point of the unit ball of H^∞ (hereafter referred to simply as an extreme point) if and only if $\log(1 - |b|^2)$ is not in L^1 of the unit circle. It is the preceding condition that enters the proof of Proposition 1.

The part of Proposition 1 involving the condition $b(w) = 0$ is immediate, because when that condition holds the kernel functions k_w^b and k_w coincide. The following criterion will be used to establish Proposition 1 in general: *The H^2 function f belongs to the range of the operator $(1 - T_b T_{\bar{b}})^{1/2}$ if and only if*

$$\lim_{r \to 1-} \|(1 - r^2 T_b T_{\bar{b}})^{-1/2} f\|_2 < \infty.$$

If the preceding limit is finite, its value is the H^2-norm of the function that $(1 - T_b T_{\bar{b}})^{1/2}$ maps to f. This is an instance of a well-known property of positive operators that is easily deduced from the spectral theorem.

For $0 < r < 1$ we let a_r denote the outer function whose modulus on ∂D equals $(1 - r^2 |b|^2)^{1/2}$ and that takes a positive value at 0. (The latter condition is included for definiteness and will play no role.) We have

$$(1 - r^2 T_b T_{\bar{b}})^{-1} = \sum_{n=0}^{\infty} (T_{rb} T_{r\bar{b}})^n = 1 + T_{rb} \left[\sum_{n=0}^{\infty} (T_{r\bar{b}} T_{rb})^n \right] T_{r\bar{b}}$$

$$= 1 + T_{rb} (1 - T_{r^2 |b|^2})^{-1} T_{r\bar{b}}.$$

As $1 - T_{r^2|b|^2} = T_{|a_r|^2} = T_{\bar{a}_r} T_{a_r}$, we have $(1 - T_{r^2|b|^2})^{-1} = T_{1/a_r} T_{1/\bar{a}_r}$, and thus

$$(1 - r^2 T_b T_{\bar{b}})^{-1} = 1 + T_{rb/a_r} T_{r\bar{b}/\bar{a}_r}.$$

Hence, because k_w is an eigenvector of $T_{r\bar{b}/\bar{a}_r}$ with eigenvalue $\overline{rb(w)/a_r(w)}$, we conclude that

$$\|(1 - r^2 T_b T_{\bar{b}})^{-1/2} k_w\|_2^2 = \|k_w\|_2^2 + \|T_{r\bar{b}/\bar{a}_r} k_w\|_2^2$$

$$= \|k_w\|_2^2 + \frac{|rb(w)|^2}{|a_r(w)|^2} \|k_w\|_2^2$$

$$= (1 - |w|^2)^{-1} \left(1 + \frac{|rb(w)|^2}{|a_r(w)|^2} \right).$$

Using the criterion above we see that, if $b(w) \neq 0$, the function k_w belongs to $\mathcal{H}(b)$ if and only if $\lim_{r \to 1} a_r(w) > 0$. The preceding condition clearly holds if and only if $\log(1 - |b|^2)$ is in L^1 of ∂D, so Proposition 1 is established.

If b is not an extreme point, we denote by a the outer function whose modulus on ∂D is $(1 - |b|^2)^{1/2}$ and, for definiteness, that takes a positive value at 0. (However, the latter condition will be ignored below in order to give certain formulas a balanced appearance.) The proof above shows that if b is not an extreme point then

$$\|k_w\|_b^2 = (1 - |w|^2)^{-1} \left(1 + \frac{|b(w)|^2}{|a(w)|^2} \right).$$

The technique used to establish Proposition 1 will be employed repeatedly.

PROPOSITION 2. *If b has a zero of order m at the origin, then the function z^{m-1} belongs to $\mathcal{H}(b)$, but z^m belongs if and only if b is not an extreme point.*

In fact, if b has a zero of order m at the origin, the equality $(1-T_bT_{\bar{b}})z^{m-1} = z^{m-1}$ shows that z^{m-1} belongs to $\mathcal{H}(b)$. Moreover, using the expression for $(1-r^2 T_b T_{\bar{b}})^{-1}$ found in the proof of Proposition 1, one easily obtains the equality

$$\|(1-r^2 T_b T_{\bar{b}})^{-1/2} z^m\|_2^2 = 1 + \frac{|r\hat{b}(m)|^2}{|a_r(0)|^2}$$

(where, in general, $\hat{b}(n)$ denotes the nth Maclaurin coefficient of b). The proof of Proposition 2 is now concluded by exactly the same reasoning as was used in the proof of Proposition 1. We see in addition that, if b has a zero of order m at the origin and is not an extreme point, then

$$\|z^m\|_b^2 = 1 + \frac{|\hat{b}(m)|^2}{|a(0)|^2}.$$

COROLLARY. *The function b belongs to $\mathcal{H}(b)$ if and only if b is not an extreme point.*

To establish this, assume b has a zero of order m (possibly 0) at the origin. The function $(1 - T_b T_{\bar{b}})z^m = z^m - \overline{\hat{b}(m)}b$ then belongs to $\mathcal{H}(b)$, but, by Proposition 2, z^m itself belongs if and only if b is not an extreme point. The corollary follows.

The preceding corollary together with the formula for X^* found earlier implies that $\mathcal{H}(b)$ is S-invariant if and only if b is not an extreme point. The spaces $\mathcal{H}(b)$ for b not an extreme point have additional interesting properties, which, however, will not be touched upon here. The author intends to explore them in a sequel to this paper.

6. Calculations. The analysis of the operator X requires the knowledge of the values of certain norms and inner products in the space $\mathcal{H}(b)$. The calculations are carried out in the present section. From the reasoning used in the proof of Proposition 1, one sees that, for f and g in $\mathcal{H}(b)$,

(3) $$\langle f, g \rangle_b = \langle f, g \rangle + \lim_{r \to 1-} \langle T_{r\bar{b}/\bar{a}_r} f, T_{r\bar{b}/\bar{a}_r} g \rangle.$$

This relation is the basis of the computations that follow.

LEMMA 1. *$\|S^*b\|_b^2$ equals $1 - |b(0)|^2$ in case b is an extreme point and $1 - |b(0)|^2 - |a(0)|^2$ in the contrary case.*

To establish this we use (3) with $f = g = S^*b$. We have, for $0 < r < 1$,

$$rT_{r\bar{b}/\bar{a}_r} S^*b = P\left(\frac{\bar{z}r^2|b|^2}{\bar{a}_r}\right) = P\left(\frac{\bar{z}(1-|a_r|^2)}{\bar{a}_r}\right) = -S^*a_r.$$

Hence
$$\|S^*b\|_b^2 = \|S^*b\|_2^2 + \lim_{r \to 1} r^{-2}\|S^*a_r\|_2^2$$
$$= \|b\|_2^2 - |b(0)|^2 + \lim_{r \to 1} r^{-2}(\|a_r\|_2^2 - |a_r(0)|^2)$$
$$= -|b(0)|^2 + \lim_{r \to 1} r^{-2}(1 - |a_r(0)|^2),$$

the last equality holding because $r^2\|b\|_2^2 + \|a_r\|_2^2 = 1$ (since $r^2|b|^2 + |a_r|^2 = 1$ on ∂D). Since $\lim_{r \to 1} a_r(0)$ is 0 in case b is an extreme point and $a(0)$ otherwise, the lemma is established.

The remaining calculations apply where b is not an extreme point.

LEMMA 2. *If b is not an extreme point, then $\|b\|_b^2 = |a(0)|^{-2} - 1$.*

We have
$$rT_{r\bar{b}/\bar{a}_r}b = P\left(\frac{r^2|b|^2}{\bar{a}_r}\right) = P\left(\frac{1 - |a_r|^2}{\bar{a}_r}\right) = \overline{a_r(0)}^{-1} - a_r.$$

Hence
$$\|b\|_b^2 = \|b\|_2^2 + \lim_{r \to 1} r^{-2}\|\overline{a_r(0)}^{-1} - a_r\|_2^2$$
$$= \lim_{r \to 1} r^{-2}(r^2\|b\|_2^2 + \|a_r\|_2^2 + |a_r(0)|^{-2} - 2)$$
$$= \lim_{r \to 1} r^{-2}(|a_r(0)|^{-2} - 1) = |a(0)|^{-2} - 1.$$

LEMMA 3. *If b is not an extreme point, then $\langle b, 1 \rangle_b = b(0)|a(0)|^{-2}$.*

By Lemma 2,
$$b(0) = \langle b, 1 - \overline{b(0)}b \rangle_b = \langle b, 1 \rangle_b - b(0)(|a(0)|^{-2} - 1),$$

from which Lemma 3 follows.

LEMMA 4. *If b is not an extreme point, then, for $n > 0$,*
$$\langle X^n b, 1 \rangle_b = \hat{b}(n) - b(0)\hat{a}(n)a(0)^{-1}.$$

If $n > 0$ then, by reasoning like that in the proof of Lemma 1,
$$rT_{r\bar{b}/\bar{a}_r}X^n b = -S^{*n}a_r.$$

From this and the equality $T_{\bar{b}/\bar{a}_r}1 = \overline{b(0)}\,\overline{a_r(0)}^{-1}$, we get
$$\langle X^n b, 1 \rangle_b = \langle S^{*n}b, 1 \rangle - \lim_{r \to 1} b(0)a_r(0)^{-1}\langle S^{*n}a_r, 1 \rangle$$
$$= \hat{b}(n) - \lim_{r \to 1} b(0)a_r(0)^{-1}\hat{a}_r(n)$$
$$= \hat{b}(n) - b(0)\hat{a}(n)a(0)^{-1}.$$

LEMMA 5. *If b is not an extreme point and if b has a zero of order m at the origin, then, for $n > 0$,*
$$\langle X^n b, z^m \rangle_b = \hat{b}(n+m) - \hat{b}(m)\hat{a}(n)a(0)^{-1}.$$

This is proved just as Lemma 4, the only change being that one replaces the equality $T_{\bar{b}/\bar{a}_r}1 = \overline{b(0)}\,\overline{a_r(0)}^{-1}$ by the equality $T_{\bar{b}/\bar{a}_r}z^m = \overline{\hat{b}(m)}\,\overline{a_r(0)}^{-1}$.

LEMMA 6. *If b is not an extreme point, then, for $n > 0$, $\langle X^n b, b \rangle_b = -\hat{a}(n)a(0)^{-1}$.*

To prove this, assume b has a zero of order m (possibly 0) at the origin. The function $z^m - \overline{\hat{b}(m)}b$ is then the image of the function z^m under the operator $1 - T_b T_{\bar{b}}$, so this function has the property that $\langle f, z^m - \overline{\hat{b}(m)}b \rangle_b = \hat{f}(m)$ for all f in $\mathcal{H}(b)$. From that and Lemma 5 we get

$$\hat{b}(n+m) = \langle X^n b, z^m - \overline{\hat{b}(m)}b \rangle_b = \hat{b}(n+m) - \hat{b}(m)\hat{a}(n)a(0)^{-1} - \hat{b}(m)\langle X^n b, b \rangle_b,$$

and Lemma 6 follows.

7. Defect operators. The analysis of the operator X to be given below hinges on the identification of the characteristic operator function of X, whose definition will be recalled in the next section. Suffice it to mention here that a knowledge of the characteristic operator function of X enables one to construct the Sz.-Nagy–Foiaş model of X and thus to relate de Branges's theory with that of Sz.-Nagy and Foiaş.

The first step in identifying the characteristic operator function of X is to compute the defect operators of X, that is, the operators $D_X = (1 - X^*X)^{1/2}$ and $D_{X^*} = (1 - XX^*)^{1/2}$. That computation is carried out in the present section.

From the formula $X^*h = Sh - \langle h, S^*b \rangle_b b$ found in §5 we obtain immediately, for h in $\mathcal{H}(b)$, the equality $XX^*h = h - \langle h, S^*b \rangle_b S^*b$, in other words

$$D_{X^*}^2 = (S^*b) \otimes (S^*b).$$

The operator D_{X^*} thus has rank 1, its range is spanned by S^*b, and its nonzero eigenvalue equals $\|S^*b\|_b$.

The structure of D_X depends upon whether or not b is an extreme point. From the formula for X^* we get, for h in $\mathcal{H}(b)$,

(4) $\qquad X^*Xh = SXh - \langle Xh, S^*b \rangle_b b = h - h(0) - \langle Xh, S^*b \rangle_b b.$

Consider first the case where b is an extreme point and $b(0) \neq 0$. Then, according to results in §5, neither b nor the constant function 1 lies in $\mathcal{H}(b)$, although $1 - \overline{b(0)}b \ (= k_0^b)$ is in $\mathcal{H}(b)$. The term $-h(0) - \langle Xh, S^*b \rangle_b b$ on the right side of (4) must therefore equal $-h(0)(1 - \overline{b(0)}b)$, which means

(5) $\qquad X^*Xh = h - \langle h, k_0^b \rangle_b k_0^b.$

On the other hand, if b is an extreme point and $b(0) = 0$, then the constant function 1 does belong to $\mathcal{H}(b)$ but b still does not, which means that the inner product $\langle Xh, S^*b \rangle_b$ from the right side of (4) must vanish for all h, so (5) remains valid. In case b is an extreme point, therefore,

$$D_X^2 = k_0^b \otimes k_0^b,$$

that is, the operator D_X has rank 1, its range is spanned by k_0^b, and its nonzero eigenvalue equals $\|k_0^b\|_b$ (which, by Lemma 1, equals the nonzero eigenvalue of D_{X^*}).

Suppose now that b is not an extreme point. Then we can rewrite (4) as

(6) $$X^*Xh = h - h(0) - \langle h, X^*Xb\rangle_b b.$$

Setting $h = b$ in (6), we get

$$X^*Xb = b - b(0) - \|Xb\|_b^2 b = b - b(0) - (1 - |b(0)|^2 - |a(0)|^2)b,$$

the last equality following by Lemma 1. Inserting the preceding expression for X^*Xb and the expression $h(0) = \langle h, 1 - \overline{b(0)}b\rangle_b$ into (6), we obtain

$$X^*Xh = h - \langle h, 1 - \overline{b(0)}b\rangle_b - \langle h, b\rangle_b b + \overline{b(0)}\langle h, 1\rangle_b b$$
$$+ (1 - |b(0)|^2 - |a(0)|^2)\langle h, b\rangle_b b$$
$$= h - \langle h, k_0^b\rangle_b k_0^b - |a(0)|^2\langle h, b\rangle_b b.$$

Thus, if b is not an extreme point, then

(7) $$D_X^2 = (k_0^b \otimes k_0^b) + |a(0)|^2 (b \otimes b).$$

The operator D_X has rank 2 in this case. Its range is spanned by the functions 1 and b.

In order to find D_X in the case where b is not an extreme point, we need to diagonalize D_X^2. It is convenient to state the relevant calculations as lemmas. The diagonalization of a 2-by-2 selfadjoint matrix is of course an elementary exercise in linear algebra, but it can involve unpleasant calculations. Fortunately, in our case one eigenvector is readily at hand.

LEMMA 7. *If b is not an extreme point, then $D_X^2 1 = 1$.*

The proof is a two-line computation based on Lemma 3. The details will be omitted.

Since D_X^2 is selfadjoint and of rank 2, we can find another eigenvector of it with a nonzero eigenvalue by finding a nontrivial linear combination of 1 and b that is orthogonal to 1. The results of this straightforward computation are stated in the next lemma.

LEMMA 8. *If b is not an extreme point, then the function $-b(0)k_0^b + |a(0)|^2 b$ is an eigenvector of D_X^2 with eigenvalue $1 - |b(0)|^2 - |a(0)|^2$.*

Indeed, that the given function is orthogonal to 1 follows immediately from Lemma 3. The corresponding eigenvalue equals the trace of D_X^2 minus 1 (since the eigenvalue of the constant function 1 is 1). By (7) and Lemma 2,

$$\operatorname{tr} D_X^2 = \|k_0^b\|_b^2 + |a(0)|^2 \|b\|_b^2$$
$$= 1 - |b(0)|^2 + |a(0)|^2(|a(0)|^{-2} - 1) = 2 - |b(0)|^2 - |a(0)|^2,$$

and Lemma 8 is established.

From Lemmas 7 and 8 one can readily write down an explicit expression for D_X in the case where b is not an extreme point.

Using the expressions for X^*X obtained above, one sees that, in case b is an extreme point,
$$\|Xh\|_b^2 = \|h\|_b^2 - |h(0)|^2,$$
and in the contrary case,
$$\|Xh\|_b^2 = \|h\|_b^2 - |h(0)|^2 - |a(0)|^2|\langle h, b\rangle_b|^2.$$

In either case one has the inequality

(8) $$\|Xh\|_b^2 \leq \|h\|_b^2 - |h(0)|^2,$$

called by de Branges the inequality for difference quotients (because X is a difference-quotient transformation: $(Xh)(z) = z^{-1}(h(z) - h(0))$). In case (8) is an equality for all h in $\mathcal{H}(b)$, de Branges says that $\mathcal{H}(b)$ satisfies the identity for difference quotients. From the expressions above, we see that happens if and only if b is an extreme point, a conclusion that should be compared with Theorem 16 of [3].

8. Characteristic operator function: extreme point case.

The characteristic operator function of X, in the theory of Sz.-Nagy and Foiaş [9], is a holomorphic operator-valued function Θ_X, defined in the unit disk, whose value at any point of the disk is a transformation from the defect space $\mathcal{D}_X = \overline{D_X \mathcal{H}(b)}$ to the defect space $\mathcal{D}_{X^*} = \overline{D_{X^*} \mathcal{H}(b)}$. It is defined by the formula
$$\Theta_X(w) = [-X + wD_{X^*}(1 - wX^*)^{-1}D_X] \mid \mathcal{D}_X.$$

The preceding definitions apply to an arbitrary Hilbert space contraction X, except that in the general case one must take $\mathcal{D}_X$ and $\mathcal{D}_{X^*}$ to be the closures of the ranges of D_X and D_{X^*} in case those ranges are not already closed. According to the theory developed in [9], a completely nonunitary contraction is uniquely determined to within unitary equivalence by its characteristic operator function. As our operator X is completely nonunitary (by the inequality for difference quotients), we shall be able to fit it into the Sz.-Nagy–Foiaş scheme once we have identified its characteristic operator function, a task that is accomplished in this and the next section.

Assume for the remainder of this section that b is an extreme point. Both defect spaces $\mathcal{D}_X$ and $\mathcal{D}_{X^*}$ then have dimension 1, so Θ_X is, essentially, a scalar-valued holomorphic function in D. To be precise, we consider the two functions k_0^b and S^*b, which have the same norm (by Lemma 1) and span $\mathcal{D}_X$ and $\mathcal{D}_{X^*}$, respectively. The scalar function θ defined by $\Theta_X(w)k_0^b = \theta(w)S^*b$ then completely determines Θ_X.

PROPOSITION 3. $\theta = b^*$.

Here, by b^* is meant the function that at z takes the value $\overline{b(\bar{z})}$. The proof of Proposition 3 is a straightforward computation. The power series expansion about the origin of the function $\langle \Theta_X(w)k_0^b, S^*b\rangle_b$ is

(9) $$\langle -Xk_0^b, S^*b\rangle_b + \sum_{n=0}^{\infty} \langle X^{*n}D_Xk_0^b, D_{X^*}S^*b\rangle_b w^{n+1},$$

and this equals $\|S^*b\|_b^2 \theta(w)$. As $Xk_0^b = -\overline{b(0)}S^*b$, the zeroth coefficient in the series (9) is $\overline{b(0)}\|S^*b\|_b^2$. Using the information about D_X and D_{X^*} obtained in §7, one easily sees that the $(n+1)$st coefficient is

$$\|k_0^b\|_b\|S^*b\|_b \langle k_0^b, S^{*(n+1)}b\rangle_b,$$

which equals $\|S^*b\|_b^2 \widehat{b}(n+1)$. Proposition 3 is established.

In particular, it follows from Proposition 3 that the operators X corresponding to two different extreme points b are unitarily equivalent if and only if the functions b in question are constant unimodular multiples of each other.

9. Characteristic operator function: nonextreme point case.

Assume in this section that b is not an extreme point. The situation is rather more complicated than in the last section, because $\mathcal{D}_X$ now has dimension 2, so that we need a pair of scalar-valued functions to describe Θ_X. To be precise, let u_1 and u_2 be a pair of orthogonal unit vectors in $\mathcal{D}_X$, and let $v = \|S^*b\|_b^{-1}S^*b$, a unit vector spanning $\mathcal{D}_{X^*}$. The operator function Θ_X is then determined by the 1-by-2 matrix function $(\theta_1 \quad \theta_2)$, where θ_j is the scalar function in D defined by $\Theta_X(w)u_j = \theta_j(w)v$. Replacing u_1, u_2 by another orthonormal basis for $\mathcal{D}_X$ amounts to multiplying the matrix function $(\theta_1 \quad \theta_2)$ from the right by a constant 2-by-2 unitary matrix.

PROPOSITION 4. *There is a choice of basis u_1, u_2 such that $\theta_1 = b^*$ and $\theta_2 = a^*$.*

Let e denote the function $-b(0)k_0^b + |a(0)|^2 b$, which according to Lemma 8 is an eigenvector of D_X, orthogonal to the eigenvector 1, with eigenvalue $(1 - |b(0)|^2 - |a(0)|^2)^{1/2}$. A straightforward calculation using Lemma 2 gives

$$\|e\|_b^2 = (|b(0)|^2 + |a(0)|^2)(1 - |b(0)|^2 - |a(0)|^2).$$

We define

$$u_1 = (|b(0)|^2 + |a(0)|^2)^{-1}[|a(0)|^2 - \overline{b(0)}(1 - |b(0)|^2 - |a(0)|^2)^{-1/2}e],$$

$$u_2 = -\overline{a(0)}(|b(0)|^2 + |a(0)|^2)^{-1}[b(0) + (1 - |b(0)|^2 - |a(0)|^2)^{-1/2}e].$$

Both u_1 and u_2 are linear combinations of the mutually orthogonal vectors 1 and e, so it is easy, using the value given above for $\|e\|_b^2$ and the value for $\|1\|_b^2$ found in §5, to show that u_1 and u_2 do form an orthonormal basis for $\mathcal{D}_X$. The details will be omitted. Moreover, using the values of $D_X 1$ and $D_X e$ given by Lemmas 7 and 8, one readily obtains the equalities $D_X u_1 = k_0^b$, $D_X u_2 = -\overline{a(0)}b$. Again, the details are omitted.

To find θ_1 and θ_2 for our choice of u_1 and u_2, we write down their power series expansions,

$$\theta_j(w) = \|S^*b\|_b^{-1} \langle \Theta_X(w) u_j, S^*b \rangle_b$$

$$= \|S^*b\|_b^{-1} \left[-\langle Xu_j, S^*b \rangle_b + \sum_{n=0}^{\infty} \langle X^{*n} D_X u_j, D_{X^*} S^*b \rangle_b w^{n+1} \right],$$

and then compute the coefficients. As $Xe = (|b(0)|^2 + |a(0)|^2) S^*b$, we have

$$Xu_1 = -\overline{b(0)}(1 - |b(0)|^2 - |a(0)|^2)^{-1/2} S^*b,$$
$$Xu_2 = -\overline{a(0)}(1 - |b(0)|^2 - |a(0)|^2)^{-1/2} S^*b.$$

Hence, the zeroth coefficient of θ_1 is

$$\|S^*b\|_b^{-1} \overline{b(0)}(1 - |b(0)|^2 - |a(0)|^2)^{-1/2} \|S^*b\|_b^2,$$

which equals $\overline{b(0)}$ by Lemma 1. A similar calculation shows that the zeroth coefficient of θ_2 is $\overline{a(0)}$.

Since $D_X u_1 = k_0^b$ and $D_{X^*} S^*b = \|S^*b\|_b S^*b$, the $(n+1)$st coefficient of θ_1 equals

$$\langle X^{*n} k_0^b, S^*b \rangle_b = \langle k_0^b, S^{*(n+1)} b \rangle_b = \hat{b}(n+1).$$

Similarly, since $D_X u_2 = -\overline{a(0)} b$, the $(n+1)$st coefficient of θ_2 equals

$$-\overline{a(0)} \langle X^{*n} b, S^*b \rangle_b = -\overline{a(0)} \langle b, X^{n+1} b \rangle_b,$$

and this equals $\hat{a}(n+1)$, by Lemma 6. Proposition 4 is now established.

By Proposition 4, the operators X corresponding to two different nonextreme points b are unitarily equivalent if and only if one can transform each of the associated 1-by-2 matrix functions $(b \; a)$ into the other by multiplying from the right with a constant 2-by-2 unitary matrix. That is the case, for instance, for the two examples $b(z) = (z-1)/2$ and $b(z) = z/2^{1/2}$. The corresponding functions a are $(z+1)/2$ and $1/2^{1/2}$, respectively, and

$$(z/2^{1/2} \; 1/2^{1/2}) = \left(\frac{z-1}{2} \; \frac{z+1}{2} \right) \begin{pmatrix} 1/2^{1/2} & -1/2^{1/2} \\ 1/2^{1/2} & 1/2^{1/2} \end{pmatrix}.$$

The situation in case $b(z) = z/2^{1/2}$ is particularly easy to understand. Then $1 - T_b T_{\bar{b}} = 1 - SS^*/2$. The powers of z form an orthogonal basis for $\mathcal{H}(b)$, with $\|z^n\|_b = 2^{1/2}$ for $n > 0$ and $\|1\|_b = 1$. The operator X is a backward weighted shift with only one weight different from 1.

Propositions 3 and 4 enable one to specify in terms of its characteristic operator function when a completely nonunitary contraction Y is unitarily equivalent to one of our operators X. That happens in the following two cases and only in those cases: (i) the defect numbers of Y (the respective dimensions of $\mathcal{D}_Y$ and $\mathcal{D}_{Y^*}$) are 1 and 1, and Θ_Y is an extreme point of the unit ball of H^∞; (ii) the defect numbers of Y are 2 and 1, and Θ_Y can be represented by a 1-by-2 inner function whose second entry is an outer function.

To illustrate, let $p(z) = (z-c)/(1-\bar{c}z)$ and $q(z) = (z+c)/(1+\bar{c}z)$, where $0 < |c| < 1$. If Θ_Y is represented by the 1-by-2 inner funtion $(2^{-1/2}p \quad 2^{-1/2}q)$, then (ii) is satisfied. Although the second entry of the preceding matrix function is not outer, right-multiplication of it by the unitary matrix

$$\begin{pmatrix} 1/2^{1/2} & -1/2^{1/2} \\ 1/2^{1/2} & 1/2^{1/2} \end{pmatrix}$$

transforms it to an equivalent function whose second entry is outer. However, if Θ_Y is represented by $(2^{-1/2}z \quad 2^{-1/2}pq)$ and $|c| < 3^{-1/2}$, then (ii) fails, for one can then show that no linear combination of the functions z and pq is outer.

10. Explicit connection with the Sz.-Nagy–Foiaş model theory.
The question now arises how to implement the unitary equivalence between the operator X and its Sz.-Nagy–Foiaş model. The answer is very simple (and perhaps should have been guessed at the outset).

Suppose first that b is not an extreme point. We form the 2-by-1 matrix inner function $B = \binom{b}{a}$, which we identify with the isometry from H^2 to $H^2 \oplus H^2$ that sends f to $bf \oplus af$. Let $\mathcal{H}(B)$ be the orthogonal complement of BH^2 in $H^2 \oplus H^2$. Then $\mathcal{H}(B)$ is invariant under $S^* \oplus S^*$, and the restriction of $S^* \oplus S^*$ to $\mathcal{H}(B)$ is the Sz.-Nagy–Foiaş model of X [9].

PROPOSITION 5. *The projection $P_+ = \begin{pmatrix} 1 & 0 \\ 0 & 0 \end{pmatrix}$ in $H^2 \oplus H^2$ sends $\mathcal{H}(B)$ isometrically onto $\mathcal{H}(b)$ and so implements the unitary equivalence between X and its Sz.-Nagy–Foiaş model.*

To establish this we observe that BH^2, being the range of the isometry

$$\begin{pmatrix} T_b \\ T_a \end{pmatrix},$$

is also the range of the projection

$$\begin{pmatrix} T_b \\ T_a \end{pmatrix} (T_{\bar{b}} \quad T_{\bar{a}}) = \begin{pmatrix} T_b T_{\bar{b}} & T_b T_{\bar{a}} \\ T_a T_{\bar{b}} & T_a T_{\bar{a}} \end{pmatrix}.$$

Hence $\mathcal{H}(B)$ is the range of the projection

$$E = \begin{pmatrix} 1 - T_b T_{\bar{b}} & -T_b T_{\bar{a}} \\ -T_a T_{\bar{b}} & 1 - T_a T_{\bar{a}} \end{pmatrix}.$$

As

$$P_+ E P_+ = \begin{pmatrix} 1 - T_b T_{\bar{b}} & 0 \\ 0 & 0 \end{pmatrix},$$

the Hilbert space contained contractively in H^2 associated with the contraction $P_+ E$ coincides with the one associated with the contraction $(1 - T_b T_{\bar{b}})^{1/2}$, that is, with $\mathcal{H}(b)$. Since E acts isometrically on its range, it follows that $\mathcal{H}(b)$ is the Hilbert space contained contractively in H^2 associated with the contraction

$P_+ \mid \mathcal{H}(B)$. Because a is an outer function, one easily verifies that the kernel of P_+ has a trivial intersection with $\mathcal{H}(B)$, so P_+ is an isometry of $\mathcal{H}(B)$ onto $\mathcal{H}(b)$.

Consider now the case where b is an extreme point. We form the function $\Delta = (1 - |b|^2)^{1/2}$ on ∂D and the 2-by-1 matrix function $B = \binom{b}{\Delta}$, which we identify with the isometry from H^2 into $H^2 \oplus \overline{\Delta L^2}$ that sends f to $bf \oplus \Delta f$. Let U denote the operator on $\overline{\Delta L^2}$ of multiplication by z, and let $\mathcal{H}(B)$ denote the orthogonal complement of BH^2 in $H^2 \oplus \overline{\Delta L^2}$. The subspace $\mathcal{H}(B)$ is invariant under $S^* \oplus U^*$ and the restriction of $S^* \oplus U^*$ to it is the Sz.-Nagy–Foiaş model of X [9].

PROPOSITION 6. *The projection* $P_+ = \begin{pmatrix} 1 & 0 \\ 0 & 0 \end{pmatrix}$ *in* $H^2 \oplus \overline{\Delta L^2}$ *sends* $\mathcal{H}(B)$ *isometrically onto* $\mathcal{H}(b)$.

The proof of this is basically the same as the proof of Proposition 5. Let M_Δ denote the operator from H^2 to $\overline{\Delta L^2}$ of multiplication by Δ. Reasoning as before, one sees that $\mathcal{H}(B)$ is the range of the projection

$$E = \begin{pmatrix} 1 - T_b T_{\bar{b}} & -T_b M_\Delta^* \\ -M_\Delta T_{\bar{b}} & 1 - M_\Delta M_\Delta^* \end{pmatrix}$$

and thus that $\mathcal{H}(b)$ is the Hilbert space contained contractively in H^2 associated with the contraction $P_+ \mid \mathcal{H}(B)$. That P_+ is an isometry of $\mathcal{H}(B)$ onto $\mathcal{H}(b)$, in other words, that the kernel of P_+ has a trivial intersection with $\mathcal{H}(B)$, follows, by virtue of the nonintegrability of $\log \Delta$, from a well-known theorem of G. Szegö [6, p. 21].

REFERENCES

1. A. Beurling, *On two problems concerning linear transformations in Hilbert space*, Acta Math. **81** (1949), 239–255.

2. L. de Branges, *Square summable power series*, (to appear).

3. L. de Branges and J. Rovnyak, *Square summable power series*, Holt, Rinehart and Winston, New York, 1966.

4. R. G. Douglas, *On majorization, factorization, and range inclusion of operators on Hilbert space*, Proc. Amer. Math. Soc. **17** (1966), 413–415.

5. P. R. Halmos, *Shifts on Hilbert spaces*, J. Reine Angew. Math. **208** (1961), 102–112.

6. H. Helson, *Lectures on invariant subspaces*, Academic Press, New York and London, 1964.

7. K. Hoffman, *Banach spaces of analytic functions*, Prentice-Hall, Engelwood Cliffs, N. J., 1962.

8. P. Lax, *Translation invariant spaces*, Acta Math. **101** (1969), 163–178.

9. B. Sz.-Nagy and C. Foiaş, *Harmonic analysis of operators on Hilbert space*, North Holland, Amsterdam, 1970.

The Cauchy Integral, Chord-Arc Curves, and Quasiconformal Mappings

STEPHEN W. SEMMES

I'd like to describe some work done in the last few years by Calderón, Coifman, David, McIntosh, Meyer, and others related to the Cauchy integral on nonsmooth curves. In some respects this area is an outgrowth of classical H^p theory, especially singular integral operators and almost-everywhere existence of boundary values. The difference is that the classical theory was concerned with the upper half-space and convolution operators, or with smooth perturbations of these that could be reduced to the flat case. For nonsmooth perturbations these reductions are not available and new methods had to be developed.

Like classical H^p theory, these questions are closely related to other problems in complex analysis, and also real analysis, partial differential equations, and operator theory. I shall concentrate on the complex plane here, and the reader is referred to [C2] and [CM5] for different perspectives.

It is important to emphasize the relationship between complex variables and these other fields, particularly real analysis. They have each been very helpful to the other. It has often been the case that real-variable methods have been important for attacking problems in complex analysis, and conversely, problems in complex analysis have often led to new real-variable methods having other applications. The Cauchy integral is an example of this. Weighted norm inequalities for maximal and area functions were used in the proof of Calderón's theorem, which solved two famous problems in complex analysis, the Denjoy conjecture and the a.e. existence of boundary values of Cauchy integrals on rectifiable Jordan curves. H^p theory, quasiconformal mappings, and the corona problem give other examples which show that this relationship is mutually beneficial.

In §1 I'll give an overview of the basic results; the papers of Calderón [C2] and Coifman and Meyer [CM5] give more information. In the second section I start over again and discuss the Cauchy integral and Riemann mapping on chord-arc curves, drawing mainly from work of Coifman and Meyer [CM2, 3, 4]. I am very fond of this setting, and I think that it is very interesting from the point of view of function theory. In many ways it is highly reminiscent of quasicircles and quasiconformal mappings, and I shall outline some of these analogies in §3.

There I'll also briefly describe some recent work of my own that attempts to reconcile these analogies by using ideas from quasiconformal mappings to give a new approach to the Cauchy integral on chord-arc curves, and which uses that to prove a version of the mapping theorem that is well adapted to chord-arc curves.

I would like to thank A. Baernstein II, D. Drasin, and R. Coifman for their very useful suggestions, and also the National Science Foundation for partial financial support in the form of a postdoctoral fellowship.

1. Basic results. Let Γ be an oriented locally rectifiable Jordan curve in the plane that passes through ∞, and let Ω_+ and Ω_- denote its complementary regions. Let f be a locally integrable function on Γ, with enough control so that our integrals converge at ∞. Define $F(z)$ on $\mathbb{C} \setminus \Gamma$ by

$$(1.1) \qquad F(z) = \frac{1}{2\pi i} \int_\Gamma \frac{f(w)\,dw}{w-z}.$$

For $w \in \Gamma$ we would like to know if $F(z)$ has nontangential boundary values $f_+(w)$ and $f_-(w)$ from Ω_+ and Ω_-, and if there are estimates, i.e., $\|f_\pm\|_p \leq C_p \|f\|_p$, $1 < p < \infty$, as there are when Γ is a line. (These L^p norms are with respect to arclength measure.)

We can convert this to a problem about singular integrals. Define the Cauchy integral $C(f)$ by

$$(1.2) \qquad C(f)(z) = \frac{1}{2\pi i} \,\text{P.V.} \int_\Gamma \frac{f(w)\,dw}{w-z}$$

for $z \in \Gamma$ such that the principal value exists. It is a classical fact that for almost all $z \in \Gamma$ the principal value exists if and only if the nontangential boundary values do, and one has the Plemelj formula:

$$(1.3) \quad f_+(z) = \tfrac{1}{2} f(z) + C(f)(z), \qquad f_-(z) = -\tfrac{1}{2} f(z) + C(f)(z), \qquad z \in \Gamma.$$

Thus $f = f_+ - f_-$ is split into two pieces, one analytic above, the other below.

Note that these limits exist if f is, say, the restriction to Γ of a C^1 function on the plane having compact support. Thus we can always define the Cauchy integral on a dense class of functions.

Everything remains simple if Γ is smooth. For example, if Γ is C^2 and agrees with the real line for z large enough, then the Cauchy integral can be compared with the Hilbert transform, leaving an error term that can be estimated brutally, by putting absolute values on its kernel.

When Γ is not smooth, this simple perturbation argument does not work, because the error term will have as bad a singularity as the Cauchy integral and cannot be estimated so easily. The cancellation needs to be taken into account. When $\Gamma = \mathbb{R}$ and the Cauchy integral is the Hilbert transform, the L^2 estimate is a consequence of the Fourier transform and Plancherel's theorem. In the general case this is again unavailable because the Cauchy integral is no longer a convolution operator.

Thus one needs methods for analyzing singular integral operators that are not convolution operators. An important tool that is available is the real-variable methods of Calderón and Zygmund which can transform an L^p estimate for some p (often L^2) into L^p estimates for all p, $1 < p < \infty$, a weak type $(1,1)$ inequality, corresponding estimates for the maximal truncated singular integral, and almost-everywhere existence of the principal value (see [**J**]). But it is still necessary to have an initial L^p estimate for some p.

In 1977 Calderón [**C1**] proved the following.

THEOREM 1. *Let $A \colon \mathbf{R} \to \mathbf{R}$ be a Lipschitz function, $A' \in L^\infty$, and let $\Gamma = \{x + iA(x) \colon x \in \mathbf{R}\}$ be its graph. There is an absolute constant $\delta > 0$ so that if $\|A'\|_\infty \leq \delta$, then the Cauchy integral on Γ is bounded on L^2.*

By real-variable methods this implies that if $f \in L^1(\Gamma)$, then the principal value in (1.2) exists a.e., and hence the boundary values of F defined in (1.1) also exist a.e. This implies the corresponding result for C^1 curves, since locally they look like Lipschitz graphs with small constant. It was already known that the result for all rectifiable Jordan curves follows from the result for C^1 curves. Thus the Cauchy integral and the boundary values of F always exist a.e., and the problem is to know when the Cauchy integral is bounded on L^p.

Another consequence of Calderón's theorem is the Denjoy conjecture, that the analytic capacity of a compact subset of a rectifiable arc is positive if its arclength measure is positive (see [**Ma**]).

Because the Cauchy integral depends nonlinearly on the function A, it is not at all clear that the hypothesis $\|A'\|_\infty \leq \delta$ in Calderón's theorem is unnecessary. This was proved five years later by Coifman, McIntosh, and Meyer [**CMM**].

THEOREM 2. *The Cauchy integral operator on any Lipschitz graph is bounded on L^2 (hence L^p, $1 < p < \infty$, etc.).*

A complete characterization of the curves for which the Cauchy integral is bounded has been obtained by Guy David [**D3**]. Call a rectifiable curve Γ regular if there is a $K > 0$ so that for all $z_0 \in \mathbf{C}$ and all $R > 0$, the arclength measure of $\{z \in \Gamma \colon |z - z_0| \leq R\}$ is at most KR. This condition also arose in [**A1**], but in a different context.

THEOREM 3. *For each p, $1 < p < \infty$, the Cauchy integral on Γ is bounded on $L^p(\Gamma)$ if and only if Γ is regular.*

For this theorem it is no longer assumed that Γ is Jordan. It is important that Γ be a curve, though. There are measures in the plane which satisfy estimates similar to those for arclength on Γ but with respect to which the Cauchy integral is not bounded [**D4**]. It is an open problem to determine the measures μ for which the Cauchy integral induces a bounded operator on $L^p(\mu)$. It is also an open problem to find the higher-dimensional version of this result.

The "only if" part of Theorem 3 is easy, but the "if" part is much deeper. David shows that a regular curve can be approximated (in a certain sense) by

Lipschitz graphs, and he uses this and a good-λ inequality to estimate the Cauchy integral. He used a similar perturbation argument to derive Theorem 2 from Theorem 1. David also proved estimates for curves that are not regular, but a weight has to be introduced.

There is the question of best estimates. Suppose A is a real-valued Lipschitz function, and define operators T_A and C_A by

$$T_A f(x) = \text{P.V.} \int_{-\infty}^{\infty} \frac{1}{(x-y) + i(A(x) - A(y))} f(y)\, dy,$$

$$C_A f(x) = \text{P.V.} \int_{-\infty}^{\infty} \frac{1 + iA'(y)}{(x-y) + i(A(x) - A(y))} f(y)\, dy.$$

Thus C_A is the Cauchy integral on the graph of A with respect to the graph parameterization, and T_A is the same but using the measure dy instead of $dz(y) = (1 + iA'(y))\, dy$, so that $\|C_A\| \leq (1 + \|A'\|_\infty)\|T_A\|$. One would like to know the growth of $\|T_A\|$ and $\|C_A\|$ as operators on L^2, say, in terms of $\|A'\|_\infty$. This is relevant in building new operators out of the Cauchy integral, as in [**CDM**], and estimates on $\|T_A\|$ and $\|C_A\|$ are important for this. The growth of $\|C_A\|$ also controls what sort of weights are needed for the Cauchy integral on nonregular curves.

The proof of Coifman, McIntosh, and Meyer gave $\|T_A\| \leq C(1 + \|A'\|_\infty)^9$. Using the perburbation method of David (which also gives polynomial estimates) as a starting point, T. Murai [**M**] and P. Tchamitchian [**T**] have improved this estimate, with Murai [**M**] obtaining $\|T_A\| \leq C(1 + \|A'\|_\infty)^{1/2}$. David [**D4**] has shown that this is best possible: for each M large enough, there is a Lipschitz function A such that $\|A'\|_\infty \leq M$ and $\|T_A\| \geq \frac{1}{10} M^{1/2}$. He also shows that $\|C_A\| \leq C(1 + \|A'\|_\infty)^{3/2}$ is best possible.

The Cauchy integral on Lipschitz graphs can also be considered from the point of view of power series. Let Ω denote the open subset of the space of complex-valued Lipschitz functions A on $\mathbb{R}$ such that $\|\operatorname{Re} A'\|_\infty < 1$. On Ω define the nonlinear operator-valued functions $\beta(A)$ and $\gamma(A)$ by

$$\beta(A) f(x) = \text{P.V.} \int_{-\infty}^{\infty} \frac{1}{(x-y) + (A(x) - A(y))} f(y)\, dy,$$

$$\gamma(A) f(x) = \text{P.V.} \int_{-\infty}^{\infty} \frac{1 + A'(y)}{(x-y) + (A(x) - A(y))} f(y)\, dy.$$

If $A = iB$ and B is real valued, then $\beta(A) = T_B$ and $\gamma(A) = C_B$. If A is real valued, then $\gamma(A) = V_A H V_A^{-1}$, where $V_A f(x) = f(x + A(x))$ and H is the Hilbert transform (except for a multiplicative constant),

$$H f(x) = \text{P.V.} \int_{-\infty}^{\infty} \frac{1}{x-y} f(y)\, dy.$$

If $A = A_0 + iA_1$, A_0 and A_1 real valued, and if $h(x) = x + A_0(x)$, $B = A_1 \circ h^{-1}$ (which is still Lipschitz), then $\gamma(A) = V_{A_0} \gamma(iB) V_{A_0}^{-1} = V_{A_0} C_B V_{A_0}^{-1}$. We know that C_B is bounded on $L^2(\mathbb{R})$ for all Lipschitz functions B, and V_{A_0} and $V_{A_0}^{-1}$ are also bounded if $\|A_0'\|_\infty < 1$, and hence $\gamma(A)$ and $\beta(A)$ are bounded operators on $L^2(\mathbf{R})$ for all $A \in \Omega$. For A such that

$$\|\operatorname{Re} A'\|_\infty \leq K < 1 \quad \text{and} \quad \|\operatorname{Im} A'\|_\infty \leq L < \infty,$$

$\|\gamma(A)\|$ and $\|\beta(A)\|$ will be uniformly bounded by some constant $C(K, L)$.

The functions $\beta(A)$ and $\gamma(A)$ are clearly formally holomorphic in A in the sense that they are holomorphic when restricted to finite-dimensional affine subspaces. Because they are locally bounded, they are holomorphic in the strong sense, which means that they have norm convergent power series representations in a neighborhood of each point in Ω. This is a general fact; see [**HP**, Chapter 3], for example.

Let us do the calculation in a specific case. Around $A = 0$, β has the formal power series expansion $\beta(A) = \sum \beta_n(A)$, where

$$(1.4) \qquad \beta_n(A)(f)(x) = (-1)^n \text{P.V.} \int_{-\infty}^\infty \frac{(A(x) - A(y))^n}{(x-y)^{n+1}} f(y)\, dy.$$

Thus $\beta_n(A)$ is a homogeneous polynomial in A of degree n. Because $\beta_n(zA) = z^n \beta_n(A)$,

$$(1.5) \qquad \beta_n(A) = \frac{1}{2\pi} \int_0^{2\pi} \beta(\|A'\|_\infty^{-1} r e^{i\theta} A) \left(\frac{\|A'\|_\infty}{r}\right)^n e^{-in\theta}\, d\theta,$$

and hence $\|\beta_n(A)\| \leq C_r \|A'\|_\infty^n r^{-n}$ for all $r < 1$, where $C_r = \sup\{\|\beta(B)\| : \|B'\|_\infty \leq r\}$. This implies that the series $\sum_{n=0}^\infty \beta_n(A)$ converges absolutely in the L^2 operator norm if $\|A'\|_\infty < 1$, and it converges to $\beta(A)$.

Conversely, a renormalization trick allows one to pass from $\|\beta(A)\| < \infty$ for all A, $\|A'\|_\infty < 1$, to the boundedness of the Cauchy integral on all Lipschitz graphs (see [**CMM**]). In fact, Coifman, McIntosh, and Meyer proved their theorem by showing that $\|\beta_n(A)\| \leq C(n+1)^4 \|A'\|_\infty^n$. Similarly, Calderón's theorem is equivalent to $\|\beta_n(A)\| \leq C^n \|A'\|_\infty^n$ for some $C > 0$, although that is not how he proved it.

Let's review this. Because the Cauchy integral on Lipschitz graphs is bounded, we can find an analytic continuation to the open set Ω, where there is a power series representation. Conversely, if the power series converges, then the Cauchy integral is bounded.

I've cheated here a little, because I have not said exactly what a power series is in this context, that is, for functions defined on an open subset of one Banach space and taking values in another. In essence, a power series is a series of homogeneous polynomials of degree n, $n = 0, 1, 2, \ldots$, and the precise definition is given in (2.2) in the next section. The expansion $\beta(A) = \sum_{n=0}^\infty \beta_n(A)$ is a nice and simple example of what a power series should be, and it should be kept in mind when reading the precise definition.

The restriction of a holomorphic function in the plane to the imaginary axis is real analytic. Similarly, $T_A = \beta(iA)$ is real analytic on the space $\mathbf{R}\operatorname{Lip} 1$ of real-valued Lipschitz functions on $\mathbf{R}$. Real analytic means that there is a norm convergent power series representation about each point, but now the power series is made up of polynomials of real variables, just as in the finite-dimensional case. It is clear what this means for the example of $\beta(A)$; see §2 for the general definition.

Since $\gamma(A)$ is holomorphic on Ω, $C_A = \gamma(iA)$ is real analytic on $\mathsf{R}\operatorname{Lip} 1$. In other words, the Cauchy integral is a real-analytic function of the graph.

When $A \in \mathsf{R}\operatorname{Lip} 1$ and $\|A'\|_\infty < 1$, we saw that $\gamma(A) = V_A H V_A^{-1}$, and so $V_A H V_A^{-1}$ is also real analytic. (However, V_A is not even continuous in the norm topology as an operator-valued function of A.) Thus the Cauchy integral C_A can be obtained by analytic continuation of $V_A H V_A^{-1}$ as a function of A. In some sense the Cauchy integral is the Hilbert transform conjugated by a change of variables that takes the real line to the given curve.

In the beginning of this section, I pointed out that the Cauchy integral is not amenable to the classical techniques for analyzing singular integral operators. However, there do exist now general criteria for the boundedness of singular integrals that have Theorems 1 and 2 as applications. I am referring to the $T(1)$ theorem of David and Journé [**DJ**] and the $T(a)$ theorem of McIntosh and Meyer [**MM**], which has been extended by David, Journé, and myself [**DJS**].

2. The Cauchy integral and Riemann mapping on chord-arc curves.

I'd like to start over now and discuss some work of Coifman and Meyer [**CM2, 3, 4**]. The point of view will be an expansion of an idea encountered in the preceding section, which is to think of the Cauchy integral as a function of the curve, instead of simply as a fixed operator associated to a fixed curve, and to study that function as we would any other function in calculus. The first question is whether this function is continuous, or, rather, what is the natural space of curves on which it is continuous.

To make this precise, I need to be more careful. For Cauchy integrals on different curves to be compared, they must act on the same space. Thus we identify the Cauchy integral defined in (1.2) with the operator C_Γ acting on $L^2(\mathsf{R})$ and defined by

$$(2.1) \qquad C_\Gamma f(t) = \frac{1}{2\pi i} \operatorname{P.V.} \int_{-\infty}^\infty \frac{f(s)}{z(s) - z(t)} z'(s)\, ds,$$

where $z(s)$ is the arclength parameterization of Γ. It is an irritating fact that $z(s)$ is not determined uniquely, and so we normalize by requiring that $0 \in \Gamma$ and $z(0) = 0$.

The first question is when is C_Γ continuous at the "origin," i.e., the real line.

THEOREM 1. $\|C_\Gamma - C_\mathsf{R}\|$ *is small if and only if* Γ *is a chord-arc curve with small constant.*

The norm is the $L^2(\mathsf{R})$ operator norm. (L^p, $1 < p < \infty$, is O.K. too.) The curve Γ is a chord-arc curve with constant $K > 0$ if $|s - t| \leq (1 + K)|z(s) - z(t)|$ for all $s, t \in \mathsf{R}$. (The length of the chord cannot be too much shorter than the length of the arc.) If $K = 0$, then Γ is a line. The precise statement of the theorem is that for each $\varepsilon > 0$ there exists $\delta > 0$ such that $\|C_\Gamma - C_\mathbf{R}\| < \delta$ implies $K < \varepsilon$ and $K < \delta$ implies that $\|C_\Gamma - C_\mathbf{R}\| < \varepsilon$. Calderón's theorem is a consequence of this.

These curves can be naturally parameterized by BMO, the space of locally integrable functions b on $\mathbf{R}$ such that for all intervals I,

$$\frac{1}{|I|} \int_I |b(x) - b_I|\, dx \leq \|b\|_* < \infty, \qquad b_I = \frac{1}{|I|} \int_I b(x)\, dx.$$

BMO functions are defined only up to an additive constant. Note that $L^\infty \subseteq$ BMO, but the inclusion is proper, because $\log|x| \in$ BMO. The John-Nirenberg theorem says that $e^{b(x)}$ is locally integrable if $\|b\|_*$ is small enough, and so $\log|x|$ is fairly typical.

There are many problems for which L^∞ is unnatural and BMO is the right substitute. The Hilbert transform is not bounded on L^∞, but it is bounded on BMO, and BMO is the smallest space containing L^∞ with this property. Another example is the following.

THEOREM 2. *Γ is a chord-arc curve with small constant if and only if $z'(t) = e^{ib(t)}$ for some real-valued $b \in$ BMO with small norm.*

The function b is unique up to adding $2\pi n$. Adding any other real number to b simply rotates Γ about the origin, which doesn't change C_Γ or the chord-arc constant of Γ, and it will be convenient to identify these rotations of Γ with Γ. If $b(t) = \alpha \log|t|$, α small, then Γ is a logarithmic spiral. If $\|b\|_\infty < \pi/2$, then Γ is a Lipschitz graph.

Suppose $b \in$ BMO has small norm, let $s, t \in \mathbf{R}$ be given, and let $I = [s, t]$. Let $\beta_I = e^{ib_I}$, so that $|\beta_I| = 1$. Thus

$$|z(s) - z(t) - \beta_I(s - t)| = \left| \int_s^t (z'(u) - \beta_I)\, du \right| = \left| \int_s^t (e^{i(b(u) - b_I)} - 1)\, du \right|$$

$$\leq \int_s^t |b(u) - b_I|\, du \leq |s - t|\, \|b\|_*,$$

which implies that $(1 + \|b\|_*)^{-1}|s - t| \leq |z(s) - z(t)|$. The other half of the theorem is harder (see [**CM3**]).

Let us come back to the proof of Theorem 1. Let $H^2_+(\Gamma)$ and $H^2_-(\Gamma)$ be the Hardy spaces of $L^2(\Gamma)$ functions that extend holomorphically to Ω_+ and Ω_-, respectively. These may be identified with subspaces of $L^2(\mathbf{R})$ by using the arclength parameterization.

The idea of Coifman and Meyer is that if Γ has small chord-arc constant, then $H^2_+(\Gamma)$ is close to $H^2_+(\mathbf{R})$, and similarly for H^2_-. The precise statement is that if $S_+(\Gamma)$ denotes the orthogonal projection onto $H^2_+(\Gamma)$, i.e., the Szegö projection, then $\|S_+(\Gamma) - S_+(\mathbf{R})\|$ is small. Since the H^2 spaces are moved only a little, the Cauchy integral will move only a little. Indeed, the Cauchy projection $CP(\Gamma) = \frac{1}{2}I + C_\Gamma$ is precisely the nonselfadjoint projection of $L^2(\Gamma)$ onto $H^2_+(\Gamma)$

with kernel $H^2_-(\Gamma)$, by the Plemelj formula (1.3). It is an exercise in abstract operator theory to show that $CP(\Gamma)$ changes slightly if $H^2_+(\Gamma)$ and $H^2_-(\Gamma)$ are changed slightly.

The estimate $S_+(\Gamma) - S_+(\mathbb{R})$ is reduced to the following real-valued problem. Let H denote the Hilbert transform, let $h: \mathbb{R} \to \mathbb{R}$ be a locally absolutely continuous homeomorphism, and let U_h be the unitary operator defined by $U_h f(x) = f(h(x))\sqrt{h'(x)}$. Then $\|U_h H U_h^{-1} - H\|$ is small if $\log h' \in \text{BMO}$ has small norm. This is proved in [**CM2, 3**].

Notice how similar this is to the Helson-Szegö theorem (see [**G**]), which says in particular that $\|e^b H e^{-b} - H\|$ is small if $b \in \text{BMO}$ has small norm. (Here we identify e^b with its multiplication operator.)

The converse part of Theorem 1 can be proved directly or derived from a result of David [**D2**]: If $H^2_+(\Gamma)$ and $H^2_-(\Gamma)$ are almost orthogonal in the sense that their angle is nearly $\pi/2$, then Γ is a chord-arc curve with small constant. This condition on $H^2_+(\Gamma)$ and $H^2_-(\Gamma)$ is equivalent to the condition that $\|S_+(\Gamma) - (I - S_-(\Gamma))\|$ is small.

The same arguments show that this operator is compact iff Γ is asymptotically smooth in the sense of Pommerenke [**P2, 3**], i.e., if the chord-arc constant tends to zero uniformly when $s, t \in \mathbb{R}$ tend to each other or to ∞. One can also characterize the curves for which this operator is trace class, for example (see [**S1**]).

There is also a version of these two theorems relative to any fixed chord-arc curve Γ_0, not just the real line. Note that David's theorem, Theorem 3 of §1, implies that the Cauchy integral on chord-arc curves is bounded.

Let us say that Γ is close to Γ_0 if $z'(t) = e^{ib(t)} z'_0(t)$, where $b \in \text{BMO}$ is real valued and has small norm. If $\|b\|_*$ is small enough, then Γ is chord-arc if Γ_0 is, with nearly the same constant. If b is a constant, then Γ is a rotation of Γ_0 about 0, and we identify Γ and Γ_0. This induces a topology on the space of chord-arc curves, which we denote by Ω.

It also gives a coordinate system about each point of Ω that turns Ω into a Banach manifold modelled on (real-valued) BMO. In fact, one can choose $\arg z'(t)$ in a canonical way that identifies Ω with an open subset of BMO (see [**D1**]). (This choice of $\arg z'(t)$ must be made carefully.)

THEOREM 3. *If Γ_0 is a chord-arc curve, then $\|C_\Gamma - C_{\Gamma_0}\|$ is small if and only if Γ is close to Γ_0 in the above sense.*

Thus the mapping $\Gamma \mapsto C_\Gamma$ is a homeomorphism of Ω onto its image in $B(L^2(\mathbb{R}))$, and so we have chosen the correct topology for Ω. This correspondence is also locally bilipschitz.

More is true: this function is smooth and, in fact, real analytic. A function Λ defined on an open subset U of a real Banach space E and taking values in a real Banach space F is called real analytic if it has a power series representation in a neighborhood of each point $e_0 \in U$. This means that

$$\Lambda(e) = \Lambda(e_0) + \sum_{k=1}^{\infty} \Lambda_k(e - e_0), \qquad \|e - e_0\| < \delta, \qquad (2.2)$$

for some $\delta > 0$, where $\|\Lambda_k(a)\|_F \leq C^k \|a\|_E^k$ for some C, and where Λ_k is a homogeneous polynomial of degree k, i.e, there is a k-linear map $\tilde{\Lambda}_k \colon E \times \cdots \times E \to F$ such that $\Lambda_k(a) = \tilde{\Lambda}_k(a, \ldots, a)$. If E and F are complex Banach spaces and $\tilde{\Lambda}_k$ is complex multilinear, then Λ is holomorphic. We saw in §1 that the Cauchy integral on Lipschitz graphs is a real-analytic operator-valued function.

THEOREM 4. *The mapping $\Gamma \mapsto C_\Gamma$ is a real-analytic mapping of Ω into $B(L^2(\mathbf{R}))$.*

The proof is similar to the argument for Lipschitz graphs, but the calculations are messier. See [**CM3**] for details. Thus the Cauchy integral is the analytic continuation of $V_a H V_a^{-1}$, where $V_a(f) = f \circ h$, $h(x) = \int_0^x e^{a(t)}\, dt$, and a is a real-valued BMO function with small norm. Also, $V_a H V_a^{-1}$ will be a real-analytic operator-valued function of a.

Note that Theorem 4 implies one half of Theorem 3. The other half can be proved directly or by arguments similar to some in [**CM4**].

Coifman and Meyer have also studied the Riemann mapping as a functional on the manifold of chord-arc curves, but the set-up is trickier.

For a given chord-arc curve Γ let Φ be a conformal mapping of the upper half-plane onto Ω_+ such that $\Phi(\infty) = \infty$. It is a theorem of Lavrentiev and Pommerenke that $\log \Phi'(z)$ lies in BMOA(**R**) and that the functions arising this way span an open subset of BMOA (see [**P1, 2, 3**, and **JK**]). (BMOA(**R**) denotes the space of holomorphic functions on the upper half-plane that are Poisson integrals of BMO functions.)

Although this is a reasonable way of parameterizing the space of chord-arc curves, it is not compatible with the structure defined above. The point is that one needs to work with Φ^{-1}. Define a homeomorphism $h \colon \mathbf{R} \to \mathbf{R}$ by $h(t) = \Phi^{-1} \circ z(t)$. By Lavrentiev's theorem and facts about weights, $\log h'(t) \in$ BMO. (See [**JK, CF**, and **J**].)

As a BMO function, $\log h'(t)$ does not depend on the choice of conformal map Φ. If Φ_1 is another, then $\Phi_1(z) = \Phi(az + b)$ for some $a, b \in \mathbf{R}$, $a > 0$. Thus $\log(\Phi_1^{-1})' = \log(\Phi^{-1})' + \log a$, and if h_1 corresponds to Φ_1, $\log h_1' = \log h' + \log a$. Thus $\log h_1' \equiv \log h'$ as BMO functions. This is an important advantage of Φ^{-1} over Φ.

Using the results of [**CM3**] it is not hard to show that the functions $\log h'$ corresponding to chord-arc curves span an open subset $\tilde{\Omega}$ of (real-valued) BMO. I mentioned earlier that David [**D1**] has shown that one can choose $b(t) = \arg z'(t)$ carefully so that the $b(t)$'s corresponding to chord-arc curves also span an open set Ω of (real-valued) BMO, which we identified with the manifold of chord-arc curves. Coifman and Meyer [**CM4**] have proved the following.

THEOREM 5. *The mapping $b \mapsto \log h'$ is a real-analytic map of Ω onto $\tilde{\Omega}$ (in the sense of (2.2)), and its inverse is also real analytic.*

Briefly, the Riemann mapping is a real-analytic function of the curve, and vice-versa.

This would not be true if we worked with h^{-1} instead of h, which means Φ instead of Φ^{-1}. In the topology we are working in, the map $h \mapsto h^{-1}$ is not even continuous: if you perturb h slightly, you may change h^{-1} a lot. It is possible to make this continuous by making smoothness assumptions on h, but this is not natural and the results are not nice. If one tries to force the mapping $h \to h^{-1}$ to be smooth, it gets even worse.

Let me describe some of the main points in the proof of Theorem 5. As in [**CM2, 3**], one observes that $b = -V_h H V_h^{-1}$, $V_h f = f \circ h$. This is a reformulation of the fact that $\log \Phi'$ is holomorphic in the upper half-plane, and it has been used in other problems concerning conformal mapping. See Chapter 10 of Pommerenke's book [**P4**], for example.

To show that $\log h' \to b$ is real analytic, it is enough to show that $\log h' \to V_h H V_h^{-1}$ is real analytic if we think of $V_h H V_h^{-1}$ as an operator on BMO. This comes from extending Theorem 4 and its proof from L^2 to BMO.

To show that $b \to \log h'$ is real analytic, it is enough to show that the differential of $\log h' \to b$ is always invertible on BMO. At the origin it is just $-H$, so $b \to \log h'$ is O.K. near there. At other points it is not clear how to invert the differential.

To overcome this difficulty Coifman and Meyer went through operator theory. It is enough to show that $\log h'$ is a locally Lipschitz function of b. By Theorem 4, the Cauchy integral (as an operator on L^2) is real analytic in b, hence Lipschitz. The same is true of the Cauchy projection $CP(\Gamma) = \frac{1}{2}I + C_\Gamma$, which (by the Plemelj formula) is the nonselfadjoint projection of $L^2(\Gamma)$ onto $H^2_+(\Gamma)$ with kernel $H^2_-(\Gamma)$.

Let $S(\Gamma)$ be again the orthogonal (Szegö) projection of $L^2(\Gamma)$ onto $H^2_+(\Gamma)$. A formula of Kerzman and Stein [**KS**] states that

$$S(\Gamma) = CP(\Gamma)\{I + (CP(\Gamma) - CP(\Gamma)^*)\}^{-1}.$$

The inverse is O.K. by spectral theory, and from this one gets that $S(\Gamma)$ is a Lipschitz function of $CP(\Gamma)$, hence of b. A theorem of David [**D2**] gives that $\log h'$ is a Lipschitz function of $S(\Gamma)$, hence of b. This implies that the differential of $\log h' \to b$ is invertible, as desired.

These five theorems of Coifman and Meyer tell us a lot about the local structure of the space of chord-arc curves and the behavior of the Cauchy integral and Riemann mappings as functionals on it. The global behavior is much less clear. For example, it is still not known if the space of chord-arc curves is connected.

Another problem is to characterize chord-arc curves in terms of conformal welding. Suppose Γ is a Jordan curve in the plane that passes through ∞. Let Φ_+ and Φ_- be conformal maps of the upper and lower half-planes onto the complementary regions of Γ that fix ∞. Define the welding homeomorphism $h\colon \mathbf{R} \to \mathbf{R}$ by $h = (\Phi_+^{-1}|_\Gamma) \circ (\Phi_-|_\mathbf{R})$. The problem is to characterize the h's that correspond to chord-arc curves.

Some partial results are known. David [**D2**] showed that Γ has small chord-arc constant if and only if $\log h'$ has small BMO norm. A consequence of Lavrentiev's theorem (see [**JK**]) is that if Γ is chord-arc, then $h'(x)$ lies in the class A_∞ of weights of Muckenhoupt. This means that h is uniformly absolutely continuous on intervals: for each $\varepsilon > 0$ there is a $\delta > 0$ such that if I is any interval, $E \subseteq I$, and $|E|/|I| < \delta$, then $|h(E)|/|h(I)| < \varepsilon$, where $|E|$ denotes the measure of E. (Ordinary absolute continuity allows δ to depend on I.) The converse is not true. In [**S2**] a nonrectifiable curve is constructed whose corresponding h is bilipschitz, i.e.,

$$C^{-1} \leq \frac{h(x) - h(y)}{x - y} \leq C \quad \text{for all } x, y \in \mathbf{R}.$$

In other words, $C^{-1} \leq h'(x) \leq C$, which is much stronger than $h' \in A_\infty$.

3. Connections with quasiconformal mappings.

There is a strong analogy between the results described in §2 and quasiconformal mappings in the plane, which is similar to the analogy between BMO and the Bloch space (see [**G**]).

Recall that a Jordan curve Γ passing through ∞ is a quasicircle if it is the image of the real line under a quasiconformal mapping of the plane to itself. Ahlfors [**A2**] proved that this is equivalent to the condition that if z_1, z_2 lie on Γ, then the diameter of the arc joining them is at most $C|z_1 - z_2|$, where C does not depend on z_1 and z_2. Like the chord-arc condition, this says that Γ does not smash into itself, but it measures in terms of diameter instead of arclength. Analogous to the first definition of quasicircle there is a theorem of Tukia [**Tu1, 2**] and Jerison and Kenig [**JK**] that a curve is a chord-arc curve iff it is the image of the line under a bilipschitz map of the plane to itself.

In the last section I mentioned that if f is a conformal mapping of the upper half-plane U onto a domain bounded by a chord-arc curve, then $\log f' \in \text{BMOA}$ and these $\log f'$'s span an open subset of BMOA. The quasicircle analogue of this is the classical universal Teichmüller space, defined as follows.

If f is a conformal mapping of U into $\mathbf{C}$, then its Schwarzian derivative $S(f) = (f''/f')' - \frac{1}{2}(f''/f')^2$ lies in the Banach space $\{g$ holomorphic in $U\colon \sup y^2|g(z)| < \infty\}$, by the distortion theorems. Ahlfors [**A2**] proved that the Schwarzian derivatives of the f's such that $f(U)$ is bounded by a quasicircle form an open subset of this Banach space. This open set (with the induced topology) is called the universal Teichmüller space.

Similarly, $\log f'$ lies in the Bloch space $\{g$ holomorphic to $U\colon \sup y|g'(z)| < \infty\}$ and the subset corresponding to the f's such that $f(U)$ is bounded by a quasicircle is again open. In fact, the natural map of $\log f' \to S(f)$ is a homeomorphism of this open set onto the universal Teichmüller space. See [**B**, §6].

The universal Teichmüller space is known to be connected, even contractible, because the mapping theorem allows it to be realized as the continuous image of the unit ball of L^∞ of the lower half-plane (see [**B**, §6]). This argument is not available in the chord-arc case.

It is not clear that there is an analogue for quasicircles of the other topology on the space of chord-arc curves, given by $\arg z'(t) \in \mathrm{BMO}$, since quasicircles do not seem to have a distinguished parameterization like the arclength parameterization.

Quasicircles have a simple characterization in terms of conformal welding (defined at the end of the last section). Using the mapping theorem and the extension theorem of Beurling and Ahlfors [**A2**], Ahlfors showed that if $h: \mathbf{R} \to \mathbf{R}$ is a homeomorphism satisfying

$$\frac{1}{C} \leq \frac{h(x+t) - h(x)}{h(x) - h(x-t)} \leq C,$$

then h corresponds to a quasicircle, unique up to an affine mapping. Conversely, an h corresponding to a quasicircle must satisfy the above doubling condition.

This doubling condition is analogous to the condition $h'(x) \in A_\infty$, stated at the end of §2. For example, doubling homeomorphisms h are characterized by the condition that V_h be bounded on the real Dirichlet space D, where $V_h f = f \circ h$ and D is the space of functions f whose Poisson extension Pf satisfies $\iint_U |\nabla Pf|^2 \, dx \, dy < \infty$. On the other hand, Jones [**Js**] has shown that V_h is bounded on BMO iff $h'(x) \in A_\infty$.

This analogy is not perfect, since $h' \in A_\infty$ does not characterize chord-arc curves by conformal welding. It is not too far off, though, since the h''s that come from chord-arc curves span an open subset of A_∞, where A_∞ is given the topology induced by the fact that $\{\log W : W \in A_\infty\}$ is an open subset of BMO.

Recently I have been able to reconcile, to some extent, this analogy between quasiconformal mappings and the results of §2 (see [**S3**]). The idea is to use some of the formalism of the former plus real-variable tools, like BMO and Carleson measures. This gives a new approach to the Cauchy integral and also a variation of the mapping theorem that is well suited to chord-arc curves. Unfortunately, these results are not strong enough to deal with global problems, like the connectedness of the manifold of chord-arc curves.

To understand how this works, it is important to understand the close connection between the mapping theorem and the Cauchy integral. (Recall that the mapping theorem says that given any $\mu \in L^\infty(\mathbf{C})$, $\|\mu\|_\infty < 1$, there is a q.c. homeomorphism $\rho = \rho_\mu$ such that $\overline{\partial}\rho = \mu \partial \rho$ (see Chapter 5 of [**A3**]).) They are both $\overline{\partial} - \mu \partial$ problems, where $\overline{\partial} = \frac{1}{2}(\partial/\partial x + i\partial/\partial y)$, $\partial = \frac{1}{2}(\partial/\partial x - i\partial/\partial y)$. The proof of the mapping theorem described in [**A3**] and [**AB**] requires estimates for $\overline{\partial}$ and $T = \partial \overline{\partial}^{-1}$, especially the facts that T is an isometry on $L^2(\mathbf{C})$, it is bounded on $L^p(\mathbf{C})$, and its norm on L^p tends to 1 as p tends to 2. Thus if $\|\mu\|_\infty < 1$, then $\|\mu T\| < 1$ on L^p for p close to 2, so that the mapping ρ_μ can be obtained by summing a Neumann series.

These ideas can be used to show that the Cauchy integral on a quasicircle Γ is a bounded operator on the Dirichlet space. Let me explain what I mean by this. To simplify matters let us assume that Γ is smooth but get estimates that depend only on the fact that it is a quasicircle.

A moment ago I defined the real Dirichlet space D on the real line. This can be extended to any quasicircle by requiring it to be preserved by (the restriction to R of the) conformal mapping. Because Γ is a quasicircle, it does not matter which side of Γ the conformal mapping maps to; both give the same space of functions. This can be verified using the Dirichlet principle and the quasi-invariance of the Dirichlet integral.

Thus we know what we mean by the Dirichlet space $D(\Gamma)$ on a quasicircle Γ. To say that the Cauchy integral is bounded on $D(\Gamma)$ means that if $f \in D(\Gamma)$, then $f = f_+ + f_-$, where $f_\pm \in D(\Gamma)$, $\|f_\pm\| \leq C\|f\|$, and f_+ and f_- extend holomorphically to Ω_+ and Ω_-, the complementary domains of Γ. If $F = f_+$ on Ω_+ and $F = -f_-$ on Ω_-, then F is holomorphic off Γ and the jump of F across Γ equals $f_+ + f_- = f$. When Γ is nice, f_+, f_-, and F are given by the Cauchy integral, as in (1.1), (1.2), and (1.3).

This can be reexpressed in terms of distributions. Define the distribution dz_Γ by $\langle g, dz_\Gamma \rangle = \int_\Gamma g \, dz_\Gamma$ for test functions g, so that $dz_\Gamma = \pm \bar\partial \chi_{\Omega_\pm}$ in the sense of distributions, by Green's formula. Hence

$$\bar\partial F = f \, dz_\Gamma.$$

This is important conceptually, but practically it is usually easier to work with the condition that f is the jump of F across Γ. Both conditions essentially characterize F, since the difference between any two such F's must be entire, and hence constant, if mild growth restrictions are assumed. (Elements of $D(\Gamma)$ are determined only up to additive constants.)

Suppose $\rho: \mathbb{C} \to \mathbb{C}$ is a quasiconformal mapping with dilatation μ such that $\rho(\mathsf{R}) = \Gamma$. If $g = f \circ \rho$ and $G = F \circ \rho$, then $(\bar\partial - \mu\partial)G = 0$ off R and G has jump g across R. Let

$$C(g)(z) = \frac{1}{2\pi i} \int_{\mathsf{R}} \frac{g(t)}{t-z} \, dt,$$

which is holomorphic off R and has jump g also. Thus $H = G - C(g)$ has no jump and therefore satisfies $(\bar\partial - \mu\partial)H = \mu C'(g)$ in the distributional sense on all of $\mathbb{C}$, not just on $\mathbb{C} \setminus \mathsf{R}$. Here $C'(g)(z) = (d/dz)C(g)(z)$.

Because Dirichlet integrals are quasi-invariant, g lies in the Dirichlet space $D(\mathsf{R})$, which means that $C'(g) \in L^2(\mathbb{C})$. Since $\|\mu\|_\infty < 1$ and $T = \partial \bar\partial^{-1}$ is an isometry on L^2,

$$\bar\partial H = (I - \mu T)^{-1}(\mu C'(g)) \in L^2(\mathbb{C}),$$

and hence $\partial H = T(\bar\partial H) \in L^2(\mathbb{C})$. This implies that the boundary values of H lie in $D(\mathsf{R})$, by the Dirichlet principle or by direct real-variable arguments.

The boundary values of $C(g)$ from above or below also lie in $D(\mathsf{R})$, since g does, and so the same is true for G. By the quasi-invariance of the Dirichlet integral again, the boundary values f_+ and f_- of $F = G \circ \rho^{-1}$ on Γ from Ω_+ and Ω_- lie in $D(\Gamma)$. Thus the Cauchy integral on Γ acting on $D(\Gamma)$ is bounded.

To use this set-up to estimate the Cauchy integral on a chord-arc curve Γ acting on $\mathrm{BMO}(\Gamma)$ or $L^p(\Gamma)$, we need an estimate more subtle than $\|\mu T\| < 1$ as an operator on $L^2(\mathbf{C})$. It turns out that right conditions are given in terms of Carleson measures.

A positive measure λ on $\mathbf{C}$ is called a Carleson measure relative to $\mathbf{R}$ if for all $x_0 \in \mathbf{R}$ and all $\mathbf{R} > 0$,
$$\lambda(\{z \in \mathbf{C}: |z - x_0| \leq \mathsf{R}\}) \leq C\mathsf{R}.$$
The smallest such C is the norm of λ. Usually Carleson measures are defined relative to the upper half-plane, but that makes little difference. See [G] for basic properties.

For a function $a(z)$ on $\mathbf{C}$ we define conditions (A) and (B) as follows:
(A) $|a(z)|^2 |y|^{-1} \, dx \, dy$ is a Carleson measure;
(B) $|y \nabla a|^2 |y|^{-1} \, dx \, dy$ is a Carleson measure.

Let (B_n) denote the condition (B) with $y \nabla a$ replaced by $y^k \nabla^k a$, $k = 1, \ldots, n$. We say that $a(z)$ satisfies (AB) if it satisfies both, (AB_∞) if it satisfies all of them, and (A + B) if $a = a_1 + a_2$, with a_1 satisfying (A) and a_2 satisfying (B). The associated norms are defined in the obvious way.

In practice these conditions will be applied to a dilatation μ, and smoothness on hyperbolic balls $\{z: |z - z_0| \leq \frac{1}{2}|y_0|\}$, $z_0 = x_0 + iy_0 \in \mathbf{C} \setminus \mathbf{R}$, will not be an issue. Thus the condition (A) should be thought of as stronger than (B), because it requires some vanishing at the boundary. Constants satisfy (B) but not (A), unless they are 0. Smooth functions with compact support always satisfy (B), but they satisfy (A) only if they are zero in $\mathbf{R}$.

There are three basic results.

THEOREM 1. *Suppose Γ is a chord-arc curve with small constant, and $z(t)$ is its arclength parameterization. Then there is a bilipschitz map $\rho: \mathbf{C} \to \mathbf{C}$ with small constant such that $\rho(\mathbf{R}) = \Gamma$, $\rho(t) = z(t)$ for $t \in \mathbf{R}$, the dilatation μ of ρ satisfies (AB_∞) with small (AB_{10}) norm (say), and $\|\mu\|_\infty$ is small.*

THEOREM 2. *Suppose $g \in \mathrm{BMO}(\mathbf{R})$, μ satisfies (A + B) with small norm, and $\|\mu\|_\infty$ is small. Then one can solve $(\overline{\partial} - \mu \partial) H = \mu C'(g)$ so that H has BMO boundary values above and below, with norm estimates.*

THEOREM 3. *Suppose $\rho: \mathbf{C} \to \mathbf{C}$ is quasiconformal, its dilatation satisfies (A + B) with small norm, and $\|\mu\|_\infty$ is also small. Then ρ restricted to $\mathbf{R}$ is absolutely continuous, $\rho(\mathbf{R}) = \Gamma$ is a chord-arc curve with small constant, and in fact $\rho(x) = \rho(0) + \int_0^x e^{\alpha(t)} \, dt$ for $x \in R$, where α is a complex-valued BMO function with small norm.*

In fact, one can generalize Theorem 1 so that it applies to any $z(t): \mathbf{R} \to \mathbf{C}$ satisfying the conclusion of Theorem 3. Theorem 2 can be extended to solve $(\overline{\partial} - \mu \partial) H = R$ for more general functions $R(z)$. For example, it is O.K. if

$|R(z)|\,dx\,dy$ is a good Carleson measure, which means almost the same thing as a Carleson measure except better L^p control is required on hyperbolic balls, which is usually not a problem. Also, there is an L^p version of Theorem 2, $1 < p < \infty$.

All three theorems can be extended to the case where R is replaced by any chord-arc curve Γ_0. For Theorems 2 and 3 it is necessary to assume that the Cauchy integral on Γ_0 is bounded.

In the case where $\rho(\mathsf{R}) = \mathsf{R}$ and μ satisfies (A), Theorem 3 (modified a little) is a consequence of a result of Dahlberg [**Dh**] on the absolute continuity of harmonic measure relative to divergence-form Laplacians. This extended a result of Carleson [**Cr**] in which one assumes that μ satisfies a square Dini condition (stronger than (A)).

Theorems 1 and 2 imply that the Cauchy integral on chord-arc curves with small constant is bounded on BMO (and hence L^p, $1 < p < \infty$, by Calderón-Zygmund theory), by the same argument as we used for the Dirichlet space on a quasicircle. The idea of the proof of Theorem 3 is to run this argument backwards. If $\Gamma = \rho(\mathsf{R})$, then Theorem 2 says that the Cauchy integral on Γ has to be O.K., and one can use this to control $\rho|_\mathsf{R}$.

These results go a long way toward characterizing the quasiconformal maps satisfying the conclusion of Theorem 3 in terms of their dilatation. In a certain sense these mappings are characterized modulo quasiconformal maps that fix each point of R. This is not too bad for two reasons. First, such mappings have to be fairly rigid, because of the distortion theorems. Second, if a complete characterization exists, it is probably a mess, because one can construct quasiconformal maps that fix each point on R but whose dilatations are awful.

Theorem 3 and a generalization of Theorem 1 give a new approach to the map $b \to \log h'$ in Theorem 5 (the Riemann mapping as a functional on the space of chord-arc curves). The idea is to use a remark of Ahlfors [**A4**], that a quasiconformal map taking one domain to another may be transformed into the conformal mapping by using the mapping theorem. These results allow one to do that with BMO estimates.

One can also give a new proof of the result of David on conformal welding of chord-arc curves with small constant, stated at the end of §2. This uses the set-up of Ahlfors [**A2**] for quasicircles, but uses these results to get the right BMO estimates.

A nice example of Theorem 3 and the conditions (A) and (B) on the dilatation is the following. Recall that the theorem of Ahlfors and Weill [**AW**] states that if f is holomorphic on the upper half-plane and its Schwarzian derivative satisfies $|y^2 S(f)| \le k < 2$, then f extends to a quasiconformal map on C. In fact, the dilatation of this extenstion is $-\frac{1}{2} y^2 \overline{S(f)(\overline{z})}$ on the lower half-plane.

If $\log f' \in$ BMOA with small norm, then $y^2 S(f)$ satisfies (A) and (B) with small norm. Conversely, if $y^2 S(f)$ satisfies (A) with small norm, then $y^2 |S(f)(z)| \le \varepsilon$ on U because $S(f)$ is holomorphic, and so the Ahlfors-Weill construction can be applied. Theorem 3 can then be applied to this extension of f, to conclude that $\log f' \in \mathrm{BMO}$, with small norm. This conclusion can be proved much more directly; Lemma 8 in [**Z1**] is a more general result, and its proof is very simple in this case. Nonetheless, it is a good example of how the condition (A) can arise. See [**S3**] for other examples.

What is missing in all of this are the global results, and it is not even clear what they should be. In particular, I do not know how to prove the boundedness of the Cauchy integral on all chord-arc curves, or even all Lipschitz graphs, using these results. It is clear that any naive global version of Theorem 3 should be false, since the naive conjecture for conformal welding is false, as I pointed out at the end of the previous section. See [**S3**] for details and further results. Also, see the papers of Zinsmeister [**Z1, 2**] for more about the BMO–chord-arc curve version of the universal Teichmüller space.

REFERENCES

[**A1**] L. V. Ahlfors, *Zur Theorie der Überlagerungsflächen*, Acta Math. **65** (1935), 157–194.

[**A2**] ____, *Quasiconformal reflections*, Acta Math. **109** (1963), 291–301.

[**A3**] ____, *Lectures on quasiconformal mappings*, Van Nostrand, New York, 1966.

[**A4**] ____, *Conformality with respect to Riemannian metrics*, Ann. Acad. Sci. Fenn. Ser. A 1 **206** (1955), 3–22.

[**AB**] L. V. Ahlfors and L. Bers, *Riemann's mapping theorem for variable metrics*, Ann. of Math. (2) **72** (1960), 385–404.

[**AW**] L. V. Ahlfors and G. Weill, *A uniqueness theorem for Beltrami equations*, Proc. Amer. Math. Soc. **13** (1962), 975–978.

[**B**] J. Becker, "Conformal mappings with quasiconformal extensions," in *Aspects of contemporary complex analysis*, edited by D. Brannan and J. Clunie, Academic Press, London, 1980, pp. 37–77.

[**C1**] A. P. Calderón, *Cauchy integrals on Lipschitz curves and related operators*, Proc. Nat. Acad. Sci. U.S.A. **74** (1977), 1324–1327.

[**C2**] ____, *Commutators, singular integrals on Lipschitz curves and applications*, Proc. Internat. Congr. Math. (Helsinki, 1978), Acad. Sci. Fennica, Helsinki, 1980, pp. 85–96.

[**Cr**] L. Carleson, *On mappings conformal at the boundary*, J. Analyse Math. **19** (1967), 1–13.

[**CF**] R. Coifman and C. Fefferman, *Weighted norm inequalities for maximal functions and singular integrals*, Studia Math. **51** (1974), 241–250.

[**CM1**] R. Coifman and Y. Meyer, "Fourier analysis of multilinear convolutions, Calderón's theorem, and analysis on Lipschitz curves," in *Euclidean harmonic analysis*, edited by J. Benedetto, Lecture Notes in Math., vol. 779, Springer-Verlag, 1979.

[**CM2**] ____, *Le théorème de Calderón par les méthodes de variable réelle*, C. R. Acad. Sci. Paris, Series A, **289** (1979), 425–428.

[**CM3**] ____, "Une généralisation du théorème de Calderón sur l'intégrale de Cauchy," in *Fourier analysis*, Proc. Sem. (El Escorial, 1979), edited by M. de Guzmán and I. Peral, Asoc. Mat. Española, Madrid, 1980.

[**CM4**] ____, *Lavrentiev's curves and conformal mapping*, Institut Mittag-Leffler, Report No. 5, 1983.

[**CM5**] ____, "Non-linear harmonic analysis, operator theory, and P.D.E.," in *Beijing lectures in harmonic analysis*, edited by E. M. Stein, Ann. of Math. Stud. no. 112, Princeton Univ. Press, Princeton, N. J., 1986.

[**CDM**] R. Coifman, G. David, and Y. Meyer, *La solution des conjectures de Calderón*, Adv. in Math. **48** (1983), 144–148.

[CMM] R. Coifman, A. McIntosh, and Y. Meyer, *L'intégrale de Cauchy définit un opérateur borné sur L^2 pour les courbes lipschitziennes*, Ann. of Math. (2) **116** (1982), 361–387.

[D1] G. David, *Thèse de troisième cycle*, Université de Paris XI, 91405 Orsay, France.

[D2] ___, *Courbes corde-arc et espaces de Hardy généralisés*, Ann. Inst. Fourier (Grenoble) **32** (1982), 227–239.

[D3] ___, *Opérateurs intégraux singuliers sur certaines courbes du plan complex*, Ann. Sci. École Norm. Sup. **17** (1984), 157–189.

[D4] ___, *A lower estimate for the norm of the Cauchy integral operator on Lipschitz curves*, preprint.

[DJ] G. David and J. L. Journé, *A boundedness criterion for generalized Calderón-Zygmund operators*, Ann. of Math. (2) **120** (1984), 371–397.

[DJS] G. David, J. L. Journé, and S. Semmes, *A generalized boundedness criterion for singular integral operators*, Rev. Mat. Hisp.-Amer. (to appear).

[Dh] B. E. J. Dahlberg, *On the absolute continuity of elliptic measures*, preprint.

[G] J. B. Garnett, *Bounded analytic functions*, Academic Press, 1981.

[HP] E. Hille and R. Phillips, *Functional analysis and semi-groups*, Colloq. Publ., vol. 31, Amer. Math. Soc., Providence, R. I., 1957.

[J] J. L. Journé, *Calderón-Zygmund operators, pseudo-differential operators, and the Cauchy integral of Calderón*, Lecture Notes in Math., vol. 994, Springer-Verlag, 1983.

[Js] P. Jones, *Homeomorphisms of the line which preserve BMO*, Ark. Mat. **21** (1983), 229–231.

[JK] D. Jerison and C. Kenig, *Hardy spaces, A_∞, and singular integrals on chord-arc domains*, Math. Scand. **50** (1982), 221–247.

[K] C. Kenig, *Weighted Hardy spaces on Lipschitz domains*, Amer. J. Math. **102** (1980), 129–163.

[KS] N. Kerzman and E. Stein, *The Cauchy integral, the Szegö kernel and the Riemann mapping functions*, Math. Ann. **236** (1978), 85–93.

[M] T. Murai, *Boundedness of singular integrals of Calderón type (VI)*, Adv. in Math. (to appear).

[Ma] D. Marshall, "Removable sets for analytic functions," in *Linear and complex analysis problem book*, Lecture Notes in Math., vol. 1043, Springer-Verlag, 485–490.

[MM] A. McIntosh and Y. Meyer, *Algèbres d'opérateurs définis par des intégrales singulières*, C. R. Acad. Sci. Paris Sér. I Math. **301** (1985), 395–397.

[P1] Ch. Pommerenke, *Schlichte Funktionen und analytische Funktionen beschränkter mittlerer Oszillation*, Comment. Math. Helv. **52** (1977), 591–602.

[P2] ___, *On univalent functions, Bloch functions, and VMOA*, Math. Ann. **236** (1978), 199–208.

[P3] ___, "Boundary behaviour of conformal mappings," in *Aspects of contemporary complex analysis*, edited by D. Brannan and J. Clunie, Academic Press, London, 1980, pp. 313–331.

[P4] ___, *Univalent functions*, Vandenhoeck and Ruprecht, Göttingen, 1975.

[S1] S. Semmes, *The Cauchy integral and related operators on smooth curves*, preprint.

[S2] ___, *A counterexample in conformal welding concerning chord-arc curves*, Ark. Mat. **24** (1986) (to appear).

[S3] ___, *BMO-Carleson measure estimates for $\overline{\partial} - \mu \partial$ and applications to the Cauchy integral and quasiconformal mappings*, Trans. Amer. Math. Soc. (to appear).

[T] P. Tchamitchian, *La norme de l'opérateur de Cauchy sur les courbes Lipschitziennes*, preprint.

[Tu1] P. Tukia, *The planar Schonflies theorem for Lipschitz maps*, Ann. Acad. Math. Sci. Fenn. Ser. A I Math. **5** (1980), 49–72.

[Tu2] ___, *Extension of quasisymmetric and Lipschitz embeddings of the real line into the plane*, Ann. Acad. Math. Sci. Fenn. Ser. A I Math. **86** (1981), 89–94.

[Z1] M. Zinsmeister, *Représentation conforme et courbes presque lipschitziennes*, Ann. Inst. Fourier (Grenoble) **34** (1984), 29–44.

[Z2] ___, *Domaines réguliers du plan*, Ann. Inst. Fourier (Grenoble) **35** (1985), 49–55.

Zippers and Univalent Functions

WILLIAM P. THURSTON

1. Introduction. Every simply connected domain **C**, with the exception of **C** itself, can be parametrized by a conformal map from the unit disk, that is, by a univalent function. Families of domains which vary continuously with parameters arise in several mathematical contexts, particularly in the theory of Kleinian groups and in the theory of other 1-dimensional complex dynamical systems. A family of domains gives rise to a family of univalent functions. There are various topologies on the set of univalent functions. One topology which has proven valuable for the theory of Kleinian groups is the topology of uniform convergence of the Schwarzian derivative (to be described presently), uniform as measured using the hyperbolic metric on the disk. In this note, we will construct simple examples of domains whose corresponding univalent functions are isolated points in this topology. These domains have complements which are arcs, in fact, quasi-intervals, and they are reminiscent of zippers.

The construction and result here are related to that of Gehring [1].

2. Preliminaries. For the benefit of people to whom the Schwarzian derivative may seem a mystery, we will set the stage by discussing the Schwarzian derivative. Actually, the Schwarzian derivative is not really essential to the current discussion. We will use it only for the definition of a topology on the set of univalent maps, and logically we really use an equivalent description of the topology (to be explained later) which makes no mention of Schwarzian derivatives. This is not meant to downplay the significance of the Schwarzian derivative; by analogy, many qualitative constructions in differential geometry can be done without ever mentioning curvature, even though curvature is central to differential geometry.

The Schwarzian derivative is very much like a kind of curvature: the various kinds of curvature in differential geometry measure deviation of curves or manifolds from being flat, while the Schwarzian derivative measures the deviation of a conformal map from being a Mœbius transformation. A Mœbius transformation is determined by its two-jet (its value and first two derivatives) at any point. The two-jet is arbitrary, except for the proviso that the first derivative be nonsingular. Thus, for any locally univalent f and any point w in the domain of f there is associated a unique Mœbius transformation $M(f, w)$ which agrees with f through second order. Note that if A is a Mœbius transformation, then

$$M(A \circ f, w) = A \circ M(f, w).$$

More generally, if f and g are two locally univalent functions, then

$$M(g \circ f, w) = M(g, f(w)) \circ M(f, w).$$

Conformal mappings up to postcomposition (composition on the left) with Mœbius transformations are therefore conveniently described by their three-jets, up to the action of Mœbius transformations on three-jets. The quotient space of this action, at any point z in the domain of the mapping, is isomorphic to **C**. To see the isomorphism, normalize a map f by postcomposing with the Mœbius transformation $M^{\circ-1}(f, w)$ to obtain a new map having the two-jet of the identity at w. The information that remains in the three-jet is the third derivative, which is an arbitrary element of **C**. Note that if the original map is a Mœbius transformation, it will have been normalized to be the identity, so its third derivative will be 0.

As z varies, this determines a complex vector bundle over the domain of the map whose fiber consists of the bundle of three-jets modulo the chain rule action of Mœbius transformations on three-jets. The linear structure in the fiber arises not because the action is linear (which it isn't), but because a linear structure is inherent in the complex structure of **C**, once the origin is determined.

What vector bundle is it? A third derivative is a cubic mapping from the tangent space of the domain to the tangent space of the range. However, for a nonsingular mapping, the tangent space of the domain is identified with the tangent space of the range via the first derivative, so we may describe the third derivative in terms of a cubic map P from the tangent space of the domain to itself. We can form the quotient $P(w)/w$ to replace this cubic map by a quadratic map from the tangent bundle of the domain to **C** which contains the same information.

It follows that this invariant for locally univalent functions up to Mœbius transformations, called the Schwarzian derivative, is a holomorphic quadratic differential form, usually referred to as a quadratic differential.

Sketching the picture of a conformal map which has the two-jet of the identity at a point (Figure 2.1: jets of locally univalent mappings), one can see that the third derivative tends to make the image of a small circle around a point in the domain look elliptic in the range. Since Mœbius transformations preserve circles, the Schwarzian derivative may be interpreted as measuring the third-order distortion of the shape of the image of a small circle.

A formula for the Schwarzian derivative can be readily determined from the information above, or it may be looked up—someplace else. Like the formula for the curvature of a curve in the plane, the formula looks somewhat mystical at first, and in a qualitative discussion, the formula tends to be a distraction from the real issues.

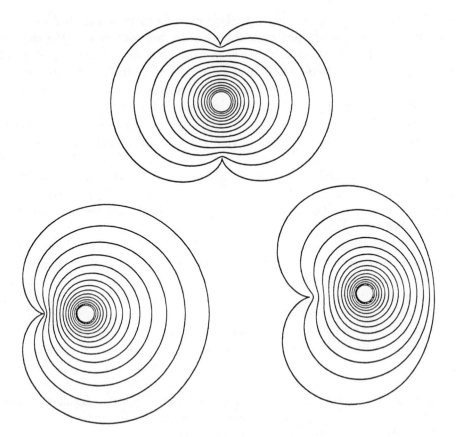

2.1. FIGURE: JETS OF LOCALLY UNIVALENT FUNCTIONS. Here are three univalent mappings illustrating the effects of second and third derivatives of a locally univalent mapping on a family of concentric circles. In (a) the mapping is $z \mapsto z + z^3/3$, and in (b), $z \mapsto z + z^2/2$. Thus, mapping (a) has the two-jet of the identity, and the third-order effect on the eccentricity of circles, as measured by the Schwarzian derivative, can be seen. Mapping (b) has the 1-jet but not the 2-jet of the identity, so even though its third derivative is 0, its Schwarzian derivative is not 0. The second derivative can be seen in the way that the centers of the images of circles are displaced toward the right. The 2-jet can be corrected by composing with the Mœbius transformation $z \mapsto z/(1-.5z)$. The composition is illustrated in (c). The third-order term in the Taylor expansion of the composition is $-z^3/4$, and again it can be seen in the third-order effect on eccentricity.

How does the Schwarzian derivative behave under composition? If f and g are two locally univalent mappings which have order two contact with the identity at a point z, then the third derivative of the composition is the sum of their third derivatives, so the Schwarzian derivative of the composition is the sum of the Schwarzian derivatives, at least in this case. The same formula holds in general, provided we are careful to treat the Schwarzian derivative as a quadratic form. This easily follows from the fact that the Schwarzian derivative is invariant under postcomposition by a Mœbius transformation, and it transforms as a quadratic form under precomposition by a Mœbius tranformation. More explicitly, for the Schwarzian derivative of an arbitrary composition $f \circ g$ evaluated at a point z, one can replace it by $(f \circ M_1) \circ (M_1^{\circ^{-1}} \circ g \circ M_2)$, where M_1 and M_2 are the Mœbius transformations chosen so that the terms in parentheses have the 2-jet of the identity at $f \circ g(z)$.

The Schwarzian derivative together with the 2-jet of a locally univalent mapping determines the 3-jet, which is to say that the mapping satisfies a third-order differential equation in terms of its Schwarzian derivative. Since it is an ordinary differential equation, the equation can be locally integrated no matter what the holomorphic quadratic differential. Thus, every quadratic differential occurs, at least locally, as the Schwarzian derivative of a conformal map. Since Mœbius tranformations act transitively on nonsingular 2-jets, the three degrees of freedom for the solution are accounted for by compositions with Mœbius transformations. If the quadratic differential is defined over a simply connected domain, local solutions can be pasted together using Mœbius transformations to give a locally univalent mapping defined from the entire domain to the Riemann sphere $\hat{\mathbf{C}}$ and having the prescribed Schwarzian derivative.

Here is a slightly different way to think of the Schwarzian derivative. As a point w varies in the domain of f, the approximating Mœbius transformation $M(f, w)$ varies. The derivative of $M(f, w)$ with respect to w associates to any tangent vector V at w an infinitesimal Mœbius transformation, which is a holomorphic vector field on the Riemann sphere

$$X(f, w, V) = \frac{\partial}{\partial t} M^{\circ^{-1}}(f, w) \circ M(f, w + tV).$$

The order of composition has been chosen so that $X(A \circ f, w, V) = X(f, w, V)$, so we may normalize f to have order 2 contact with the identity at w without changing X. Then it is clear that $X(f, w, V)$ vanishes through order 1 at w. X is thus determined by its second derivative, which is an element of $\hom(T \otimes T, T)$, where T is the tangent space to $\hat{\mathbf{C}}$ at w. Therefore we have as an invariant of f at w an element of $\hom(T, (T \otimes T, T))$, which is canonically isomorphic to $\hom(T \otimes T \otimes T, T)$, which in turn is isomorphic to $\hom(T \otimes T, \mathbf{C})$ (taking into account the fact that T is 1-dimensional).

This invariant is, of course, the same as the Schwarzian derivative of f, but it has perhaps a clearer interpretation as measuring the rate of change of the best approximating Mœbius transformation.

Consider a univalent mapping f, that is, a conformal embedding of the open unit disk in $\mathbf{C}$. We can normalize f by postcomposition with an affine transformation until it has the 1-jet of the identity at the origin. Once this is done, it is easy to see that each derivative of f is bounded in a way which is independent of f. It follows that the Schwarzian derivatives of univalent functions are uniformly bounded with respect to the hyperbolic metric on the domain disk.

One useful topology on the set of univalent functions is the topology of uniform convergence on compact subsets of the open disk; this topology is used in the theory of normal families, and in particular, in the standard proof of the Riemann mapping theorem. The group of complex affine transformations acts on univalent functions, and the quotient space is contractible with respect to this topology: if f is normalized to have derivative 1 at the origin, then the isotopy

$$f_t(z) = \frac{1}{t} f(tz) \qquad [0 < t \leq 1]$$

connects f to the identity map (which is the limit as t approaches 0).

This topology is much too weak for many other purposes, however. Since Schwarzian derivatives are uniformly bounded, another natural topology comes from uniform convergence on quadratic differentials in the hyperbolic plane. This works best on the set of univalent maps up to postcomposition with Mœbius transformations. Denote this space by S. An element of S may be represented by any conformal embedding of the unit disk on the Riemann sphere $\hat{\mathbf{C}}$—up to Mœbius transformations; it is pointless to require that infinity is in the complement of the image.

Because of the chain rule above for Schwarzian derivatives, an ε neighborhood of a univalent mapping f consists precisely of mappings which have the form $g \circ f$, where g is a univalent mapping defined on the image of f whose Schwarzian derivative has norm less than ε as measured in the Poincaré metric of the image of f.

Here is a macroscopic (rather than infinitesimal) description of the topology of S. If f_1 and f_2 are two univalent mappings, we can compare them by considering the disk of radius R with respect to the hyperbolic metric about a point z in the domain of the mappings. Change coordinates to put z at the origin, and normalize the mappings to have the 2-jet of the identity at 0, i.e., let B be a conformal automorphism of the unit disk sending the origin to z, and let the normalized transformation be $g_i = M^{\circ^{-1}}(f_i \circ B, 0) \circ f_i \circ B$. Then define $d_{z,R}(f_1, f_2)$ to be the maximum distance between g_1 and g_2 in the hyperbolic disk of radius R about the origin, measured with respect to the spherical metric in the range. We can define f_1 and f_2 to be (ε, R) close if $d_{z,R} < \varepsilon$ for all z in the open unit disk. This determines a neighborhood basis for a topology. Actually, we would arrive at a neighborhood basis for the same topology if we picked any particular R and held it fixed.

In fact, the resulting topology would be the same if the normalized maps were prescribed to be not only pointwise close, but C^r close, since derivatives of holomorphic maps are estimated in terms of values, on account of the Cauchy integral formula. Clearly, this implies the Schwarzian derivatives are close. On the other hand, solutions of differential equations depend continuously on the data, so if the Schwarzian derivatives of f_1 and f_2 are uniformly close, then f_1 and f_2 are (ε, R) close. Thus, the topology is the same as that of S.

What is the macroscopic analogue of the description of neighborhoods of univalent functions in terms of compositions with functions having small Schwarzian derivatives? To construct such a description, we must modify the definition of $d_{z,R}$ so that it can be applied to functions defined on a domain U other than the unit disk. The difficulty is that we need a way to measure the distance between the values of two mappings which is not affected by postcomposition with a Mœbius transformation. In the infinitesimal setting, we used the Poincaré metric of the domain, but this will not do in the macroscopic setting since the mapping is unlikely to have image contained in U.

The solution is to measure using the unique round metric $R(z, U)$ on $\hat{\mathbf{C}}$ which has first-order contact to the Poincaré metric at a point z. By round, we mean the metric is equivalent to the standard spherical metric by a Mœbius transformation. If B is a conformal mapping of the unit disk to U which sends 0 to z, then this round metric is the image of the standard spherical metric under $M(B, 0)$. More generally, the round metric associated to a point z in a domain U transforms by $M(g, z)$ if g is a conformal embedding of U in $\hat{\mathbf{C}}$. (Note that the transformation of the 1-jet of a metric depends on the 2-jet of a mapping.)

One way to visualize the associated round metric is to observe that a choice of a round metric is equvialent to a choice of a point p in the $\mathbf{H}^3$ bounded by $\hat{\mathbf{C}}$. The round metric is the visual metric at p, that is, it is induced from the metric on the unit tangent space to $\mathbf{H}^3$ at p via the homeomorphism which sends a tangent vector to the endpoint of its geodesic on the sphere at infinity. Thus, to a point z in a domain U there is canonically associated a point $p(z, U)$ in $\mathbf{H}^3$. In the case of a round disk, the associated point is the perpendicular projection to the hyperbolic plane with the same bounding circle. The point $p(z, U)$ is transformed by the formula $p(f(z), f(U)) = M(f, z) \circ p(z, U)$ whenever U is mapped to another domain by an embedding (or even a covering map).

Now it is clear that another neighborhood basis for an element f of S consists of the sets of maps of the form $g \circ f$, where g is a conformal embedding of the image U of the unit disk which has the property that for each $z \in U$, $M^{\circ^{-1}}(g, z) \circ g$ is within ε of the identity when restricted to the Poincaré ball of radius R about z and measured using $R(z, U)$.

We shall prove that S has uncountably many isolated points. This contrasts strongly with the contractibility of the set S when it is equipped with the topology of uniform convergence on compact subsets.

In passing, we point out that there is a close connection developed by Lipman Bers between the space S and Teichmüller spaces. In fact, denote by U the space of locally univalent functions with the topology of uniform convergence of

Schwarzian derivatives, so that $S \subset U$ has the topology induced from U. The interior T of S is called the universal Teichmüller space. All ordinary Teichmüller spaces are embedded in the universal Teichmüller space in a natural way. T is also characterized as consisting of those univalent functions whose image is bounded by a quasicircle (see definition below). The main result of Gehring [1] is that S is not the closure of T.

3. Zippers. The image of a univalent mapping determines the mapping, up to precomposition with an isometry of the domain hyperbolic plane. The isometries of $\mathbf{H}^2$ do not act continously on S. Still, we will analyze univalent functions in terms of the shapes of their images.

We will say that an open set U is *rigid* if there is some ε such that any conformal embedding U in $\hat{\mathbf{C}}$ with Schwarzian derivative less than ε is a Mœbius transformation. When U is simply connected, this is equivalent to the condition that *every* Riemann mapping from the open disk to the domain represents an isolated point in S. Note that the condition of rigidity makes sense even when the domain is not simply connected.

A *quasi-interval* in $\hat{\mathbf{C}}$ is an arc α in $\hat{\mathbf{C}}$ with the property that there is a constant K such that for any two points $a, b \in \alpha$, the subarc between a and b lies in a disk of radius $K\, d(a,b)$ about a. Ahlfors showed that quasi-intervals are exactly the quasiconformal images of a unit interval in the real line. (Note: this follows from a similar characterization of the quasiconformal images of circles, using a square root construction.) The main point of this paper will be to prove

3.1. THEOREM: RIGID QUASI-INTERVALS. *There are many quasi-intervals with rigid complement.*

A more precise formulation of the statement will be given in Theorem 4.3: incorrigible implies rigid, where "many" will be replaced by a specific definition.

The proof of this theorem may make it sound like S has such a strong topology that almost no domains should be nonrigid, except for cases such as Jordan domains whose complement contains an open set, and geometrically simple cases such as domains with piecewise-smooth frontier. Perhaps so. On the other hand, there are many interesting nonrigid examples which arise in the theory of Kleinian groups: the complements of the limit sets of all known degenerate groups are nonrigid. Maybe the technique here can be turned around to help give information about the geometry and topology of these limit sets.

We will now give an intuitive description of a construction for rigid quasi-intervals. Rather than making the specific estimates needed to demonstrate the particular example rigorously, we will in the next section give a general proof that a general class of quasi-intervals (including the ones we construct) have rigid complements.

The idea is based on zippers. Zippers have the property that they cannot be pulled apart in the middle without a lot of distortion. In the same way, the

3.2. FIGURE. A simple zipper curve.

simple zipper curve of Figure 3.2 has the property that it separates the plane into two components which cannot be moved apart disjointly without first making a definite amount of distortion, because the zipper teeth on the two sides of the curve interlock. The usual custom in the physical world is to make zippers with rectilinear teeth, but these smooth teeth will serve the same purpose while being more convenient for constructing higher-order zippers.

The curve of Figure 3.2 is part of an embedded line in $\mathbf{C}$ invariant by a translation, which is a parabolic Mœbius transformation. Similarly, we can construct a simple zipper arc Z_1 which is invariant by a hyperbolic Mœbius transformation. Any conformal embedding of the complement of Z_1 with small Schwarzian derivative (measured as a quadratic differential using the Poincaré metric of the complement of Z_1) will have zipper teeth of approximately the same shape. If ε is small enough that a few of the teeth are interlocked, there is a cascade effect: all the other teeth will be forced to be interlocked.

The complement of Z_1 is not rigid. For one thing, the Schwarzian derivative varies continuously if the curve is wiggled in the C^1 topology. To take care of this, we iterate the zipper construction. Inductively replace the curve Z_k by a new curve Z_{k+1} lying in a thin band about the original and having teeth of a smaller scale. There is a limit curve Z_∞.

Any conformal embedding of the complement of Z_∞ with Schwarzian derivative uniformly near 0 must have the largest of the top level teeth interlocked. Because of the cascade effect, all the top level teeth are interlocked. Also, this forces the level 2 teeth to be approximately interlocked; hence they too are interlocked. The cascade effect proceeds to finer and finer levels, forcing all teeth at all levels to be interlocked.

This implies that the new conformal embedding extends continuously over Z_∞, giving a homeomorphism of $\hat{\mathbf{C}}$ which is conformal at least in the complement of Z_∞.

3.3. PROPOSITION: EXTENSION CONFORMAL. *Any homeomorphism of $\hat{\mathbf{C}}$ which is conformal in the complement of a quasi-interval α is a Mœbius transformation.*

PROOF. By the theorem of Ahlfors mentioned above, α can be straightened to a line segment by a quasiconformal homeomorphism. The proposition is easy to prove in the case of a line segment, using the Cauchy integral formula. This implies that the original map f is at least quasiconformal. Its quasiconformal distortion is zero almost everywhere; therefore f is conformal. Consequently, it is a Mœbius transformation. □

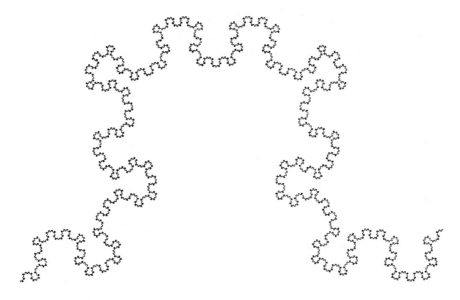

3.4. FIGURE: ITERATED ZIPPER CURVE. A detail of the iterated zipper arc Z_∞. The picture will look much the same if any small tooth is enlarged to the size of this large tooth. On a large scale, the entire zipper arc fits inside a banana-shaped region of the $\mathbf{C}$, and it is invariant by a hyperbolic transformation.

This gives a demonstration, at least on the intuitive level, that the complement of Z_∞ is rigid.

4. Compactness. We will make a more rigorous and more general analysis by considering all possible quasi-intervals microscopically. For this, it is convenient to consider quasi-intervals in $\mathbf{C}$, so that a K-quasi-interval when magnified by any constant is still a K-quasi-interval. (On the 2-sphere, the magnification would bring the two endpoints close together, so that the constant would have to go to infinity.) As a limiting case, there are also quasilines and quasihalf-lines, which are properly embedded copies of $(-\infty, \infty)$ or $[0, \infty)$.

The Hausdorff metric is a metric on the set of closed subsets of a metric space which says that two sets are within ε if every point in either set is within ε of some point of the other set. Since $\mathbf{C}$ is noncompact, it is convenient to extend closed subsets of $\mathbf{C}$ to subsets of $\hat{\mathbf{C}}$ by forming the closure, and to measure the Hausdorff distance of the resulting closed subsets of $\hat{\mathbf{C}}$ using the spherical metric. The set of closed subsets of $\mathbf{C}$ with this topology forms a compact metric space.

Let $Q(K)$ be the set of closed sets consisting of the empty set and all singletons, together with all K-quasi-intervals, K-quasilines, and K-quasihalf-lines.

4.1. PROPOSITION: CLOSURE QUASIARCS. *The set $Q(K)$ is closed and hence compact.*

PROOF. This is left as a nice exercise for the reader. □

Given any K-quasi-interval α, let $E(\alpha)$ consist of all enlargements of α, that is, the set of all images of α under transformations $z \mapsto az + b$, where $|a| \geq 1$.

4.2. DEFINITION. A K-quasi-interval α is *incorrigible* if the closure of $E(\alpha)$ contains no straight lines.

It is easy to see that incorrigibility is equivalent to the condition that the closure of $E(\alpha)$ contains no circles.

4.3. THEOREM: INCORRIGIBLE IMPLIES RIGID. *The complement of any incorrigible K-quasiarc is rigid.*

4.4. COROLLARY: RIGID DOMAINS EXIST. *There are domains which are rigid.*

PROOF. For one example, use the complement of Z_∞. There will be no lines in the closure of $E(Z_\infty)$ provided the construction is carried out in such a way that the ratio of the size of the teeth of the kth level to the teeth of the $(k+1)$st level is bounded.

A more interesting example comes from the limit set of any quasi-Fuchsian group which has no parabolic elements and which is not Fuchsian. Let α be any arc on such a limit set. The limit set has the property that the set of subsets of $\mathbf{C}$ obtained by conjugating the limit set by any Mœbius transformation which makes it pass through 0 and ∞ is already a closed subset of $Q(K)$, for some K. In fact, the set of such conjugating transformations, up to the action of the quasi-Fuchsian group, is compact. This implies that $E(\alpha)$ contains no lines. □

Note that if a quasi-Fuchsian group contains parabolics, the closure of $E(\alpha)$ in fact does contain lines, although a refinement of the proof of the theorem would show that the complement of any subarc of such a limit set is rigid anyway.

PROOF OF THEOREM 4.3. We will prove the theorem by proving a more positive form of the statement: If α is a K-quasi-interval and f_i a sequence of conformal embeddings which are not Mœbius transformations but whose Schwarzian derivatives converge uniformly to 0, then there are lines (in fact, many lines) in the closure of $E(\alpha)$.

Note that the closure of $E(\alpha)$ is invariant under the process of enlargement. What we will do is pass to a limit of enlargements of α, then a limit of enlargements of this, and so on until we find a line.

We may assume that f_i is normalized at some base point on $\hat{\mathbf{C}}$, to have the 2-jet of the identity transformation at that point. It is convenient to take as base point the point at infinity. The f_i converge uniformly on compact sets of the domain $U = \hat{\mathbf{C}} - \alpha$ to the identity.

We need a bit of information about the qualitative properties of a quasi-interval and the behavior of maps with small Schwarzian derivatives.

4.5. LEMMA. *Let t be any point in the interior of α, and let D be the disk of radius $3Kr$ about t. There must be points within D which are at distance exactly r from α.*

PROOF. Indeed, t separates α into two subarcs α_1 and α_2. Let A be the annulus about t with inner radius $2Kr$ and outer radius $3Kr$. The r-neighborhoods of the α_i intersected with A must be disjoint, since α is a K-quasiarc.

In the typical case that these intersections are nonempty, since the annulus is connected we can find points in the frontier of their union, which is the r-neighborhood of α, as claimed. If each α_i intersects the circle of radius $3Kr$, we can even find a point as claimed on both homological sides of the intersection of α with the $3Kr$-disk.

Even near the ends of α, where only one of the pieces (say α_1) of α intersects the annulus, a related argument still works. To see this, consider the double cover of $\mathbf{C}$ branched at t. The inverse image of the r-neighborhood of α_1 intersect A must have two components, so there are points in their frontier. □

The lemma implies that the collection of disks centered about points in the complement of α whose radius is $3K$ times the distance to α cover $\mathbf{C}$. In fact, the subcollection consisting only of those disks of any fixed radius cover some neighborhood of α.

We will define a quantity $D(f,z)$ which is a measure of how far f is distorted from a Mœbius transformation in the vicinity of $z \in U$. D will measure distortion roughly on the scale of the distance $r(z)$ of z from α. The definition is that $D(f,z)$ is the maximum distance of $M^{\circ^{-1}}(f,z) \circ M(f,w)$ from the identity, measured with respect to the round metric $R(z,U)$ on $\hat{\mathbf{C}}$, as w ranges over the set of points a distance at least $r(z)/2$ from α but distance at most $10Kr(z)$ from z.

Note that $D(f,z)$ goes to 0 as z goes to infinity. What happens for z near α?

4.6. PROPOSITION: SMALL DISTORTION IMPLIES CONTINUOUS. *There is a constant $a > 0$ such that if $D(f,z)$ is less than a in some neighborhood of α, then f extends continuously to α, and hence f is a Mœbius transformation.*

PROOF. The condition that a point w is within the disk of radius $10Kr(z)$ about z but has distance more than $r(z)/2$ from α implies that the round metrics $R(z,U)$ and $R(w,U)$ are roughly compatible, or in other words, there is an a priori bound for $d(p(z,U), p(w,U))$. In fact, among points z such that $r(z)$ is less than a bounded fraction of the diameter of α, the metric $R(z,U)$ deviates by a bounded amount from the metric obtained by enlarging the picture until $r(z) = 1$, then mapping to a sphere via the stereographic projection which sends the unit circle about z to the equator of the sphere. To see this, consider the case that $r(z) = 1$. Make a parabolic transformation fixing z so that α goes through ∞. The first two derivatives of a univalent function which sends the origin to z are bounded, which means that the 1-jet of the Poincaré metric deviates by a bounded amount from the 1-jet of the Euclidean metric, which is the same as the 1-jet of the round metric defined by stereographic projection.

The proof of the proposition has to do with the nesting properties of the disks of radius $10Kr(z)$ about points z of distance $r(z)$ from α.

For any point z' with $r(z') = r(z)/2$ and such that the $3Kr$ disks for z and z' intersect, then they are at distance no more than $3Kr(z) + 3Kr(z') = 4.5Kr(z)$ from each other. Then the $10Kr(z') = 5Kr(z)$ disk about z' is contained in the $10Kr(z)$ disk about z, leaving a $r(z)/2$ margin of safety.

Under the hypothesis that $D(f,z)$ is small, the transformation $M^{\circ^{-1}}(f,z) \circ M(f,z')$ sends the $10Kr(z')$ disk safely inside the $9.9Kr(z)$ disk about z.

Consider a pair $A \supset B$ of nested disks and all possible images of them under Mœbius transformations taking A to the finite complex plane. There is a maximum for the ratio of the radii of these images; this upper bound is attained when the two disks are concentric. It follows that there is an upper bound $\beta < 1$ to the ratio of the radius of the image of the $10Kr(z')$ disk about z' by $M(f,z')$ to that of the $10Kr(z)$ disk about z under $M(f,z)$, where z and z' range over pairs of points chosen as above. In fact, β can be found as a function only of $D(f,z)$; in the circumstances above, certainly $\beta < .99$.

Renormalize f for convenience to make sure that $M(f,z)$ maps the $10Kr(z)$ disk about z to a disk E in the finite complex plane. Let R be the radius of E. The image of the $10Kr(z')$ disk about z' by $M(f,z')$ is contained in E and has a radius less than βR.

Inductively, it follows that for any point w such that $r(w) = 2^{-k}r(z)$ whose $3Kr(w)$ disk intersects the $3Kr(z)$ disk about z, the diameter of its image under $M(f,w)$ is a disk of radius less than $R\beta^k$ contained in the original disk E. In particular, the image by f of the entire $3Kr(z)$ disk about z is contained in the image by $M(f,z)$ of the $10Kr(z)$ disk.

Consequently, for any point t on α, there are neighborhoods of t in **C** whose images under f (where f is defined) have arbitrarily small diameter. Therefore f converges along α to a continuous map. □

It is interesting to remark here that when f is a mapping with small Schwarzian derivative, the hypothesis of the proposition would be satisfied if the measure of distortion $D(f,z)$ were modified so that only points w are considered which are on the same side of α as z. The proof of the proposition therefore shows that f has a continuous limit on each side of α—but in general, the limits from the two sides will be different.

PROOF OF THEOREM 4.3 (CONTINUED). Now back to our sequence f_i. By assumption, the f_i are not Mœbius transformations, so there is a constant a such that $D(f_i,z) > a$ for points z arbitrarily close to α.

By the intermediate value theorem, for each sufficiently large integer $n > 0$, there is a point $x_{i,n}$ where D_i takes the value $1/n$. Let $f_{i,n}$ be a rescaling of f_i so that $x_{i,n}$ is at the origin, and α becomes a curve $\alpha_{i,n}$ at distance one from the origin. We renormalize $f_{i,n}$ to have the 2-jet of the identity at the origin.

If we fix n and let i tend to infinity, there is a subsequence of the i's so that $\alpha_{i,n}$ converges to a quasiline or quasihalf-line $\alpha_{\infty,n}$, and so that the $f_{i,n}$ converge to a transformation $f_{\infty,n}$ which is the identity near 0 and is locally Mœbius (since its Schwarzian derivative is 0) but not Mœbius, since $D(f_{i,n}, 0)$ is constant. Obviously, the complement of $\alpha_{\infty,n}$ cannot be connected. Therefore, $\alpha_{\infty,n}$ must be a quasiline which separates $\mathbf{C}$ into two components, U_0 which contains the origin and U_1 which does not. The transformation $f_{\infty,n}$ is the identity on U_0, and on U_1 it is a nontrivial Mœbius transformation taking U_1 into itself.

Now we let n go to infinity. There is some subsequence such that the $\alpha_{\infty,n}$ converge to a quasiline α_∞. The transformations $f_{\infty,n}$ converge to the identity on both sides of the limiting quasiline, since $D(f_{\infty,n},0)$ converges to 0.

Let X be any vector field which represents a limiting direction for the Mœbius transformations $f_{\infty,n}$. By "direction," we refer to the direction from the identity in the Lie group of Mœbius transformations. More explicitly, let

$$X(z) = \left[\lim a_{i_j}(f_{\infty,i_j}(z) - z)\right] \frac{\partial}{\partial z},$$

where the subsequence $\{i_j\}$ and sequence of constants a_{i_j} are chosen to make the limit exist and be nonzero.

We claim that X carries U_1 into itself. To see this, consider any flow line of X through a point x in U_1. For any point y downstream from x on the flow line, there is some $f_{\infty,n}$ and some integer q such that its qth iterate carries x approximately to y. It follows that y is at least in the closure of U_1. Since the image of an open set by a time t map of a flow is open, y must be in U_1 itself.

Choose a point t on α where the flow is not zero, and conjugate by a sequence of enlargements centered about t so that in the limit X becomes an infinitesimal translation in the i direction. Passing to the limit of a subsequence, we obtain a new quasiline β, which intersects each line parallel to the y-axis in an interval or a point.

If any of these intersections are intervals, we immediately obtain a line in the closure of $E(\alpha)$. In fact, if the slopes of secant lines connecting pairs of points of β are unbounded (either above or below), there is a sequence of enlargements carrying such pairs of points to pairs at distance n from each other. The limit is a vertical line, so we are done in this case also.

If the slopes of the secant lines are bounded, then β is the graph of a Lipschitz function. A Lipschitz function is differentiable almost everywhere. A sequence of enlargements centered at any point where the function is differentiable converges to a line whose slope is the derivative. □

References

F. W. Gehring, *Spirals and the universal Teichmüller space*, Acta Math. **141** (1978), 99–113.

The Story of the Verification of the Bieberbach Conjecture

LOUIS DE BRANGES

You do not have to know much about the Bieberbach conjecture to be able to read this story. All you have to know is that it was, until recently, one of the most famous unsolved problems of analysis. It was proposed in 1916 by the German mathematician Ludwig Bieberbach.

Assume that a power series $z + a_2 z^2 + a_3 z^3 + \cdots$ represents a function which has distinct values at distinct points of the unit disk. Bieberbach conjectured that the inequality

$$|a_n| \leq n$$

holds for every $n > 1$.

The Bieberbach conjecture was verified for the second coefficient in 1916, for the third coefficient in 1923, for the fourth coefficient in 1955, for the sixth coefficient in 1968, and for the fifth coefficient in 1972.

No other case of the Bieberbach conjecture was known when I distributed a manuscript, on the first of March 1984, claiming to contain a proof of the Bieberbach conjecture for all coefficients.

At that time I had been working on the Bieberbach conjecture for nearly seven years, but I was not an acknowledged expert on that subject. Most of my research, over a period of twenty-five years, had been on other problems.

The final assault on the Bieberbach conjecture was made in a round-about way. I spent the year 1983 writing a book on *Square Summable Power Series*. Only one of the six chapters is concerned with the Bieberbach conjecture. The other five are about apparently unrelated problems on invariant subspaces, perturbation theory, polynomial approximation, and substitution transformations.

So the chapter on the Bieberbach conjecture was only a part of a much larger project. I was delayed in writing the chapter because my wife Tatiana and I translated into English a Russian manuscript related to the invariant subspace theory in the second chapter. We accepted the translation because its author was a student of M. S. Livšic, a personal friend and one of the creative geniuses of invariant subspace theory.

The chapter on the Bieberbach conjecture was completed only at the end of January 1984. The chapter improves on an estimation theory which I had obtained in the fall of 1982. I was very much surprised by the outcome of my calculations, because they were very close to a general proof of the Bieberbach conjecture.

But to obtain the Bieberbach conjecture, I needed to know that certain polynomials are positive in the unit interval. For the $(n + 1)$st coefficient, there are n polynomials, of degrees 1 through n. The polynomials are hypergeometric series with rational coefficients. So they are easily computable, and by examining them I could verify the Bieberbach conjecture for the second, third, and fourth coefficients. But the fifth coefficient made me work and I nearly did not make my way through the calculations for the sixth coefficient. So I was very happy when Walter Gautschi, seemingly without effort, ran off the calculations on the Purdue computer to verify the Bieberbach conjecture up to the twenty-fifth coefficient.

I leave it to him to tell his own story of how that happened. In fact he went farther than twenty-five, but I was already impressed at twenty-five because the Bieberbach conjecture was doubted by experts for odd coefficients, beginning around the nineteenth. I concluded that these doubts were unfounded and began searching for a general proof of the inequality.

At this point there occurred one of two remarkable coincidences connected with the proof of the Bieberbach conjecture. Walter Gautschi consulted Richard Askey at the University of Wisconsin about the calculations. It turned out that the inequality being tested on the computer was a theorem which Askey had obtained in 1976 in joint work with George Gasper of Northwestern University. That completed the proof of the Bieberbach conjecture.

That allowed me to complete my manuscript on *Square Summable Power Series*. I mailed more than a dozen copies to experts for verification. That was in the beginning of March. To my disappointment every one of them wrote back giving me a good reason why he could not check the proof at that time.

It was at this point that a second remarkable coincidence occurred. This happened because I was scheduled for a three-month visit to the Soviet Union under the exchange agreement between the National Academy of Sciences and the Academy of Sciences of the USSR.

Here I need to explain that I had proved a stronger conjecture than the Bieberbach conjecture. I had proved a conjecture of I. M. Milin, which implies the Bieberbach conjecture. I made a telephone call to my hosts at the Steklov Mathematical Institute in Leningrad. Sergei Kisliakov told me that Milin was in Leningrad.

On the twentieth of April, I took Lufthansa flights from Chicago to Frankfurt and from Frankfurt to Leningrad. In one hand I carried my manuscript on *Square Summable Power Series*, neatly bound in a black folder. A bleak winter landscape greeted me on arrival, and I felt unable to speak a word of Russian despite an excellent reading knowledge of the language and the ability to understand what was said when it was spoken. My host Sergei Khruschev greeted me at customs and took me through the city to a three-room apartment which was to be mine.

Then we went by subway to the apartment of N. K. Nikol'skiĭ, the leader of the seminar in functional analysis. Present also were V. P. Havin and Mrs. Nikol'skiĭ. After a meal which she served, we scheduled my talks to the functional analysis seminar in April and May. There was just enough time to cover the five chapters of my book on *Square Summable Power Series* which were not related to the Bieberbach conjecture. That part was left to the seminar in geometric function theory.

My stay in Leningrad was one of the happiest periods of my life. A typical day began with a walk through a huge park nearby. Then after bathing, eating, and shopping, I was free to do whatever I wanted. Tuesdays and Thursdays were seminar days. On those days I took the subway downtown and walked along Nevskiĭ prospect until I reached the Mathematical Institute, which is located along a bank of one of the canals passing through the city.

Functional analysis has an impressive seminar room which must hold fifty people. A meeting of the seminar ordinarily consists of two hour-and-a-half sessions interrupted by a break for tea. I spoke in English, doing much writing on the blackboard. My work was received with interest and appropriate questions in Russian and in English. At the end of the five lectures it seemed to me that every member of the seminar had spoken to me of some work of his own which was related to my lectures.

The verification of the proof of the Bieberbach conjecture was made in a smaller room with a smaller audience whose members were less familar with English. I was introduced to Professor Milin by Asya Greenshpan, a professional translator and the wife of Arkady Greenshpan. Milin had an amused look which I interpreted as doubts about the soundness of my undertaking. This seemed to be confirmed a few minutes later when a question was asked about my work on the Peterson conjecture, which contains an uncorrected error. I had to explain that I did not know whether my argument could be saved and that I had not tried to do so because the Peterson conjecture was contained in the Weil conjecture which had subsequently been proved by Deligne.

No one seemed to believe a word of my lecture, but the situation was better than it seemed. The reason was that Khrushchev had visited Purdue University during the fall semester 1982 and had brought back to Leningrad a manuscript which I had written at that time. After my telephone call, the results of the manuscript were presented to the seminar in geometric function theory by a student named Emel'ianov. This occurred some three weeks before my arrival. Already in my second meeting with the seminar, Genia Emel'ianov was saying that the proof was correct. A meeting of the seminar then occurred in which I was not present. In my third meeting with the seminar, Professor Milin shook my hand, without losing the glint of humor in his eye, and said that I was a very talented mathematician. He also said that my argument was elegant.

Both Emel'ianov and Milin submitted written reports confirming the proof of the Bieberbach conjecture and presenting variants of the argument considered to be advantageous. The reports were accepted by the other members of the seminar, including Arkady Greenshpan and the seminar leader, G. V. Kuz'mina, as correct.

Spring had come to Leningrad and it was now June. Every day was an ideal twenty degrees Celsius. No more seminars met. Everyone went on sightseeing expeditions to the Winter Palace and to other palaces outside the city such as Peterhof and Pushkin. But during that time I worked for long hours with Professor Kuz'mina to prepare the preprint, in Russian and in English, which contains the findings of the seminar.

She was extremely conscientious in her capacity as an editor. The manuscript underwent constant revisions not only for mathematical content but also for details of expressions and for historical accuracy. She was, for example, concerned that the cases of equality be included in the preprint. I assumed at the time that this work was done by Emel'ianov, but this may have reflected modesty on her part.

Our working sessions began at two in the afternoon and ended at nine or ten in the evening. The only break was for tea. When I left, she complained about the adjustments to living without the Bieberbach conjecture, which had become such an absorbing part of her life.

During the second week of June, I had made a visit to the Gonchar seminar at the Steklov Mathematical Institute in Moscow. Afterwards Professor Gonchar let it be known that he thought the proof should be published in *Mat. Sbornik.* I agreed that was appropriate in view of the contribution of the Leningrad seminar in geometric function theory to the verification of the proof. As I left, Professor Kuz'mina and I promised to keep working together to bring the manuscript into suitable form for such publication.

The Soviet Academy of Sciences granted me permission to leave Leningrad two weeks earlier than scheduled so that I would have a chance to present my proof of the Bieberbach conjecture in Western Europe before catching my return flight from Frankfurt to Chicago. It also paid me for those two weeks, which allowed me to take an Aeroflot flight from Leningrad to East Berlin.

Sergei Khrushchev accompanied me to the Leningrad airport on the morning of July first. One of the things which was different on leaving Leningrad was that by that time I could speak Russian.

My first reaction on arriving in West Berlin was to have a normal meal. I then went by train to Heidelberg and rested for several days at my father's house. I learned from his scales that I had lost five kilos during the ten weeks in Leningrad.

In the first two weeks of July, I presented the proof of the Bieberbach conjecture at the University of Würzburg, the University of Hannover, and the Free University in Amsterdam. The proof was also presented by Sergei Khrushchev at an international symposium on operator theory in Romania in the second week of June. As I was catching my return flight to Chicago on July fifteenth, Volodya Peller of the Leningrad seminar in functional analysis was about to present the proof of the Bieberbach conjecture at an international symposium on operator theory in Lancaster, England.

Everywhere that the proof of the Bieberbach conjecture was presented, it was received with great excitement. With all that had happened, it was several weeks after my return to Lafayette that I could resume my work.

As soon as I was back to my typewriter and to my usual means of producing manuscripts, I became dissatisfied with the publication arrangements made in the Soviet Union. My main concern in Leningrad had been to get the seminar to say yes. I tried to do nothing which would prevent or delay that important affirmation. However, during the month of June, I found improvements in technique which could not have been incorporated in the preprint during the remaining time of my visit. The mails to Leningrad were too slow and the rules of *Mat. Sbornik* were too strict to make and check out all the needed changes later.

I decided to begin a new and final version of the proof of the Bieberbach conjecture. As I was working on it, I had the feeling that I was producing a historical document. I put other concerns aside and gave myself entirely to the writing of the paper, without any hurry.

I asked friends and colleagues about the best journal for publication. The unanimous answer was *Acta Mathematica*.

The paper was completed at the end of August, just as the new semester was beginning at Purdue University. The manuscript was mailed to Stockholm in the middle of September. Professor Hörmander accepted it for publication in the middle of October. Publication occurred within four months of acceptance.

Note: Other accounts of the verification have been written by I. M. Milin in these proceedings and by G. V. Kuz'mina (in a joint article with O. M. Fomenko) in the *Mathematical Intelligencer*, **8** (1986), No. 1, pp. 40–47.

Reminiscences of My Involvement in de Branges's Proof of the Bieberbach Conjecture

WALTER GAUTSCHI

Around February 3, 1984 (I can't remember the exact date), Louis de Branges came to my office and asked whether he could talk to me for a minute about some work he was doing; perhaps I could be of help. I distinctly remember the first thought that ran through my mind: "Me? Helping de Branges?" We hardly knew each other, never engaged in any mathematical conversation in all the 20 or so years we were at Purdue, and—so I believed—had interests diametrically apart. He sat down and told me that he had a way of proving the Bieberbach conjecture, but needed to establish certain inequalities involving hypergeometric functions. He felt it would be worthwhile, as a first step, to check as many of these inequalities as possible on the computer. Could I do this for him?

Now this was a time when I happened to be under all sorts of pressures. I was expected (and very much wanted) to write a paper for *BIT* to honor Germund Dahlquist on his 60th birthday. Through some mix-up the invitation had reached me just a few days earlier (on February 1)—way past the deadline of December 31, 1983—but I was graciously given an extension through February 29. So I had less than four weeks in which to produce worthwhile results and a publishable paper. At the same time I was in the midst of rewriting a chapter of a survey article for the *MAA Studies in Numerical Analysis* in order to be ready to incorporate the new version on the galley proofs that were to arrive at any time. Also, I was scheduled to leave for Europe on March 7 for lectures in Italy, Yugoslavia, and Germany. As if this were not enough, I had, as the newly appointed managing editor of *Mathematics of Computation*, to deal with a constant stream of manuscripts for this journal. And classes had to be taught also, department committee meetings attended to, etc., etc.

I didn't, of course, tell Louis all these things, but they weighed heavily on my mind when I replied that I would probably not have the time to do anything for him right away. He then told me that he was soon going to give a seminar

---

The work described in this article was supported, in part, by the National Science Foundation under grant DCR-8320561.

on the subject and asked me to at least attend the seminar and in that way get some more concrete ideas of what was involved.

The seminar took place on February 7, and I managed to attend. I was immediately struck by the clarity, freshness, and elegance of Louis' talk and began to appreciate how those inequalities came about. To my delight, they could be written in terms of orthogonal polynomials—currently a subject very much on my mind. What was needed was to show that for any positive integer n the set of n inequalities

(1) $$F_{n,k}(x) := \int_0^1 t^{n-k-1/2} P_k^{(2n-2k,1)}(1-2tx)\,dt > 0,$$
$$0 < x < 1, \ k = 0, 1, 2, \ldots, n-1$$

is valid, where $P_k^{(\alpha,\beta)}$ is the Jacobi polynomial of degree k with parameters α, β. (For $k = 0$, the inequality is trivially true.) Louis' theory in fact states that the validity of (1) for some n implies the Bieberbach conjecture for the $(n+1)$st coefficient (but not vice versa). Louis concluded the lecture by showing how he evaluated $F_{n,k}$—a polynomial of degree k—explicitly for the first few values of n (for $n \leq 4$, I believe) and how he could verify the correctness of (1) in these cases. Unfortunately, they did not include the largest value of n for which Bieberbach's conjecture had already been proven.

I saw right away how (1) could be verified computationally using Gauss-Jacobi quadrature (with weight function $t^{-1/2}$ on $[0,1]$; but see (2) below), and I pointed this out during the discussion, claiming, with zest, that it would be easy for me to go as far up with n as $n = 100$, if that should be necessary. I was clearly fired up by now and was determined to carry out the computations immediately, no matter what. Having developed reliable software for orthogonal polynomials and Gaussian quadrature during the past few years, I knew that it shouldn't take too much time for me to write the necessary programs.

To begin with, I noted by a simple symmetry argument that one needed only the classical Gauss-Legendre quadrature rule on $[-1,1]$. If $\tau_\nu^{(2m)}, \lambda_\nu^{(2m)}, \nu = 1, 2, \ldots, 2m$, are the nodes and weights, respectively, of the $2m$-point Gaussian quadrature rule, with $1 > \tau_1^{(2m)} > \tau_2^{(2m)} > \cdots > \tau_{2m}^{(2m)} > -1$, then in fact

(2) $$\int_0^1 t^{-1/2} p(t)\,dt = 2 \sum_{\nu=1}^m \lambda_\nu^{(2m)} p([\tau_\nu^{(2m)}]^2)$$

for any $p \in \mathsf{P}_{2m-1}$, $m = 1, 2, 3, \ldots$. Since the integral in (1) is of the form (2), with p a polynomial of degree n, it suffices to take $2m - 1 \geq n$ in (2), for example, $m = [n/2] + 1$. The Gauss formula involved in (2) can easily be generated for any value of m (this indeed is done by one of the easier parts of my software package), and the polynomial $P_k^{(\alpha,\beta)}$ in (1) is readily and accurately generated by the well-known three-term recurrence relation. I actually found it slightly more convenient to compute

(3) $$f_{n,k}(x) = \int_0^1 t^{n-k-1/2} \pi_k^{(2n-2k,1)}(1-2tx)\,dt,$$

for $0 \leq x \leq 1$, $k = 1, 2, \ldots, n-1$, where $\pi_k^{(\alpha,\beta)}$ is the monic Jacobi polynomial.

My first program ran the next day, on February 8, and "verified" (1) for all $n \leq 18$. It cost me \$3.69 in computer time on the CDC 6500. The program, of course, was still fairly primitive; I simply evaluated $f_{n,k}$ for up to 400 equally spaced points on the interval $[0, 1]$ and printed the minimum value and corresponding x-value to see whether the minimum was positive (or a "machine zero" when $x = 1$ and k is odd). I took this simple-minded approach because I was fairly sure that I was going to hit a negative minimum for some early value of n, which would render Louis' argument inconclusive for that value of n, and I could quit and go on with my own work. It didn't work out that way!

After this first piece of positive evidence, I began to improve the program, incorporated error-monitoring devices, compared double precision with single precision results, determined all minima and maxima of $f_{n,k}$ on $[0, 1]$ to full machine precision using Newton's method, and checked between any two extrema for possible additional oscillations that I may have missed. I then pushed this improved version of the program up to $n = 30$ and found the validity of (1) confirmed in every case. The most expensive run (for $27 \leq n \leq 30$) still cost me only \$10.84.

At this point I was convinced that (1) is true for all n. I began to play with the idea of writing up this work in a short note entitled, tentatively, "Numerical evidence in support of a conjecture of L. de Branges." (I didn't dare yet to bring Bieberbach into the title!) I even prepared neat photoready printout on our Diablo printer that could be reproduced, together with the program listing, in the microfiche or supplements section of the journal. (I had in mind my own *Mathematics of Computation*.) A brief excerpt from these tables is shown on the next page.

Happy about this encouraging development, I called on February 13 my good friend Luigi Gatteschi at the University of Turin and asked him whether I could possibly give a second lecture in Turin (one had already been scheduled); the title: "La congettura di Bieberbach è (probabilmente) vera." He readily agreed (though I seemed to detect a skeptical tone in his voice) and subsequently arranged the first lecture to be given at the University and the second at the Polytechnic.

Still not completely satisfied with the strictly computational nature of my work, I began to develop complicated analytic criteria that would insure, mathematically, that $f_{n,k}$ could not have any zeros on any given sufficiently small subinterval of $[0, 1]$. By applying these criteria in a judicious manner, one could then in principle prove (again with the help of the computer) that the whole interval $[0, 1)$ is free of zeros. I spent about a week on efforts along these lines, but did not get very far, since the program I wrote turned out to be extremely slow. About this time, on February 20, Professor Jack Schwartz from the Courant Institute at NYU visited our Computer Sciences Department. I requested beforehand ten minutes of his time to talk to him briefly about this computational

n	k	x	f(x)	df(x)	ddf(x)
25	23	0	2.992383247e-04	-2.85e-02	2.439457320e+00
		3.205597026e-01	2.341983715e-08	-7.58e-20	7.210458286e-06
		3.417860015e-01	2.391757633e-08	9.68e-20	-5.828107204e-06
		4.340919006e-01	1.072194076e-08	-1.27e-20	3.934615048e-06
		4.662672133e-01	1.136217104e-08	2.70e-20	-3.184816919e-06
		5.522705925e-01	5.785863359e-09	2.36e-20	2.457239610e-06
		5.912254258e-01	6.401691048e-09	-1.98e-20	-2.091412963e-06
		6.679768905e-01	3.545491509e-09	1.17e-20	1.891163295e-06
		7.096583223e-01	4.124040146e-09	-2.57e-21	-1.744348249e-06
		7.744958058e-01	2.406779734e-09	-3.54e-21	1.841974435e-06
		8.146887759e-01	2.974169399e-09	-1.11e-20	-1.903521612e-06
		8.656751077e-01	1.775935986e-09	-1.01e-20	2.376104557e-06
		9.002022063e-01	2.375351261e-09	8.51e-21	-2.907485610e-06
		9.362182846e-01	1.395724928e-09	-7.17e-21	4.540772788e-06
		9.612703112e-01	2.103525418e-09	-1.40e-20	-7.509062965e-06
		9.819556336e-01	1.112364664e-09	-3.14e-20	1.838703966e-05
		9.944763566e-01	2.166255736e-09	1.27e-19	-7.519729386e-05
		1.000000000e+00	4.784809710e-19	-1.02e-06	-3.376022124e-04
25	24	0	1.583271559e-05	-2.13e-03	2.534501112e-01
		8.821620539e-02	8.890073589e-08	6.86e-19	9.519682273e-05
		1.007789679e-01	9.102976075e-08	1.47e-18	-6.626872232e-05
		1.654376237e-01	3.432305315e-08	-4.90e-20	2.621243315e-05
		1.895837701e-01	3.655406944e-08	4.55e-19	-1.841404435e-05
		2.621430762e-01	1.721658961e-08	-6.88e-21	9.824593756e-06
		2.961071425e-01	1.896011080e-08	5.88e-21	-7.338968761e-06
		3.726369664e-01	1.015730680e-08	-9.74e-21	4.916658213e-06
		4.142485505e-01	1.154094862e-08	-6.56e-21	-3.940854645e-06
		4.905874739e-01	6.707130829e-09	-4.35e-20	3.131842409e-06
		5.370536552e-01	7.869709830e-09	1.27e-20	-2.707663255e-06
		6.092538964e-01	4.817924151e-09	-2.91e-21	2.475361689e-06
		6.572727170e-01	5.859388113e-09	-2.44e-21	-2.328490771e-06
		7.218483349e-01	3.696588620e-09	-2.38e-21	2.418952894e-06
		7.678171738e-01	4.693289289e-09	-1.50e-20	-2.516816547e-06
		8.219130195e-01	2.988555425e-09	-4.24e-22	2.987795528e-06
		8.621912336e-01	4.012828437e-09	2.54e-21	-3.552522979e-06
		9.036814275e-01	2.511845675e-09	1.49e-20	4.988334162e-06
		9.348846218e-01	3.660165547e-09	2.51e-20	-7.304448233e-06
		9.624003854e-01	2.141929923e-09	-4.80e-20	1.360061023e-05
		9.817059823e-01	3.622132535e-09	1.46e-19	-3.126465152e-05
		9.945898427e-01	1.627629589e-09	-4.94e-19	1.385542984e-04
		1.000000000e+00	5.412894218e-09	1.82e-06	6.092753732e-04

Extrema of $f_{n,k}(x)$ on $[0,1]$, with first and second derivatives, for $n = 25$ and $k = 23, 24$. (The computer printout is in floating-point E-format, so that $e - xx$ is to be read as 10^{-xx}. Note also that the derivatives at the interior extrema are not exactly zero, but approximately $\varepsilon \cdot f_{n,k}(0)$, where $\varepsilon = 3.55 \times 10^{-15}$ is the machine precision of the CDC computer.)

work on the Bieberbach conjecture. He showed polite interest in the matter, but didn't say much. Only at the end of our brief meeting he casually asked why not use Sturm sequences. I remember how this question caught me by surprise and how I wondered why I hadn't thought of it myself. After all, I used Sturm sequences in a similar setting some eight years ago in connection with Chebyshev-type quadrature rules. On second thought, however, I could understand why Sturm didn't enter my mind: The polynomial $F_{n,k}$ in (1) is given in the form of an integral, and it was not immediately obvious how to generate Sturm sequences in rational form.

It soon occurred to me, however, that the explicit power representation of $F_{n,k}$ can be obtained (in rational form) rather easily by substituting the representation

$$(4) \quad P_k^{(\alpha,\beta)}(u) = \frac{\Gamma(\alpha+k+1)}{k!\Gamma(\alpha+\beta+k+1)} \sum_{m=0}^{k} \binom{k}{m} \frac{\Gamma(\alpha+\beta+k+m+1)}{2^m \Gamma(\alpha+m+1)} (u-1)^m$$

into (1). Indeed, with $u = 1 - 2tx$, one gets $u - 1 = -2tx$, and the integral in (1), using (4), is easily evaluated, yielding a polynomial with coefficients in the form of ratios of factorials (in fact, a $_3F_2$). So the work to be done from now on was clearly mapped out for me: Apply the Sturm sequence algorithm to (1) [or alternatively, to (3)] on the interval $[0,1]$ in rational arithmetic—for example, using the MACSYMA system—and in this way show compellingly, once and for all, and for as many n as desired, that $F_{n,k}$ cannot vanish on $[0,1)$ and therefore, since $F_{n,k}(0) > 0$, that it remains positive on $(0,1)$. Time, however, was getting short, and I decided to postpone this work until after my return from Europe. In the meantime, I programmed the Sturm sequence algorithm for (1) in double precision FORTRAN in order to check out the algorithmic details and to make it easier for me, upon my return, to transcribe the program into the MACSYMA language (a system I still had to get better acquainted with). By February 26 (a Sunday) I had the program running satisfactorily on the CDC computer and producing results as expected.

A week earlier, incidentally, I managed to complete my paper for Dahlquist and sent it off to the editors.

My departure for Europe was becoming imminent. Since my verification work seemed well under way, and in good shape now, I let it rest for a while and turned my attention to the lectures I was to give in Europe. Just to set my mind at ease, I wanted to make sure, however, that the inequality (1) was not by chance already known in the literature. Actually, looking at the rather delicate behavior of $f_{n,k}(x)$ for x near 1 and k near n, as exemplified in the short table above, I rather doubted that analytical techniques could be sharp enough to provide a proof of (1). But it didn't hurt to check. I knew there was only a handful of mathematicians in the world who could possibly be familiar with a result of the type (1) and even come up with a proof of it, among them Dick Askey at the University of Wisconsin, whom I knew best. So I called him on February 29 and told him of the inequality (1) and what it implied according to de Branges. He immediately interrupted me with an emphatic: "I don't believe it!" and recounted some rather outrageous claims that had been made in the past by a number of people. I countered that de Branges was a serious mathematician and that we were dealing here with first-rate work. Even if Louis' implication should not hold tight, I said, the inequalities (1) were quite interesting in their own right and ought to be scrutinized further. Besides, I was fully convinced of their validity. Dick now agreed to look into it.

I was working late at home on my lectures, that same night, when the phone rang and I heard Dick Askey's triumphant voice on the other end of the line: "The inequality is not a conjecture—it's a theorem!" He then pointed out a result in a joint 1976 paper with George Gasper that contains (1) as a special case. I was, of course, delighted to hear this incredible news, but also disappointed, realizing that all my hectic work had been in vain. After I checked and confirmed the result myself, I saw Louis the next morning and told him the good news. He replied, rather matter-of-factly: "Well, that proves Bieberbach's conjecture."

Immediately after I talked to Louis, I called Luigi Gatteschi in Turin and asked him to change the title of my second lecture. There was no point anymore to talk about numerical evidence for a conjecture that had turned into a theorem, and I proposed, instead, to talk about the work I did in the paper for Dahlquist. I told him that I would explain everything when I was in Turin. He agreed to send out a change of title notice, and he scheduled this second lecture to be held also at the University on March 13.

It was during the first ten minutes of this lecture that I first apologized for the change of subject and then briefly announced the validity of the Bieberbach conjecture subject to verification of de Branges's work. This was probably the first time that the word got out in Europe, but it was a small audience, consisting largely of graduate students and only a few faculty members. A week later, I attended a conference in Munich celebrating the 25th anniversary of the journal *Numerische Mathematik*. There I saw another good friend of mine, Dick Varga, and told him informally of de Branges's proof of the Bieberbach conjecture. I knew he was going to give a talk himself about a number of conjectures, including the Riemann hypothesis. At the end of the discussion period he turned towards me and put me on the spot with: "Speaking of conjectures, how about Bieberbach's conjecture, Professor Gautschi?" So I went to the blackboard and announced again de Branges's proof of the conjecture and the role played by the inequality of Askey and Gasper. But this time, it was before a large international audience of experts, and I felt the enormous impact of my brief presentation. The word now spread to different parts of Europe.

Looking back at this episode, I cannot help concluding with a few philosophical remarks. 1. *The computer is an important aid in theorem proving.* In our case, the computer could have been used to prove Bieberbach's conjecture (using Sturm sequences in MACSYMA) if not for all n, then at least for as many n as desirable and practicable. Equally importantly, my computer results gave Louis confidence in his overall proof strategy; his approach indeed seemed capable of proving the complete Bieberbach conjecture. 2. *The availability of high-quality mathematical software is of the utmost importance in scientific research.* While this statement is undisputed among computer scientists, it deserves to be better understood and appreciated by the mathematical community. Had I not had available my own software package on orthogonal polynomials, I would probably not have undertaken these computations, given the severe time constraints under which I was operating. 3. *One should never underestimate the usefulness of a result in pure mathematics.* No one in his wildest dreams, least of all the authors, could have imagined that the Askey/Gasper nonnegativity result would provide a critical link in the proof of the Bieberbach conjecture. *Inequalities,* in particular, *are always potentially useful.* I am fortunate to have inherited a love for inequalities from my teacher, Professor Alexander Ostrowski, who was a master at them, and from Professor Mauro Picone, who openly confessed to me his fondness for inequalities. Perhaps it is an auspicious omen that the Bieberbach conjecture itself—now de Branges's theorem—consists of a set of inequalities.

My Reaction to de Branges's Proof of the Bieberbach Conjecture

RICHARD ASKEY

On February 29, 1984, Walter Gautschi called to ask about an inequality. He prefaced his request with the statement that Louis de Branges had reduced the Bieberbach conjecture to showing that an integral of a class of Jacobi polynomials was positive. I said that was preposterous, since the Bieberbach conjecture was a complex inequality, and it could not be proved from a real inequality for a real function. Gautschi gave me the inequality and said it seemed to be true, but that it was probably very hard to prove. He had put some special cases on a computer and they seemed to be true, but there were times when the polynomials were very small. I said I knew something about these questions and would look at it. That evening I computed the integral to see what $_3F_2$ it was and then looked at a paper George Gasper and I wrote about ten years earlier. It appeared in the *American Journal of Mathematics* in 1976. To my surprise, the inequality de Branges wanted was proved in this paper. I called Gautschi to tell him that the inequality was true, but that I was almost sure it would not prove the Bieberbach conjecture. In addition to my earlier reservation, there was another good reason why I felt this way. The $_3F_2$ inequality is equivalent to the positivity of a sum of Jacobi polynomials, and Gasper had proven the positivity of another sum of Jacobi polynomials that is a fractional derivative of order one-half of the inequality de Branges needed. Thus while this inequality is sharp in some senses, as Gautschi realized from his computations, in other ways there were much stronger inequalities. I did not think the Bieberbach conjecture would follow from anything except a very sharp inequality.

Two days later de Branges called to ask me if I knew the consequences of this inequality. I said I knew what Gautschi had told me, but that I had serious doubts and told him what they were. In less than two weeks de Branges sent his manuscript, *Square Summable Power Series*. Before that I called Peter Duren, and he said he had been informed and had suggested that de Branges circulate his proof widely. At that time I felt there was something like a 2 percent chance that de Branges had proven the Bieberbach conjecture. The reason this was greater than zero was that he had discovered a deep inequality without knowing it before, so he clearly had something. It seemed likely he would be able to prove the Bieberbach conjecture for a subclass of univalent functions, since weaker inequalities for trigonometric polynomials had been used for that purpose before.

I tried to read the last chapter of de Branges's book, but was unable to follow his arguments. The next thing I heard was from Walter Rudin, when he returned in the summer from a meeting in England. He called to say that de Branges had proved the Bieberbach conjecture. After I told Rudin of my doubts, he told me there was a new version of de Branges's proof, with all the operator theory removed, and that the details had been checked by a number of very good mathematicians. He gave me a Russian version from Leningrad. After a week I had reached the point of getting out a Russian dictionary when David Drasin called to tell me the latest and to ask if the inequality Gasper and I had published was really true. He said that Pommerenke had checked the rest of the proof. I assured him that our inequality was true, and that there were two proofs. The first appeared in our paper, and I had earlier included a proof of

$$(1) \qquad \sum_{k=0}^{n} P_k^{(\alpha,0)}(x) > 0, \qquad -1 < x \le 1, \ a > -2$$

in my Regional Conference Lectures: *Orthogonal Polynomials and Special Functions*, SIAM, 1975. When $\alpha = 2, 4, \ldots$, this is equivalent to the inequality de Branges needed. This is an appropriate place to recall something I first mentioned in these Lectures. The proof of (1) given there was found by Gasper. Gasper has found a second proof of (1) which is shorter than his first proof, but not as well motivated. Actually there are now five proofs, four due to Gasper and one found by Koornwinder.

Drasin said there was another version of de Branges's proof written by Fitz-Gerald and Pommerenke. He sent a copy, and after reading it, I was convinced that my original reservations had come because I did not know enough. It was also clear that de Branges's marvelous proof should lead to deeper inequalities than Milin's conjecture. Koornwinder has found a deeper inequality in one direction, and there should be more in other directions.

There are a number of things to learn from this work. First, on the technical level, de Branges has shown the true depth of Loewner's work. It was always known to be deep, but I doubt if anyone except de Branges really appreciated how deep. Second, there is a lot to be said for doing two things de Branges did. He worked on a hard problem, has worked on other hard problems in the past, and was not discouraged by past failures. Also it often pays not to talk with experts in an area, or at least not to take their pessimism too seriously. Experts in univalent functions would have told de Branges that his method could not work, and they would have been wrong. My reaction was similar. I knew too much in one direction, where my optimism had led to a number of deep inequalities, and I did not know enough in another area. These two led to a skepticism that was unwarranted. Do not take the pessimistic views of an expert very seriously, but optimistic views should be taken very seriously, for they often lead to something important. Third, if a newspaper reporter calls, say you want ten minutes to think and ask them to call back, or offer to return their call. I have been disappointed by the press coverage of this work. The important fact is de Branges's proof, not whether he had published false proofs of other conjectures. The fact that de Branges was an outsider in univalent functions is important, but no one who knows his early work can say he was on the fringe of the research community. If one does not know, then it is best to say nothing and refer the reporter to someone who does. Finally, the polynomial inequalities Gasper and I proved (both jointly and separately) were primarily conjectured while trying to understand papers of Fejér that I was reading in his *Gesammelte Arbeiten*. Collected papers of great mathematicians are very valuable, and every mathematician should own some and read seriously in at least one. The time and money spent is a very good investment.

Poem

There was a professor in Berlin.
With him my story does begin.
Be careful that his name you catch:
He was called Ludwig Bieberbach,
Which makes me wonder whether it might
Be that his father had second sight,
That in mysterious ways knew he
Of the future coming of the great Louis
And that he called his child after him
Who later his son's fame would dim.
As the years passed young Ludwig acquired
Mathematical insight much to be admired.
He raised a question, then quite new:
What do the coefficients do
Of functions that are schlichtly mapping
The disk on regions non-overlapping?
He normalized to $a_1 = 1$,
And then after he had carefully gone
Through all the functions in his ken,
He found a_n absolute at most n.
Herr Bieberbach was very clever,
But despite desperate endeavor
He never managed more to do
Than the case $n = 2$.

The learned world was full of glee
When Löwner cracked $n = 3$.
Then fell 4, 5, 6 right up to 9,
Some done by wizards who here with us dine.
Each step took years, and it was clear
That the final solution was far from near.

Some tried to get it in ways very devious
By making conjectures more abstruse than the previous.

But all of them labored to no avail.
We saw Robertson, Milin, Lebedev fail.

But just as we started to feel quite blue
And worked as if under a pall,
There came de Branges out of Purdue
And solved the problems all.

This is no secret, all over it chimes,
It is in *Pravda* and *The New York Times*.
Their millions of readers are happy to know,
But we quite special thanks do owe.
For we belong to the fortunate set
Who for this feast are in Louis' debt
And for what a conference, unsurpassed
With lots to learn from a stellar cast.
I can only feel an icy pity
For those who frown on work by committee.
Ours here is of a marvelous sort:
It got wonderful people and financial support,
Fine lectures, excellent drink and food,
Through its kindness we've never had it so good.
But perhaps it is not kindness at all,
Instead an urgent emergency call:
Do they fear future mathematicians bereft
Because there is nothing for them to do left?
And have they called us to do our bit
To make conjectures so that the wit
Of coming Louis, Enricos, and Walters
Can show its brilliance long after ours falters?
Rest at peace, committee, no need to moan:
Eight new problems from Peter Jones alone.
Well, it is eight well-nigh,
One or two will be solved by Murai.
There is just one more thing I want to say:
To the present Louis a rousing hurrah!

Wolfgang Fuchs

*Read at the banquet of the Bieberbach conference,
March 12, 1985*